Natural Phenolic Compounds for Health, Food and Cosmetic Applications

Natural Phenolic Compounds for Health, Food and Cosmetic Applications

Editor

Lucia Panzella

MDPI • Basel • Beijing • Wuhan • Barcelona • Belgrade • Manchester • Tokyo • Cluj • Tianjin

Editor
Lucia Panzella
Department of Chemical Sciences,
University of Naples "Federico II"
Italy

Editorial Office
MDPI
St. Alban-Anlage 66
4052 Basel, Switzerland

This is a reprint of articles from the Special Issue published online in the open access journal *Antioxidants* (ISSN 2076-3921) (available at: https://www.mdpi.com/journal/antioxidants/special_issues/aatural_phenolic_compounds).

For citation purposes, cite each article independently as indicated on the article page online and as indicated below:

LastName, A.A.; LastName, B.B.; LastName, C.C. Article Title. *Journal Name* **Year**, *Article Number*, Page Range.

ISBN 978-3-03936-734-4 (Hbk)
ISBN 978-3-03936-735-1 (PDF)

© 2020 by the authors. Articles in this book are Open Access and distributed under the Creative Commons Attribution (CC BY) license, which allows users to download, copy and build upon published articles, as long as the author and publisher are properly credited, which ensures maximum dissemination and a wider impact of our publications.

The book as a whole is distributed by MDPI under the terms and conditions of the Creative Commons license CC BY-NC-ND.

Contents

About the Editor . . . . **vii**

Lucia Panzella
Natural Phenolic Compounds for Health, Food and Cosmetic Applications
Reprinted from: *Antioxidants* **2020**, *9*, 427, doi:10.3390/antiox9050427 . . . . **1**

**J. Fausto Rivero-Cruz, Jessica Granados-Pineda, José Pedraza-Chaverri, Jazmin Marlen
Pérez-Rojas, Ajit Kumar-Passari, Gloria Diaz-Ruiz and Blanca Estela Rivero-Cruz**
Phytochemical Constituents, Antioxidant, Cytotoxic, and Antimicrobial Activities of the
Ethanolic Extract of Mexican Brown Propolis
Reprinted from: *Antioxidants* **2020**, *9*, 70, doi:10.3390/antiox9010070 . . . . **7**

**Po-Han Lin, Yi-Fen Chiang, Tzong-Ming Shieh, Hsin-Yuan Chen, Chun-Kuang Shih,
Tong-Hong Wang, Kai-Lee Wang, Tsui-Chin Huang, Yong-Han Hong, Sing-Chung Li and
Shih-Min Hsia**
Dietary Compound Isoliquiritigenin, an Antioxidant from Licorice, Suppresses Triple-Negative
Breast Tumor Growth via Apoptotic Death Program Activation in Cell and Xenograft
Animal Models
Reprinted from: *Antioxidants* **2020**, *9*, 228, doi:10.3390/antiox9030228 . . . . **19**

**Fabiana Tortora, Rosaria Notariale, Viviana Maresca, Katrina Vanessa Good, Sergio Sorbo,
Adriana Basile, Marina Piscopo and Caterina Manna**
Phenol-Rich *Feijoa sellowiana* (Pineapple Guava) Extracts Protect Human Red Blood Cells from
Mercury-Induced Cellular Toxicity
Reprinted from: *Antioxidants* **2019**, *9*, 220, doi:10.3390/antiox9070220 . . . . **35**

**Andréa A. M. Shimojo, Ana Rita V. Fernandes, Nuno R. E. Ferreira, Elena Sanchez-Lopez,
Maria H. A. Santana and Eliana B. Souto**
Evaluation of the Influence of Process Parameters on the Properties of Resveratrol-Loaded NLC
Using 2^2 Full Factorial Design
Reprinted from: *Antioxidants* **2019**, *8*, 272, doi:10.3390/antiox8080272 . . . . **49**

**Eun-Sol Ha, Woo-Yong Sim, Seon-Kwang Lee, Ji-Su Jeong, Jeong-Soo Kim, In-hwan Baek,
Du Hyung Choi, Heejun Park, Sung-Joo Hwang and Min-Soo Kim**
Preparation and Evaluation of Resveratrol-Loaded Composite Nanoparticles Using
a Supercritical Fluid Technology for Enhanced Oral and Skin Delivery
Reprinted from: *Antioxidants* **2019**, *8*, 554, doi:10.3390/antiox8110554 . . . . **63**

**Maria Laura Alfieri, Giovanni Pilotta, Lucia Panzella, Laura Cipolla and Alessandra
Napolitano**
Gelatin-Based Hydrogels for the Controlled Release of 5,6-Dihydroxyindole-2-Carboxylic Acid,
a Melanin-Related Metabolite with Potent Antioxidant Activity
Reprinted from: *Antioxidants* **2020**, *9*, 245, doi:10.3390/antiox9030245 . . . . **81**

**Davide Liberti, Maria Laura Alfieri, Daria Maria Monti, Lucia Panzella and Alessandra
Napolitano**
A Melanin-Related Phenolic Polymer with Potent Photoprotective and Antioxidant Activities
for Dermo-Cosmetic Applications
Reprinted from: *Antioxidants* **2020**, *9*, 270, doi:10.3390/antiox9040270 . . . . **97**

César Leyva-Porras, María Zenaida Saavedra-Leos, Elsa Cervantes-González, Patricia Aguirre-Bañuelos, Macrina B. Silva-Cázarez and Claudia Álvarez-Salas
Spray Drying of Blueberry Juice-Maltodextrin Mixtures: Evaluation of Processing Conditions on Content of Resveratrol
Reprinted from: *Antioxidants* **2019**, *8*, 437, doi:10.3390/antiox8100437 . . . . **111**

Carol López de Dicastillo, Constanza Piña, Luan Garrido, Carla Arancibia and María José Galotto
Enhancing Thermal Stability and Bioaccesibility of Açaí Fruit Polyphenols through Electrohydrodynamic Encapsulation into Zein Electrosprayed Particles
Reprinted from: *Antioxidants* **2019**, *8*, 464, doi:10.3390/antiox8100464 . . . . **123**

Antonio de Jesus Cenobio-Galindo, Juan Ocampo-López, Abigail Reyes-Munguía, María Luisa Carrillo-Inungaray, Maria Cawood, Gabriela Medina-Pérez, Fabián Fernández-Luqueño and Rafael Germán Campos-Montiel
Influence of Bioactive Compounds Incorporated in a Nanoemulsion as Coating on Avocado Fruits (*Persea americana*) during Postharvest Storage: Antioxidant Activity, Physicochemical Changes and Structural Evaluation
Reprinted from: *Antioxidants* **2019**, *8*, 500, doi:10.3390/antiox8100500 . . . . **139**

Min-Soo Kim, Sung-Joon Jeon, So Jung Youn, Hyungjae Lee, Young-Joon Park, Dae-Ok Kim, Byung-Yong Kim, Wooki Kim and Moo-Yeol Baik
Enhancement of Minor Ginsenosides Contents and Antioxidant Capacity of American and Canadian Ginsengs (*Panax quinquefolius*) by Puffing
Reprinted from: *Antioxidants* **2019**, *8*, 527, doi:10.3390/antiox8110527 . . . . **151**

Milad Haydari, Viviana Maresca, Daniela Rigano, Alireza Taleei, Ali Akbar Shahnejat-Bushehri, Javad Hadian, Sergio Sorbo, Marco Guida, Caterina Manna, Marina Piscopo, Rosaria Notariale, Francesca De Ruberto, Lina Fusaro and Adriana Basile
Salicylic Acid and Melatonin Alleviate the Effects of Heat Stress on Essential Oil Composition and Antioxidant Enzyme Activity in *Mentha* × *piperita* and *Mentha arvensis* L.
Reprinted from: *Antioxidants* **2019**, *8*, 547, doi:10.3390/antiox8110547 . . . . **165**

Lucia Panzella, Federica Moccia, Maria Toscanesi, Marco Trifuoggi, Samuele Giovando and Alessandra Napolitano
Exhausted Woods from Tannin Extraction as an Unexplored Waste Biomass: Evaluation of the Antioxidant and Pollutant Adsorption Properties and Activating Effects of Hydrolytic Treatments
Reprinted from: *Antioxidants* **2019**, *8*, 84, doi:10.3390/antiox8040084 . . . . **177**

Roberta Bernini, Isabella Carastro, Francesca Santoni and Mariangela Clemente
Synthesis of Lipophilic Esters of Tyrosol, Homovanillyl Alcohol and Hydroxytyrosol
Reprinted from: *Antioxidants* **2019**, *8*, 174, doi:10.3390/antiox8060174 . . . . **191**

Sinemyiz Atalay, Iwona Jarocka-Karpowicz and Elzbieta Skrzydlewska
Antioxidative and Anti-Inflammatory Properties of Cannabidiol
Reprinted from: *Antioxidants* **2020**, *9*, 21, doi:10.3390/antiox9010021 . . . . **201**

About the Editor

Lucia Panzella is Associate Professor in Organic Chemistry at the Department of Chemical Sciences of Naples University "Federico II". Her research interests focus on the exploitation of antioxidant phenolic polymers for the development of functional materials to be used in cosmetics, food packaging, and biomedicine, with particular relevance to those deriving from agri-food byproducts. She is coauthor of more than 110 publications in international peer-reviewed journals.

Editorial

Natural Phenolic Compounds for Health, Food and Cosmetic Applications

Lucia Panzella

Department of Chemical Sciences, University of Naples "Federico II", Via Cintia 4, I-80126 Naples, Italy; panzella@unina.it; Tel.: +39-081-674-131

Received: 12 May 2020; Accepted: 13 May 2020; Published: 15 May 2020

Based on their potent antioxidant properties, natural phenolic compounds have gained more and more attention for their possible exploitation as food supplements, as well as functional ingredients in food and in the cosmetic industry [1–9]. This Special Issue, concerning innovative applications of natural phenolic compounds from edible or non-edible natural sources in the field of nutrition, biomedicine, food, and the cosmetic sector, contains fifteen contributions, of which fourteen are research articles and one review.

The phytochemical constituents, as well as the antioxidant, cytotoxic, and antimicrobial activities of the ethanolic extract of Mexican brown propolis have been reported by Rivero-Cruz et al. [10]. Twelve known compounds have been isolated and identified by nuclear magnetic resonance spectroscopy (NMR). Additionally, 40 volatile compounds, including nonanal, α-pinene, and neryl alcohol, have been identified by means of headspace-solid phase microextraction with gas chromatography and mass spectrometry time of flight analysis (HS-SPME/GC-MS-TOF). The extract showed anti-proliferative effects on glioma cells and was able to decrease the proliferation and viability of cervical cancer cells.

Lin et al. [11] investigated the effect of isoliquiritigenin (ISL) on the proliferation of triple-negative breast cancer cells. The authors found that treatment with ISL inhibited triple-negative breast cancer cell line (MDA-MB-231) cell growth and increased cytotoxicity. ISL was able to reduce cell cycle progression through the reduction of cyclin D1 protein expression and increased the sub-G1 phase population. The expression of Bcl-2 protein was reduced by ISL treatment, whereas the Bax protein level increased; subsequently, the downstream signaling molecules caspase-3 and poly ADP-ribose polymerase (PARP) were activated. Moreover, ISL reduced the expression of total and phosphorylated mammalian target of rapamycin (mTOR), ULK1, and cathepsin B, whereas the expression of autophagic-associated proteins p62, Beclin1, and LC3 was increased. In vivo studies further confirmed that preventive treatment with ISL could inhibit breast cancer growth and induce apoptotic and autophagic-mediated apoptosis cell death.

Natural phenolic compound rich-extracts have also been recently described as effective agents against environmental oxidative stressors, such as mercury. In particular, Tortora et al. [12] reported the beneficial properties of *Feijoa sellowiana* extracts against mercury toxicity in human red blood cells (RBCs). Both peel and pulp extracts were able to counteract the oxidative stress and thiol decrease induced in RBCs by mercury treatment, although the peel extract had a greater protective effect due in part to the amount and kind of phenolic compounds. Furthermore, *Feijoa sellowiana* extracts also prevented mercury-induced morphological changes, which are known to enhance the pro-coagulant activity of RBCs.

Increasing attention has also been recently devoted to the development of formulations allowing for higher stability and bioavailability of bioactive phenols. In particular, Shimojo et al. [13] have reported optimization of the production process of nanostructured lipid carriers (NLCs) for resveratrol. NLCs were produced by a high shear homogenization and ultrasound method using Compritol® ATO C888 as a solid lipid and Miglyol 812® as a liquid lipid. Based on the factorial design, which was used to optimize the variables of the NLCs production process from a small number of experiments, it was concluded that a shear rate of 19,000 rpm and a shear time of 6 min was the optimal parameters for resveratrol-loaded NLC production.

Along the same line, Ha et al. [14] created composite nanoparticles containing hydrophilic additives using a supercritical antisolvent (SAS) process to increase the solubility and dissolution properties of resveratrol for application in oral and skin delivery. In particular, resveratrol/hydroxylpropylmethyl cellulose (HPMC)/poloxamer 407 (1:4:1) nanoparticles with the highest flux (0.792 $\mu g/min/cm^2$) exhibited rapid absorption and showed significantly higher exposure 4 h after oral administration, compared to micronized resveratrol. Good correlations were observed between in vitro flux and in vivo pharmacokinetic data. The increased solubility and flux of resveratrol generated by the HPMC/surfactant nanoparticles increased the driving force on the gastrointestinal epithelial membrane and rat skin, resulting in enhanced oral and skin delivery of the compound.

Gelatin-based hydrogels have instead been reported for the controlled release of 5,6-dihydroxyindole-2-carboxylic acid (DHICA), a melanin-related metabolite with potent antioxidant activity. In particular, a paper by Alfieri et al. [15] describes the preparation of three types of gelatin-based hydrogels, that is a pristine porcine skin type A gelatin (HGel-A), a pristine gelatin cross-linked by amide coupling of lysines and glutamic/aspartic acids (HGel-B), and a gelatin/chitosan blend (HGel-C). The extent of incorporation into all the gelatins tested using a 10% *w/w* indole to gelatin ratio was very satisfactory, ranging from 60 to 90%, and an appreciable release under conditions of physiological relevance was observed, reaching 30% and 40% at 6 h for HGel-B and HGel-C, respectively. Moreover, DHICA incorporated into HGel-B proved fairly stable over 6 h, whereas the free compound at the same concentration was almost completely oxidized. The antioxidant power of the indole loaded gelatins was also monitored by chemical assays and proved unaltered even after prolonged storage in air.

The potent photoprotective and antioxidant activities of a DHICA-related phenolic polymer have also been reported by Liberti et al. [16]. In particular, the protective effect of a polymer obtained starting from the methyl ester of DHICA (MeDHICA-melanin) against ultraviolet A (UVA)-induced oxidative stress in immortalized human keratinocytes (HaCaTs) was described. At concentrations as low as 10 $\mu g/mL$, MeDHICA-melanin prevented reactive oxygen species accumulation and partially reduced glutathione oxidation in UVA-irradiated keratinocytes. Western blot experiments revealed that the polymer was able to induce the translocation of nuclear factor erythroid 2-related factor 2 (Nrf-2) to the nucleus with the activation of the transcription of antioxidant enzymes, such as heme-oxygenase 1. Spectrophotometric and HPLC analysis of cell lysate allowed the conclusion that a significant fraction (ca. 7%) of the polymer was internalized in the cells.

Instability issues concerning resveratrol were the object of investigation in a paper by Leyva-Porras et al. [17], who studied the effect of the spray drying processing conditions of blueberry juice and maltodextrin (MX) mixtures on the content and retention of resveratrol in the final product. Analysis of variance (ANOVA) showed that the concentration of MX was the main variable influencing resveratrol content. Response surface plots (RSP) confirmed the application limits of maltodextrins based on their molecular weight, where low molecular weight MXs showed better performance as carrying agents.

López de Dicastillo et al. [18] instead reported the encapsulation of açaí fruit antioxidants, especially anthocyanins, into electrosprayed zein, a heat-resistant protein, to improve their bioavailability and thermal resistance. In particular, a hydroalcoholic açaí extract was selected due to its high polyphenolic content and antioxidant capacities. Encapsulation efficiency was approximately 70%. Results demonstrated the effectiveness of the encapsulation on protecting polyphenolic content after high-temperature treatments, such as sterilization and baking. Bioaccessibility studies also indicated an increase of polyphenol levels after in vitro digestion stages of encapsulated açaí fruit extract in contrast with the unprotected extract.

Bioactive compound nanoemulsions have been evaluated as coatings to improve avocado fruit quality during postharvest by Cenobio-Galindo et al. [19]. Nanoemulsions made of orange essential oil and *Opuntia oligacantha* extract were applied as a coating in whole avocado fruits, and the following treatments were assessed: concentrated nanoemulsion (CN), 50% nanoemulsion (N50), 25% nanoemulsion (N25) and control (C). The best results were obtained with the N50 and N25 treatments not only for firmness and weight loss but also for the activity of polyphenol oxidase since a delay in browning was observed in the coated fruits. Furthermore, the nanoemulsion treatments maintained the total phenol and total flavonoid content and improved the antioxidant activity as determined by 2,2'-azino-bis(3-ethylbenzothiazoline-6-sulfonic acid) (ABTS) and 2,2-diphenyl-1-picrylhydrazyl (DPPH) assays at 60 days compared to controls. A delaying effect on the maturation of the epicarp was also observed when the nanoemulsion was applied.

Puffing has instead been proposed to enhance ginsenoside content and antioxidant activities of ginseng (*Panax quinquefolius*). In particular, Kim et al. [20] analyzed American and Canadian ginsengs puffed at different pressures and extracted with 70% ethanol. Puffing formed a porous structure, inducing an efficient elution of internal compounds that resulted in significant increases in extraction yields. The content of minor ginsenosides such as Rg2, Rg3, and compound K increased with increasing puffing pressure, whereas that of the major ginsenosides Rg1, Re, Rf, Rb1, Rc, Rd decreased, likely as a consequence of deglycosylation and pyrolysis processes. ABTS radical scavenging activity, total phenolic content, and total flavonoid content increased with increasing puffing pressure, although no improvement in DPPH reducing activity was observed.

Haydari et al. [21] reported that the natural additives salicylic acid and melatonin were able to alleviate the effects of heat stress on essential oil composition and antioxidant enzyme activity (catalase, superoxide dismutase, glutathione S-transferase, and peroxidase) in *Mentha piperita* and *Mentha arvensis* L.

Natural phenolic compounds are widely present not only in foods but also in non-edible, easily-accessible sources, such as waste materials from agri-food industries [22]. In this context, Panzella et al. [23] focused their attention on exhausted woods, which represent a by-product of the tannin industrial production processes. In particular, the authors reported the characterization of the antioxidant and other properties of practical interest of exhausted chestnut and quebracho wood, together with those of a chestnut wood fiber, produced from steamed exhausted chestnut wood. All the materials investigated exhibited good antioxidant properties in DPPH and ferric reducing/antioxidant power (FRAP) assays, as well as in a superoxide scavenging assay. An increase of the antioxidant potency was observed for both exhausted woods and chestnut wood fiber following activation by hydrolytic treatment. The three materials also proved able to adsorb organic pollutants and to remove toxic heavy metal ions from aqueous solutions.

In recent years, particular attention has also been devoted to modulation of the solubility properties of natural phenolic compounds to broaden their applications, e.g., as dietary supplements or stabilizers of foods and cosmetics in non-aqueous media. In this regard, Bernini et al. [24] reported a simple and low-cost procedure for the synthesis of lipophilic esters of tyrosol, homovanillyl alcohol, and hydroxytyrosol. The reactions were carried out under mild and green chemistry conditions, which also allowed obtaining the desired products in good yields. Notably, the procedure was also applied to hydroxytyrosol-enriched extracts obtained from olive by-products.

Finally, the cellular, antioxidant, and anti-inflammatory properties of cannabidiol and its synthetic derivatives have been reviewed by Atalay et al. [25].

In conclusion, the contributions published in this Special Issue open new perspectives toward the exploitation of phenol-rich natural extracts or pure phenolic compounds as functional ingredients in the health sector, e.g., in preventing and combating mercury-related illnesses or as alternative therapeutic agents for clinical trials against breast cancer. The growing importance of agri-food wastes as sources of phenolic compounds, as well as of synthetic derivatives of natural compounds with improved antioxidant properties have also been highlighted. Finally, novel technologies have been described to improve extraction yields, stability, bioavailability, and delivery of antioxidant compounds, e.g., for healthcare products or for skin applications.

Conflicts of Interest: The author declares no conflict of interest.

References

1. Panzella, L.; Napolitano, A. Natural phenol polymers: Recent advances in food and health applications. *Antioxidants* **2017**, *6*, 30. [CrossRef]
2. Cory, H.; Passarelli, S.; Szeto, J.; Tamez, M.; Mattei, J. The role of polyphenols in human health and food systems: A mini-review. *Front. Nutr.* **2018**, *5*, 87. [CrossRef]
3. Piccolella, S.; Crescente, G.; Candela, L.; Pacifico, S. Nutraceutical polyphenols: New analytical challenges and opportunities. *J. Pharm. Biomed. Anal.* **2019**, *175*, 112774. [CrossRef]
4. Silva, R.F.M.; Pogačnik, L. Polyphenols from food and natural products: Neuroprotection and safety. *Antioxidants* **2020**, *9*, 61. [CrossRef] [PubMed]
5. Zillich, O.V.; Schweiggert-Weisz, U.; Eisner, P.; Kerscher, M. Polyphenols as active ingredients for cosmetic products. *Int. J. Cosmet. Sci.* **2015**, *37*, 455–464. [CrossRef] [PubMed]
6. Działo, M.; Mierziak, J.; Korzun, U.; Preisner, M.; Szopa, J.; Kulma, A. The potential of plant phenolics in prevention and therapy of skin disorders. *Int. J. Mol. Sci.* **2016**, *17*, 160.
7. Panzella, L.; Napolitano, A. Natural and bioinspired phenolic compounds as tyrosinase inhibitors for the treatment of skin hyperpigmentation: Recent advances. *Cosmetics* **2019**, *6*, 57. [CrossRef]
8. Bouarab-Chibane, L.; Degraeve, P.; Ferhout, H.; Bouajila, J.; Oulahal, N. Plant antimicrobial polyphenols as potential natural food preservatives. *J. Sci. Food Agric.* **2019**, *99*, 1457–1474. [CrossRef] [PubMed]
9. Milinčić, D.D.; Levi, S.M.; Kosti, A.Ž. Application of polyphenol-loaded nanoparticles in food industry. *Nanomaterials* **2019**, *9*, 1629.
10. Rivero-Cruz, J.; Granados-Pineda, J.; Pedraza-Chaverri, J.; Pérez-Rojas, J.; Kumar-Passari, A.; Diaz-Ruiz, G.; Rivero-Cruz, B. Phytochemical constituents, antioxidant, cytotoxic, and antimicrobial activities of the ethanolic extract of Mexican brown propolis. *Antioxidants* **2020**, *9*, 70. [CrossRef]
11. Lin, P.; Chiang, Y.; Shieh, T.; Chen, H.; Shih, C.; Wang, T.; Wang, K.; Huang, T.; Hong, Y.; Li, S.; et al. Dietary compound isoliquiritigenin, an antioxidant from licorice, suppresses triple-negative breast tumor growth via apoptotic death program activation in cell and xenograft animal models. *Antioxidants* **2020**, *9*, 228. [CrossRef] [PubMed]
12. Tortora, F.; Notariale, R.; Maresca, V.; Good, K.; Sorbo, S.; Basile, A.; Piscopo, M.; Manna, C. Phenol-rich *Feijoa sellowiana* (pineapple guava) extracts protect human red blood cells from mercury-induced cellular toxicity. *Antioxidants* **2019**, *8*, 220. [CrossRef] [PubMed]
13. Shimojo, A.; Fernandes, A.; Ferreira, N.; Sanchez-Lopez, E.; Santana, M.; Souto, E. Evaluation of the influence of process parameters on the properties of resveratrol-loaded NLC using 2^2 full factorial design. *Antioxidants* **2019**, *8*, 272. [CrossRef] [PubMed]
14. Ha, E.; Sim, W.; Lee, S.; Jeong, J.; Kim, J.; Baek, I.; Choi, D.; Park, H.; Hwang, S.; Kim, M. Preparation and evaluation of resveratrol-loaded composite nanoparticles using a supercritical fluid technology for enhanced oral and skin delivery. *Antioxidants* **2019**, *8*, 554. [CrossRef] [PubMed]
15. Alfieri, M.; Pilotta, G.; Panzella, L.; Cipolla, L.; Napolitano, A. Gelatin-based hydrogels for the controlled release of 5,6-dihydroxyindole-2-carboxylic acid, a melanin-related metabolite with potent antioxidant activity. *Antioxidants* **2020**, *9*, 245. [CrossRef]
16. Liberti, D.; Alfieri, M.; Monti, D.; Panzella, L.; Napolitano, A. A melanin-related phenolic polymer with potent photoprotective and antioxidant activities for dermo-cosmetic applications. *Antioxidants* **2020**, *9*, 270. [CrossRef]
17. Leyva-Porras, C.; Saavedra-Leos, M.; Cervantes-González, E.; Aguirre-Bañuelos, P.; Silva-Cázarez, M.; Álvarez-Salas, C. Spray drying of blueberry juice-maltodextrin mixtures: Evaluation of processing conditions on content of resveratrol. *Antioxidants* **2019**, *8*, 437. [CrossRef]
18. López de Dicastillo, C.; Piña, C.; Garrido, L.; Arancibia, C.; Galotto, M. Enhancing thermal stability and bioaccesibility of açaí fruit polyphenols through electrohydrodynamic encapsulation into zein electrosprayed particles. *Antioxidants* **2019**, *8*, 464. [CrossRef]

19. Cenobio-Galindo, A.; Ocampo-López, J.; Reyes-Munguía, A.; Carrillo-Inungaray, M.; Cawood, M.; Medina-Pérez, G.; Fernández-Luqueño, F.; Campos-Montiel, R. Influence of bioactive compounds incorporated in a nanoemulsion as coating on avocado fruits (*Persea americana*) during postharvest storage: Antioxidant activity, physicochemical changes and structural evaluation. *Antioxidants* **2019**, *8*, 500. [CrossRef]
20. Kim, M.; Jeon, S.; Youn, S.; Lee, H.; Park, Y.; Kim, D.; Kim, B.; Kim, W.; Baik, M. Enhancement of minor ginsenosides contents and antioxidant capacity of american and canadian ginsengs (*Panax quinquefolius*) by puffing. *Antioxidants* **2019**, *8*, 527. [CrossRef]
21. Haydari, M.; Maresca, V.; Rigano, D.; Taleei, A.; Shahnejat-Bushehri, A.; Hadian, J.; Sorbo, S.; Guida, M.; Manna, C.; Piscopo, M.; et al. Salicylic acid and melatonin alleviate the effects of heat stress on essential oil composition and antioxidant enzyme activity in *Mentha* × *piperita* and *Mentha arvensis* L. *Antioxidants* **2019**, *8*, 547. [CrossRef] [PubMed]
22. Panzella, L.; Moccia, F.; Nasti, R.; Marzorati, S.; Verotta, L.; Napolitano, A. Bioactive phenolic compounds from agri-food wastes: An update on green and sustainable extraction methodologies. *Front. Nutr.* **2020**, *7*, 60. [CrossRef]
23. Panzella, L.; Moccia, F.; Toscanesi, M.; Trifuoggi, M.; Giovando, S.; Napolitano, A. Exhausted woods from tannin extraction as an unexplored waste biomass: Evaluation of the antioxidant and pollutant adsorption properties and activating effects of hydrolytic treatments. *Antioxidants* **2019**, *8*, 84. [CrossRef] [PubMed]
24. Bernini, R.; Carastro, I.; Santoni, F.; Clemente, M. Synthesis of lipophilic esters of tyrosol, homovanillyl alcohol and hydroxytyrosol. *Antioxidants* **2019**, *8*, 174. [CrossRef]
25. Atalay, S.; Jarocka-Karpowicz, I.; Skrzydlewska, E. Antioxidative and anti-inflammatory properties of cannabidiol. *Antioxidants* **2020**, *9*, 21. [CrossRef] [PubMed]

© 2020 by the author. Licensee MDPI, Basel, Switzerland. This article is an open access article distributed under the terms and conditions of the Creative Commons Attribution (CC BY) license (http://creativecommons.org/licenses/by/4.0/).

Article

Phytochemical Constituents, Antioxidant, Cytotoxic, and Antimicrobial Activities of the Ethanolic Extract of Mexican Brown Propolis

J. Fausto Rivero-Cruz [1],*, Jessica Granados-Pineda [1], José Pedraza-Chaverri [1], Jazmin Marlen Pérez-Rojas [2], Ajit Kumar-Passari [3], Gloria Diaz-Ruiz [1] and Blanca Estela Rivero-Cruz [1]

[1] Facultad de Química, Universidad Nacional Autónoma de México, Ciudad Universitaria, Ciudad de México 04510, Mexico; jessygpin@hotmail.com (J.G.-P.); pedraza@unam.mx (J.P.-C.); gloriadr@unam.mx (G.D.-R.); blancariv@unam.mx (B.E.R.-C.)
[2] División de Investigación Básica, Instituto Nacional de Cancerología, Av. San Fernando 22, Apartado Postal 22026, Tlalpan 14000, Mexico; jazminmarlen@gmail.com
[3] Instituto de Investigaciones Biomédicas, Universidad Nacional Autónoma de México, Ciudad Universitaria, Ciudad de México 04510, Mexico; ajit.passari22@iibiomedicas.unam.mx
* Correspondence: joserc@unam.mx; Tel.: +52-55-5622-5281

Received: 30 November 2019; Accepted: 4 January 2020; Published: 13 January 2020

Abstract: Propolis is a complex mixture of natural sticky and resinous components produced by honeybees from living plant exudates. Globally, research has been dedicated to studying the biological properties and chemical composition of propolis from various geographical and climatic regions. However, the chemical data and biological properties of Mexican brown propolis are scant. The antioxidant activity of the ethanolic extract of propolis (EEP) sample collected in México and the isolated compounds is described. Cytotoxic activity was evaluated in a central nervous system and cervical cancer cell lines. Cytotoxicity of EEP was evaluated in a C6 cell line and cervical cancer (HeLa, SiHa, and CasKi) measured by the 3-(3,5-dimethylthiazol-2-yl)2,5-diphenyltetrazolium (MTT) assay. The antibacterial activity was tested using the minimum inhibitory concentration (MIC) assay. Twelve known compounds were isolated and identified by nuclear magnetic resonance spectroscopy (NMR). Additionally, forty volatile compounds were identified by means of headspace-solid phase microextraction with gas chromatography and mass spectrometry time of flight analysis (HS-SPME/GC-MS-TOF). The main volatile compounds detected include nonanal (18.82%), α-pinene (12.45%), neryl alcohol (10.13%), and α-pinene (8.04%). EEP showed an anti-proliferative effect on glioma cells better than temozolomide, also decreased proliferation and viability in cervical cancer cells, but its effectiveness was lower compared to cisplatin.

Keywords: propolis; antioxidant activity; cytotoxic; antibacterial; México; HS-SPME/GC-MS-TOF; NMR; volatile compounds; flavonoids; phenolic acids

---

1. Introduction

Propolis also known as "bee glue" is a nontoxic hive product accumulated by bees from diverse plants containing compounds such as flavonoid aglycones, phenolic acids and their esters, phenolic aldehydes, alcohols, ketones, sesquiterpenes, coumarins, steroids, amino acids, and inorganic compounds. It functions in sealing holes, cracks, reconstruction, and smothering the inner surfaces of the beehive. Propolis and its extracts have application in treating diseases due its anti-inflammatory, antioxidant, antibacterial, antimycotic, antifungal, antiulcer, anticancer, and immunomodulatory properties [1,2]. Egyptians, Greeks, Romans, Chinese, Arabs, and Incas have traditionally used it as an antiseptic to treat wounds and as an anti-pyretic agent [1–3]. The term propolis was described in the 16th Century in France and was considered as an official drug by London Pharmacopoeia [4].

Propolis contains mainly hydrophobic terpenes and phenolics compounds and can be classified in five types. The chemogeographic patterns of propolis types reflects the geographical distribution of plants species [2]. According to this distribution, the propolis produced in North America belongs to the poplar propolis and contains mainly flavonoids without B-ring substituents, such as pinocembrin, pinobanksin, galangin, and chrysin [2,5].

In recent years, it has gained wide acceptance as traditional medicine in various parts of the world where it is claimed to improve human health and to prevent diseases such as diabetes and cancer [6,7]. Nowadays, in several countries, it is possible to find propolis as liquid extracts in bottles, vaporizers, syrups capsules, tablets, candies, creams, among others [8,9].

Data about the chemical composition and biological activity of propolis from México is limited [10–12]. Propolis in Atliplano region is prepared in several forms, including syrups, tinctures, and creams as an alternative to improve health and prevent diseases. In view of the importance of propolis in Mexican traditional medicine the chemical composition and biological activities (antioxidant, antibacterial, and cytotoxic) of the ethanolic extract of Altiplano propolis (EEP) were investigated.

The aims of this study were: (a) To isolate the major compounds of brown Mexican propolis, and (b) to evaluate the antibacterial, antioxidant, and cytotoxic activities of the EEP.

2. Materials and Methods

2.1. Chemicals and Reagents

All the chemicals to investigate the antioxidant, antibacterial, and cytotoxic activities were supplied by Sigma-Aldrich (St. Louis, MO, USA). All the solvents and chromatographic supports were purchased from Merck (Darmstadt, Germany). Deuterated solvents and TMS were provided by Cambridge Isotope Laboratories (Tewksbury, MA, USA).

2.2. Instrumentation

Nuclear magnetic resonance (NMR) spectra were taken on a Bruker AVANCE III 400 (400 MHz) spectrometer, using tetramethylsilane (TMS) as an internal standard.

2.3. Propolis Samples

Beekeepers kindly provided four raw propolis samples from Silao and Irapuato, Guanajuato, México (Table 1). The samples were harvested using traps in 2018 and 2019 and were frozen and stored at −20 °C until analysis.

Table 1. Location of recollection, harvesting method, and weight recollected.

Code	Location of Recollection	Year of Recollection	Harvesting Method	Weight (g)
GUA-1	Silao	2018	plastic nest	6.0
GUA-2	Irapuato	2018	plastic nest	8.50
GUA-3	Irapuato	2019	plastic nest	6.95
GUA-4	Silao	2019	plastic nest	32.0

2.4. Antioxidant Activity

The EEP antioxidant activity was assessed using two different assays in vitro: DPPH and ABTS. Both methods were modified and translated into 96-well plates. Each test was done in three replicates.

Radical scavenging activity (RSA) for DPPH was evaluated according to the method described in [11]. Briefly, an ethanolic solution of 0.208 mM was added to 0.1 mL of different concentrations of extracts and pure compounds. The 96-well plate was maintained in a dark at room temperature for 20 min and the absorbance was recorded at 540 nm. The RSA was calculated as: $RSA = 100 \times (A_{control} - A_{sample})/A_{control}$, where $A_{control}$ and A_{sample} are the absorbance. The IC_{50} values were calculated from the relationship curve of RSA versus concentrations of the respective sample curve.

The ABTS test was performed according to the methodology previously reported in [13,14] and slightly modified. The RSA of the ABTS radical was calculated using the following equation: % inhibition = 100 ($A_{control} - A_{sample}$)/$A_{control}$. The IC_{50} was calculated from the scavenging activities (%) versus concentrations of the respective sample curve.

2.5. Total Phenol and Total Flavonoid Content

In this paper, we used spectrometric procedures for the quantification of the total phenolic and flavonoids content in propolis. The total phenol content in extracts was adapted from the method described by Singleton and Rossi [15]. The total flavonoid content was determined using the aluminum chloride reagent and the method described by Marquele et al. [16]. The total phenol content was expressed as mg equivalents of gallic acid/g of dry extract of propolis (EEP). The total flavonoid content was expressed as mg equivalents of quercetin/g of dry extract of propolis (EEP).

2.6. Extraction and Isolation of Compounds 1–12 from EEP GUA-4

The air-dried and powdered propolis samples GUA-4 were extracted with ethanol for up to one week and the resultant extract was concentrated in vacuo. A portion of ethanol-soluble extract (10.3 g) was subjected to a silica gel vacuum column chromatography (VLC) and eluted with a gradient mixture of dichloromethane–acetone (1:0 → 0:1) to give eight pooled fractions (F2–F8). Fractions F2, F4, and F7 showed the best antioxidant activity. Fraction F2 was chromatographed over a Sephadex LH-20 column and eluted with methanol to yield **1** (40 mg). Fraction F4, eluted with dichloromethane-acetone (9:1), was chromatographed over a Sephadex LH-20 column, using methanol as eluent, to give six fractions. Fraction F4-3 (100 mg) was separated by TLC with dichloromethane–acetone (99:1), followed by TLC with dichloromethane–acetone (98:2), to give **2** (20 mg) and **3** (65 mg). Fraction F5 was rechromatographed on a Sephadex LH-20 column using methanol as solvent to give six subfractions (F5-1: 10 mg; F5-2: 12.3 mg; F5-3: 17.5 mg; F5-4: 15.7 mg; F5-5: 25.3 mg; F5-6: 18.1 mg). Subfraction F5-1 yielded crystals of 4. F5-5 (25.3 mg) yielded crystals of 5 (9.0 mg). Mother liquor was subjected to a preparative thin-layer chromatography (PTLC) with acetone-toluene (5:95) to give 6. Subfraction F5-6 (18.1 mg) yielded crystals of 7 (11.0 mg). Fraction F7 was chromatographed on silica gel using a hexane–EtOAc gradient system to give five fractions (F7-1: 13.9 mg; F7-2: 517.7 mg; F7-3: 28.4 mg; F7-4: 16.7 mg; F7-5: 32.8 mg; F7-6: 56.8 mg; F7-7: 45.1 mg). Subfraction F7-1 was identified as 8. Subfraction F7-2 was chromatographed by RP-MPLC, using H_2O–CH_3CN (1:0 → 0:1) to afford five subfractions (F7-2.1: 125.3 mg; F7-2.2: 275.1 mg; F7-2.3: 67.8 mg; F7-2.4: 37.0 mg; F7-2.5: 12.3 mg). Subfraction F7-2.3 was identified as 9. Subfractions F7-2.4, F7-2.5, and F7-2.6 were dissolved in CH_2Cl_2–MeOH and left overnight to give crystals of 10, 11, and 12, respectively.

2.7. Head Space Solid Phase Microextraction (HS-SPME), GC-MS Analysis, and Identification of Volatile Components

The HS-SPME was performed using a Stableflex® fiber 50/30 μm DVB/CAR/PDMS (1 cm) as described in our previous work [17].

The analysis of volatile compounds was carried out on a GC Agilent 6890 N (Agilent Technology, Santa Clara, CA, USA) series gas chromatograph coupled to a LECO time of flight mass spectrometer (LECO Corporation, St. Joseph, MI, USA). The volatile compounds were separated on a 5% diphenyl-95% dimethyl polysiloxane (30 m × 0.18 mm i.d.; 0.18 μm film thickness) capillary column (Bellefonte, PA, USA). The carrier gas was helium with a flow rate of 1 mL/min and split ratio of 1:50. The column initial temperature was 40 °C. It was then raised to 300 °C with a rate of 20 °C/min and was held for 5 min. The ionization electron energy was 70 eV and the mass range scanned was 40–400 *m/z*. The injector and MS transfer were set at 300 and 250 °C, respectively. The volatile constituents of propolis were identified by co-injection of the sample with standard samples when available; based

2.8. Cell Culture

Rat C6 glioma cell and human cervical cancer cell lines (HeLa, SiHa, and CaSki) were obtained from American Type Culture Collection (Manassas, VA, USA). Cells were routinely maintained in Dulbecco's modified Eagle's medium (DMEM) supplemented with fetal bovine serum (Gibco BRL) with 5% for C6 cells and 10% for cervical cancer cells, and incubated at 37 °C in an atmosphere comprising 5% CO_2 and 95% air at high humidity. Cells were harvested with 0.025% trypsin and 0.01% EDTA (Gibco BRL).

The effect of EEP, cisplatin, and temozolomide on the proliferation of cells were evaluated using the MTT assay (3-(3,5-dimethylthiazol-2-yl)2,5-diphenyltetrazolium), which is based on the reduction of a tetrazolium salt in metabolically active cells. The procedure was as follows. Viable cells were seed into 96-well plates in 100 μL per well of DMEM culture medium at a density of 3×10^3 for C6 cells, 2×10^4 for CaSki cells, and 1×10^4 cells for HeLa and SiHa cells. After treatment, the medium was removed and the MTT solution was added to each well, followed by 1 to 2 h in a humidified atmosphere containing 5% CO_2 at 37 °C. The absorbance of the samples was measured spectrophotometrically at λ 570 nm using a microtiter plate ELISA reader. Results are expressed as the percentage of MMT reduction.

C6 cells were exposed for 72 h with 2 to 200 μg/mL of EEP, since it is the time used to make the exposure with temozolomide (first line treatment used for glioblastoma), whereas HeLa, SiHa, and CaSki cells were exposed for 24 h with 15 to 500 μg/mL of EEP; after that time, cell viability was quantified using the MTT assay. Temozolomide (250 μM) and cisplatin (5–320 μM) were used as a control. The concentration of drugs to reach 50% growth inhibition (IC_{50}) was obtained from the survival curves. The experiments were conducted in triplicate in independent experiments. Values are expressed as the mean $\pm$ SEM of at least three independent experiments. SigmaPlot 12.3 software (Systat Software, Santa Clara, CA, USA) was used.

2.9. Antibacterial Activity

The in vitro antibacterial activity of EEP and compounds **1–4**, **10–13**, and **15–17** were determined using a broth microdilution test as recommended by Clinical and Laboratory Standards Institute (New York, NY, USA) M7-A11 for bacteria [20]. The minimum inhibitory concentration (MIC) was defined as the lowest concentration of the test agent that had restricted growth to a level <0.05 at 660 nm after incubation at 37 °C for 16–24 h.

3. Results and Discussion

3.1. Total Phenol and Flavonoid Content

The total polyphenol contents for the selected propolis samples were found to be 178.9 ± 5.7, 198.4 ± 4.1, 167.6 ± 7.8, and 246.3 ± 3.2 mg GAE/g of dry extract for GUA-1, GUA-2, GUA-3, and GUA-4, respectively. It should be noted that the Folin-Ciolcateau reagent was reported in the literature as not specific to only phenols and could react with other reducing compounds that could be oxidized by the Folin reagent [21]. The total flavonoid contents were 64.32 ± 5.75, 58.34 ± 2.8, 77.45 ± 6.9, and 87.5 ± 1.9 mg QE/g of dry extract for GUA-1, GUA-2, GUA-3, and GUA-4, respectively. Flavonoids can form complexes with aluminum chloride to yield a yellow solution. Valencia et al. [22] has been reported values of 629.6 ± 9.9 mg GAE/g of dry extract for TPC and 185.9 ± 3.2 mg QE/g of dry extract and TF in a propolis from Sonora, México. The most important flavonoids isolated in this propolis sample were pinocembrin, pinobanksin 3-acetate, and chrysin. The results also confirm the influence of the geographic region and the season of collection for the quality and properties of the propolis [2].

3.2. Antioxidant Activity Assays

The EEP was evaluated for its ability to quench the DPPH·, which is one of the few stable organic nitrogen radicals and bears a purple colour. This assay is based on the measurement of the loss of $DPPH^{\bullet}$ after reaction with samples. It is considered as the prior mechanism involved in the electron transfer. The IC_{50} value is a parameter widely used to measure the antioxidant activity of test samples. It is calculated as the concentration of antioxidants needed to decrease the initial DPPH concentration by 50% [23]. Thus, the lower IC_{50} value the higher antioxidant activity. The EEP GUA-4 showed an antioxidant activity (IC_{50} = 67.9 µg/mL) comparable to the reference ascorbic acid (IC_{50} = 43.2 µg/mL) and tenfold lower than Trolox (IC_{50} = 6.3 µg/mL) and seven-fold lower than quercetin (IC_{50} = 9.9 µg/mL). While caffeic acid (12) showed the highest activity (IC_{50} = 5.9 µg/mL) along with ferulic acid (10) (IC_{50} = 9.9 µg/mL) and syringic acid (11) (IC_{50} = 9.8 µg/mL). The lowest antioxidant activity corresponds to 5-methylchrysin ether (9) (IC_{50} = 112.9 µg/mL) and 5-methyl-pinobanksin ether (5) (IC_{50} = 98.4 µg/mL).

There are reports of the DPPH scavenging capacity, in terms of IC_{50}, for pinocembrin (1), chrysin (2), galangin (3), isorhamnetin (7), ferulic acid (10), syringic acid, (11) and caffeic acid (12) [23–26]. For alpinetin (4) only the percentage of scavenging activity of DPPH is reported [27]. Nevertheless, the information available for dillenetin (6), 5-methyl-pinobanksin ether (5), 5-methylchrysin ether (9), and 5-methylgalangin ether (8) is scarce. Thus, the present work stands for the first report of the IC_{50} as a measurement of their antioxidant activity (Table 2).

In the ABTS assay, the antioxidant activity is measured as the ability of test compounds to decrease the color reacting directly with the radical $ABTS^{\bullet+}$ [28]. Ferulic acid (10), syringic acid (11), caffeic acid (12), and chrysin (2) showed the lowest IC_{50} (Table 2). The EEP, 5-methylchrysin ether (9), and 5-methyl-pinobanksin ether (5) showed weak antioxidant activity (Table 2).

The analysis of their chemical structures can explain the weak antioxidant activity of the methylated flavonoids. According to Procházková et al. [29] the catechol structure in the B-ring, 2,3-double bond in conjugation with a 4-oxo function in the C-ring and the hydroxyl groups in meta position in ring-A. On the other hand, the activity of phenolic acids lies on the presence of two o-hydroxyl groups in the aromatic ring.

Table 2. Scavenging ability of ethanol extract of propolis (EEP) GUA-4 and compounds 1–12.

Compounds	IC_{50} (µg/mL)	
	DPPH	ABTS
EEP GUA-4	67.9 ± 0.1	98.7 ± 0.5
pinocembrin (1)	23.5 ± 1.8	44.8 ± 2.1
chrysin (2)	10.7 ± 0.2	16.3 ± 0.8
galangin (3)	15.3 ± 0.5	26.8 ± 0.1
alpinetin (4)	47.3 ± 1.9	69.5 ± 0.7
5-methylpinobanksin ether (5)	98.4 ± 2.3	126.9 ± 4.5
dillenetin (6)	35.7 ± 3.5	48.9 ± 2.9
isorhamnetin (7)	21.4 ± 3.1	36.7 ± 4.6
5-methylgalangin ether (8)	63.2 ± 2.6	94.2 ± 6.2
5-methylchrysin ether (9)	112.9 ± 1.9	158.4 ± 4.9
ferulic acid (10)	9.9 ± 0.7	16.7 ± 0.2
syringic acid (11)	9.8 ± 0.4	13.0 ± 0.9
caffeic acid (12)	5.9 ± 0.4	9.7 ± 0.5
Quercetin	9.9 ± 2.5	16.1 ± 2.1
Trolox	6.3 ± 1.4	3.8 ± 1.2
ascorbic acid	43.2 ± 10.3	36.8 ± 2.5

3.3. Cytotoxicity of EEP on Cancer Cells

Cytotoxicity was expressed as the percentage grow inhibition of C6, HeLa, SiHa, and CaSki cells treated with EEP. In all the cases, EEP shows a cytotoxicity concentration-depended manner. In Table 3, we show the IC_{50} value of EEP over four cancer cell lines. Those results show that EEP restricts glioblastoma cells (C6 cell cancer line) proliferation in vitro as efficiently as temozolomide (reference drug), whereas, for cervical cancer cell lines, it requires a higher concentration of the EEP compared to cisplatin.

There are a few studies of beneficial properties of Mexican propolis. Li et al. [11] reported that three of the 39 compounds isolated from the methanolic extract of Mexican propolis exhibited a potent cytotoxic effect in a colon, melanoma, lung, cervix, and fibrosarcoma cancer cell lines. Li et al. reported the isolation of flavonoids from methanolic extract of Mexican propolis, and one of them revealed significant cytotoxic effect against pancreatic human cancer cell line with IC_{50} values of 4 μM [10]. Other studies described the potent cytotoxic activity of galangin (3); ferulic acid (10); syringic acid (11); and caffeic acid (12) against different cancer cell lines [22]. Interestingly, a few researchers reported that these compounds could be useful for therapeutic treatments. For example, Benguedouar et al. [30] reported that ethanolic extract of Algerian propolis (EEP) and galangin (3) decreased the number of B16F1 melanoma cells in vitro compared to control. Celinska-Janowicz et al. [31] state that the ethanolic extract of propolis isolated abundant polyphenolic compounds such as ferulic acid (10) and caffeic acid (12) revealed pro-apoptotic activity on human tongue squamous carcinoma cells (CAL-27). From this information, we emphasized that EEP and compounds possesses anti-cancer effects against cancer cell lines.

Table 3. Cytotoxic effect of EEP on several cancer cell lines.

Cell Line	IC_{50}	Reference Drug	Reference Concentration
C6	92.2 μg/mL	Temozolamide	IC_{30} 250 μM (50 μg/mL)
HeLa	>100 μg/mL (357 μg/mL)	Cisplatin	IC_{50} 46 μM (14 μg/mL)
SiHa	>100 μg/mL (500 μg/mL)		IC_{50} 121 μM (36 μg/mL)
CaSki	>100 μg/mL (538 μg/mL)		IC_{50} 163 μM (50 μg/mL)

3.4. Antibacterial Activity

As shown in Table 4, antimicrobial screening against four oral pathogens revealed that compounds **1**, **3**, **4**, **12**, and **16** were inhibitory to the growth of *Streptococcus mutans*, *Streptococcus oralis*, *Streptococcus sanguinis*, and *Phorphyromonas gingivalis*. Among these, compounds **1**, **3**, **4**, and **12**, were either equally or more potent than their respective crude extract of origin (Table 4). Earlier in vitro studies have shown that the Sonoran ethanolic extract of propolis exhibited antibacterial activity against *E. coli* (ATCC 25922) and *S. aureus* (ATCC 6538P). The propolis constituents CAPE, pinocembrin, pinobanksin 3-O-acetate, and naringenin exhibited significant inhibitory activity on the growth of *S. aureus*. CAPE exhibited the maximum inhibitory effect on the bacterial growth (CAPE (MIC 0.1 mmol/L), pinocembrin (MIC 0.4 mmol/L), pinobanksin 3-O-acetate (MIC 0.8 mmol/mL), and naringenin (0.8 mmol/L)). None of the propolis constituents influenced the growth of *E. coli* at any of the tested concentrations [32].

Table 4. Antimicrobial activity of ethanolic extract and compounds 1–4, 10–13, 15–17, and chlorhexidine digluconate (CHX) [a].

Compounds	MIC [b] (µg/mL)			
	S. mutans	*S. oralis*	*S. sanguinis*	*P. gingivalis*
EEP	250	125	125	500
EOP	500	500	500	1000
pinocembrin (**1**)	256	128	128	512
chrysin (**2**)	512	512	512	1024
galangin (**3**)	256	256	256	1024
alpinetin (**4**)	128	256	128	516
ferulic acid (**10**)	500	250	250	500
syringic acid (**11**)	250	250	250	500
caffeic acid (**12**)	128	128	128	256
nonanal (**13**)	256	256	128	512
neryl alcohol (**15**)	1024	512	512	1024
α-pinene (**16**)	250	250	250	500
α-pinene (**17**)	500	500	500	500
[a] CHX	0.02	0.02	0.02	0.12

[a] positive control; [b] minimum inhibitory concentration.

3.5. Chemical Composition of EEP GUA-4

The EtOH-soluble extract of sample GUA-4 was fractionated by chromatography on a VLC column and by Sephadex LH-20, giving 12 known compounds (**1–12**). The compounds were identified as pinocembrin (**1**, 405.4 mg) [33], chrysin (**2**, 126.3 mg) [34], galangin (**3**, 56.7 mg) [33], alpinetin (**4**, 15.6 mg) [35], 5-methyl-pinobanksin ether (**5**, 16.8 mg) [36], dillenetin (**6**, 12.3 mg) [37], isorhamnetin (**7**, 24.6 mg) [38], 5-methylgalangin ether (**8**, 9.8 mg) [39], 5-methylchrysin ether (**9**, 16.3 mg) [40], ferulic acid (**10**, 24.9 mg) [33], syringic acid (**11**, 6.7 mg), and caffeic acid (**12**, 8.9 mg) [33] (Figure S1) by means of 1D (Table S1) and 2D NMR spectral analysis. The presence of pinocembrin (**1**), chrysin (**2**), galangin (**3**), ferulic acid (**10**), syringic acid (**11**), and caffeic acid (**12**) suggest that its main botanical source are a species of *Populus* typical of the country, such as *Populus mexicana* Wesmael, *Populus guzmanantlensis* Vazques and Cuevas and *Populus simaroa* Rzedowski. Although flavonoids without B-ring substituents appear common in temperate propolis from both the northern and southern hemispheres, in this study we report for the first time the presence of dillenetin (**6**) and isorhamnetin (**7**) with B-ring substitution as constituents of Mexican poplar propolis. To the best of our knowledge, this is the first report of dillenetin (**6**) as a propolis constituent. 5-methylpinobanksin (**5**), alpinetin (**4**), isorhamnetin (**7**), 5-methylgalangin ether (**8**), and 5-methylchrysin ether (**9**) were previously identified by diode array detection and ESI mass spectrometry (LC-DAD-ESI-MS) as constituents of Italian, Portuguese, and Czech propolis [41,42].

3.6. Volatile Compounds

Forty volatile constituents were identified as shown in Table 5. The typical analytical ion current (AIC) chromatogram is shown in Figure S2. The chemical composition of volatiles was found in agreement with previous reports [43]. The main volatile compounds identified were nonanal (18.829%, **13**), β-pinene (12.179%, **14**), 1-octen-3-ol (12.129%, **15**), neryl alcohol (10.135%, **16**), α-pinene (8.046%, **17**), 6-methyl-3,5-heptadiene-2-one (6.803%, **18**), *p*-cymen-9-ol (3.108%, **19**), and sylvestrene (3.022 %, **20**) (Figure S3). It is important to note that the content of the compounds differed from a previous study carried out on the volatile compounds in propolis samples from Yucatan (México). In that sample, hexadecanoic acid (10.9%) and trans-verbenol (7.0%) were the predominant compounds [44].

Antioxidants **2020**, *9*, 70

Table 5. Volatile components from Mexican propolis GUA-4.

	Name	Retention Index	Area %	Method of Identification
1	α-pinene	939	8.046	a, b, c
2	β-pinene	979	12.179	a, b, c
3	1-octen-3-ol	982	12.129	a, b
4	ethyl hexanoate	1003	0.042	a, b, c
5	Octanal	1006	1.273	a, b
6	Sylvestrene	1030	3.022	a, b
7	acetophenone	1076	0.486	a, b
8	1-octanol	1078	0.107	a, b
9	Thujone	1102	0.276	a, b
10	*p*-cymenene	1082	0.291	a, b
11	Nonanal	1100	18.829	a, b, c
12	6-methyl-3,5-heptadiene-2-one	1105	6.803	a, b
13	Eucalyptol	1039	0.472	a, b, c
14	Camphor	1146	0.583	a, b, c
15	*trans*-pinocamphone	1162	2.386	a, b
16	*trans*-terpineol	1163	0.102	a, b
17	*p*-menth-1,5-dien-8-ol	1167	1.097	a, b
18	(2*E*)-nonenal	1168	1.097	a, b
19	*m*-cymen-8-ol	1180	1.024	a, b, c
20	Unknown 1	1185	0.064	-
21	α-terpineol	1189	1.571	a, b
22	Myrtenol	1193	1.571	a, b
23	ᴅ-fenchone	1187	1.828	a, b
24	*p*-cymen-9-ol	1207	3.108	a, b, c
25	neryl alcohol	1217	10.135	a, b
26	trans-carveol	1219	2.953	a, b
27	Citronellol	1228	1.916	a, b
28	Ocimenone	1230	0.727	a, b
29	*cis*-chrysanthenyl acetate	1253	0.243	a, b
30	Geraniol	1278	0.314	a, b
31	neodehydro carveol acetate	1307	1.843	a, b
32	ethyl nonanoate	1319	1.283	a, b
33	trans-carvyl acetate	1337	0.357	a, b
34	α-longipinene	1352	0.181	a, b
35	β-cububene	1388	0.248	a, b
36	β-bourbonene	1398	0.217	a, b
37	n-decyl acetate	1408	0.109	a, b
38	trans-geranylacetone	1455	0.429	a, b
39	γ-gelinene	1485	0.103	a, b
40	β-bisabolene	1509	0.290	a, b

a: Retention time; b: Retention index; c: Mass spectrum.

4. Conclusions

In the present study, we have isolated twelve (**1–12**) components and identified 40 volatile compounds in Mexican propolis. The current study revealed the presence of antioxidant, antimicrobial, and cytotoxic phytochemicals [galangin (**3**); ferulic acid (**10**); syringic acid (**11**); and caffeic acid (**12**)] in EEP. It is concluded that EEP can be a potential addition in pharmaceutical products for the improvement of human health by contributing in the antioxidant defense system fighting against the production of free radicals. México, being a megadiverse country, has numerous numbers of propolis differing in chemical composition. However, unfortunately it is still unexplored.

Supplementary Materials: The following are available online at http://www.mdpi.com/2076-3921/9/1/70/s1, Figure S1: Flavonoids isolated from EEP GUA-4; Figure S2: Analytical ion current chromatogram (AIC) of propolis

GUA-4; Figure S3: Major volatile compounds of EEP GUA-4; Table S1: ^{1}H-RMN data of the flavonoids isolated from Mexican propolis.

Author Contributions: J.G.-P. performed the phytochemical study; J.F.R.-C. conceived and designed the phytochemical experiments and wrote the paper; J.M.P.-R. conceived and designed the in vitro experiments; A.K.-P.reviewed of the manuscript and contributed with in vitro experiments; G.D.-R. contributed with the antimicrobial assays; J.P.-C. reviewed the manuscript and contributed with in vitro experiments; B.E.R.-C. reviewed of the manuscript and contributed with in vitro experiments. All authors have read and agreed to the published version of the manuscript.

Funding: This research was funded by CONACyT CB-252006 and College of Chemistry, National Autonomous University of Mexico under the PAIP 5000-9138 Fellowship Scheme.

Acknowledgments: We are in debt to Médico Veterinario Zootecnista. Ángel López-Ramírez for technical assistance in propolis recolection. Jessica Granados Pineda is a doctoral student from the Programa de Doctorado en Ciencias Químicas, Universidad Nacional Autónoma de México (UNAM), Facultad de Química, and recieved a CONACyT fellowship # 273461. The authors wish to thank the technical assistance of Georgina Duarte-Lisci and Juan Rojas-Moreno.

Conflicts of Interest: The authors declare no conflict of interest.

References

1. Bankova, V.; Bertelli, D.; Borba, R.; Conti, B.J.; da Silva Cunha, I.B.; Danert, C.; Eberlin, M.N.; I Falcão, S.; Isla, M.I.; Moreno, M.I.N.; et al. Standard methods for *Apis mellifera* propolis research. *J. Apic. Res.* **2016**, *58*, 1–49. [CrossRef]
2. Salatino, A.; Fernandes-Silva, C.C.; Righi, A.A.; Salatino, M.L. Propolis research and the chemistry of plant products. *Nat. Prod. Res.* **2011**, *28*, 925–936. [CrossRef] [PubMed]
3. Kuropatnicki, A.K.; Szliszka, E.; Krol, W. Historical aspects of propolis research in modern times. *Evid. Based Cmnplement. Altern. Med. eCAM* **2013**, *2013*, 964149. [CrossRef] [PubMed]
4. Toreti, V.C.; Sato, H.H.; Pastore, G.M.; Park, Y.K. Recent progress of propolis for its biological and chemical compositions and its botanical origin. *Evid. Based Cmnplement. Altern. Med. eCAM* **2013**, *2013*, 697390. [CrossRef]
5. Christov, R.; Trusheva, B.; Popova, M.; Bankova, V.; Bertrand, M. Chemical composition of propolis from canada, its antiradical activity and plant origin. *Nat. Prod. Res.* **2006**, *20*, 531–536. [CrossRef]
6. Karapetsas, A.; Voulgaridou, G.-P.; Konialis, M.; Tsochantaridis, I.; Kynigopoulos, S.; Lambropoulou, M.; Stavropoulou, M.-I.; Stathopoulou, K.; Aligiannis, N.; Bozidis, P.; et al. Propolis extracts inhibit UV-induced photodamage in human experimental in vitro skin models. *Antioxidants* **2019**, *8*, 125. [CrossRef]
7. Nna, V.U.; Abu Bakar, A.B.; Ahmad, A.; Eleazu, C.O.; Mohamed, M. Oxidative stress, NF-κb-mediated inflammation and apoptosis in the testes of streptozotocin–induced diabetic rats: Combined protective effects of malaysian propolis and metformin. *Antioxidants* **2019**, *8*, 465. [CrossRef]
8. Papotti, G.; Bertelli, D.; Bortolotti, L.; Plessi, M. Chemical and functional characterization of italian propolis obtained by different harvesting methods. *J. Agric. Food. Chem.* **2012**, *60*, 2852–2862. [CrossRef]
9. Navarro-Navarro, M.; Ruiz-Bustos, P.; Valencia, D.; Robles-Zepeda, R.; Ruiz-Bustos, E.; Virues, C.; Hernandez, J.; Dominguez, Z.; Velazquez, C. Antibacterial activity of sonoran propolis and some of its constituents against clinically significant vibrio species. *Foodborne Pathog. Dis.* **2013**, *10*, 150–158. [CrossRef]
10. Li, F.; Awale, S.; Tezuka, Y.; Esumi, H.; Kadota, S. Study on the constituents of mexican propolis and their cytotoxic activity against panc-1 human pancreatic cancer cells. *J. Nat. Prod.* **2010**, *73*, 623–627. [CrossRef]
11. Li, F.; Awale, S.; Tezuka, Y.; Kadota, S. Cytotoxicity of constituents from mexican propolis against a panel of six different cancer cell lines. *Nat. Prod. Commun.* **2010**, *5*, 1601–1606. [CrossRef] [PubMed]
12. Lotti, C.; Campo Fernandez, M.; Piccinelli, A.L.; Cuesta-Rubio, O.; Marquez Hernandez, I.; Rastrelli, L. Chemical constituents of red mexican propolis. *J. Agric. Food Chem.* **2010**, *58*, 2209–2213. [CrossRef] [PubMed]
13. Cheng, Z.; Moore, J.; Yu, L. High-throughput relative DPPH radical scavenging capacity assay. *J. Agric. Food Chem.* **2006**, *54*, 7429–7436. [CrossRef] [PubMed]
14. Zhao, H.; Fan, W.; Dong, J.; Lu, J.; Chen, J.; Shan, L.; Lin, Y.; Kong, W. Evaluation of antioxidant activities and total phenolic contents of typical malting barley varieties. *Food Chem.* **2007**, *107*, 296–304. [CrossRef]

15. Singleton, V.L.; Rossi, J.A. Colorimetry of total phenolics with phosphomolybdic-phosphotungstic acid reagents. *Am. J. Enol. Vit.* **1965**, *16*, 144–158.
16. Marquele, F.D.; Di Mambro, V.M.; Georgetti, S.R.; Casagrande, R.; Valim, Y.M.L.; Fonseca, M.J.V. Assessment of the antioxidant activities of brazilian extracts of propolis alone and in topical pharmaceutical formulations. *J. Pharm. Biomed. Anal.* **2005**, *39*, 455–462. [CrossRef]
17. Torres-González, A.; López-Rivera, P.; Duarte-Lisci, G.; López-Ramírez, Á.; Correa-Benítez, A.; Rivero-Cruz, J.F. Analysis of volatile components from *Melipona beecheii* geopropolis from Southeast Mexico by headspace solid-phase microextraction. *Nat. Prod. Res.* **2016**, *30*, 237–240. [CrossRef]
18. Adams, R.P. *Identification of Essential Oil Components by Gas Chromatography/Mass Spectrometry*; Allured Publishing Corporation: Carol Stream, IL, USA, 2007.
19. Linstrom, P.J. *NIST Chemistry Webbook, NIST Standard Reference Database Number 69*; National Institute of Standards and Technology: Gaithersburg, MD, USA, 2005.
20. Clinical and Laboratory Standards Institute. *Methods for Dilution Antimicrobial Susceptibility Tests for Bacteria that Grow Aerobically, Approved Standard*, 10th ed.; M07-A11; Clinical and Laboratory Standards Institute: Wayne, PA, USA, 2018.
21. Escarpa, A.; González, M. Approach to the content of total extractable phenolic compounds from different food samples by comparison of chromatographic and spectrophotometric methods. *Anal. Chim. Acta* **2001**, *427*, 119–127. [CrossRef]
22. Valencia, D.; Alday, E.; Robles-Zepeda, R.; Garibay-Escobar, A.; Galvez-Ruiz, J.C.; Salas-Reyes, M.; Jimenez-Estrada, M.; Velazquez-Contreras, E.; Hernandez, J.; Velazquez, C. Seasonal effect on chemical composition and biological activities of sonoran propolis. *Food Chem.* **2012**, *131*, 645–651. [CrossRef]
23. Sánchez-Moreno, C.; Larrauri, J.A.; Saura-Calixto, F. A procedure to measure the antiradical efficiency of polyphenols. *J. Sci. Food Chem.* **1998**, *76*, 270–276. [CrossRef]
24. Lima, B.; Tapia, A.; Luna, L.; Fabani, M.P.; Schmeda-Hirschmann, G.; Podio, N.S.; Wunderlin, D.A.; Feresin, G.E. Main flavonoids, dpph activity, and metal content allow determination of the geographical origin of propolis from the province of San Juan (Argentina). *J. Agric. Food Chem.* **2009**, *57*, 2691–2698. [CrossRef] [PubMed]
25. Tominaga, H.; Kobayashi, Y.; Goto, T.; Kasemura, K.; Nomura, M. DPPH radical-scavenging effect of several phenylpropanoid compounds and their glycoside derivatives. *Yakugaku Zasshi J. Pharm. Soc. Jpn.* **2005**, *125*, 371–375. [CrossRef] [PubMed]
26. Pengfei, L.; Tiansheng, D.; Xianglin, H.; Jianguo, W. Antioxidant properties of isolated isorhamnetin from the sea buckthorn marc. *Plant Foods Hum. Nutr.* **2009**, *64*, 141–145. [CrossRef] [PubMed]
27. Lu, H.T.; Zou, Y.L.; Deng, R.; Shan, H. Extraction, purification and antiradical activities of alpinetin and cardamomin from alpinia katsumadai hayata. *Asian J. Chem.* **2013**, *25*, 9503–9507. [CrossRef]
28. Prior, R.L.; Wu, X.; Schaich, K. Standardized methods for the determination of antioxidant capacity and phenolics in foods and dietary supplements. *J. Agric. Food Chem.* **2005**, *53*, 4290–4302. [CrossRef]
29. Procházková, D.; Boušová, I.; Wilhelmová, N. Antioxidant and prooxidant properties of flavonoids. *Fitoterapia* **2011**, *82*, 513–523. [CrossRef]
30. Benguedouar, L.; Lahouel, M.; Gangloff, S.C.; Durlach, A.; Grange, F.; Bernard, P.; Antonicelli, F. Ethanolic extract of Algerian propolis and galangin decreased murine melanoma T. *Anticancer Agents Med. Chem.* **2016**, *16*, 1172–1183. [CrossRef]
31. Celińska-Janowicz, K.; Zaręba, I.; Lazarek, U.; Teul, J.; Tomczyk, M.; Pałka, J.; Miltyk, W. Constituents of Propolis: Chrysin, Caffeic Acid, p-Coumaric Acid, and Ferulic Acid Induce PRODH/POX-Dependent Apoptosis in Human Tongue Squamous Cell Carcinoma Cell (CAL-27). *Front. Pharmacol.* **2018**, *9*, 336. [CrossRef]
32. Velazquez, C.; Navarro, M.; Acosta, A.; Angulo, A.; Dominguez, Z.; Robles, R.; Robles-Zepeda, R.; Lugo, E.; Goycoolea, F.M.; Velazquez, E.F.; et al. Antibacterial and free-radical scavenging activities of Sonoran propolis. *J. Appl. Microbiol.* **2007**, *103*, 1747–1756. [CrossRef]
33. Bertelli, D.; Papotti, G.; Bortolotti, L.; Marcazzan, G.L.; Plessi, M. ^{1}H-NMR simultaneous identification of health-relevant compounds in propolis extracts. *Phytochem. Anal.* **2012**, *23*, 260–266. [CrossRef]
34. Wawer, I.; Zielinska, A. 13c cp/mas nmr studies of flavonoids. *Magn. Reson. Chem.* **2001**, *39*, 374–380. [CrossRef]

35. Dominguez, X.A.; Franco, R.; Zamudio, A.; Barradas, D.D.M.; Watson, W.H.; Zabel, V.; Merijanian, A. Mexican medicinal plants. Part 38. Flavonoids from *Dalea scandens* var. Paucifolia and *Dalea thyrsiflora*. *Phytochemistry* **1980**, *19*, 1262–1263. [CrossRef]
36. Rossi, M.H.; Yoshida, M.; Soares Maia, J.G. Neolignans, styrylpyrones and flavonoids from an aniba species. *Phytochemistry* **1997**, *45*, 1263–1269. [CrossRef]
37. Hosny, M.; Dhar, K.; Rosazza, J.P.N. Hydroxylations and methylations of quercetin, fisetin, and catechin by *Streptomyces griseus*. *J. Nat. Prod.* **2001**, *64*, 462–465. [CrossRef] [PubMed]
38. Cao, X.; Wei, Y.; Ito, Y. Preparative isolation of isorhamnetin from *Stigma maydis* using high-speed countercurrent chromatography. *J. Liq. Chromatogr. Relat. Technol.* **2009**, *32*, 273–280. [CrossRef] [PubMed]
39. Nagy, M.; Suchy, V.; Uhrin, D.; Ubik, K.; Budesinsky, M.; Grancai, D. Constituents of propolis of Czechoslovak origin. V. *Chem. Pap.* **1988**, *42*, 691–696.
40. Falcão, S.; Vilas-Boas, M.; Estevinho, L.; Barros, C.; Domingues, M.M.; Cardoso, S. Phenolic characterization of northeast portuguese propolis: Usual and unusual compounds. *Anal. Bioanal. Chem.* **2010**, *396*, 887–897. [CrossRef]
41. Gardana, C.; Scaglianti, M.; Pietta, P.; Simonetti, P. Analysis of the polyphenolic fraction of propolis from different sources by liquid chromatography–tandem mass spectrometry. *J. Pharm. Biomed. Anal.* **2007**, *45*, 390–399. [CrossRef]
42. Falcao, S.I.; Vale, N.; Gomes, P.; Domingues, M.R.M.; Freire, C.; Cardoso, S.M.; Vilas-Boas, M. Phenolic profiling of portuguese propolis by LC-MS spectrometry: Uncommon propolis rich in flavonoid glycosides. *Phytochem. Anal.* **2013**, *24*, 309–318. [CrossRef]
43. Pellati, F.; Prencipe, F.P.; Benvenuti, S. Headspace solid-phase microextraction-gas chromatography-mass spectrometry characterization of propolis volatile compounds. *J. Pharm. Biomed. Anal.* **2013**, *84*, 103–111. [CrossRef]
44. Pino, J.A.; Marbot, R.; Delgado, A.; Zumarraga, C.; Sauri, E. Volatile constituents of propolis from honey bees and stingless bees from Yucatan. *J. Essent. Oil Res.* **2006**, *18*, 53–56. [CrossRef]

 © 2020 by the authors. Licensee MDPI, Basel, Switzerland. This article is an open access article distributed under the terms and conditions of the Creative Commons Attribution (CC BY) license (http://creativecommons.org/licenses/by/4.0/).

Article

Dietary Compound Isoliquiritigenin, an Antioxidant from Licorice, Suppresses Triple-Negative Breast Tumor Growth via Apoptotic Death Program Activation in Cell and Xenograft Animal Models

Po-Han Lin [1], Yi-Fen Chiang [1,†], Tzong-Ming Shieh [2,3,†], Hsin-Yuan Chen [1], Chun-Kuang Shih [1], Tong-Hong Wang [4,5], Kai-Lee Wang [6], Tsui-Chin Huang [7], Yong-Han Hong [8], Sing-Chung Li [1] and Shih-Min Hsia [1,9,10,11,*]

1. School of Nutrition and Health Sciences, College of Nutrition, Taipei Medical University, Taipei 11031, Taiwan; phlin@tmu.edu.tw (P.-H.L.); yvonne840828@gmail.com (Y.-F.C.); hsin246@gmail.com (H.-Y.C.); ckshih@tmu.edu.tw (C.-K.S.); sinchung@tmu.edu.tw (S.-C.L.)
2. School of Dentistry, College of Dentistry, China Medical University, Taichung 40402, Taiwan; tmshieh@mail.cmu.edu.tw
3. Department of Dental Hygiene, College of Health Care, China Medical University, Taichung 40402, Taiwan
4. Tissue Bank, Chang Gung Memorial Hospital, Tao-Yuan 33305, Taiwan; cellww@adm.cgmh.org.tw
5. Graduate Institute of Health Industry Technology, Chang Gung University of Science and Technology, Tao-Yuan 33305, Taiwan
6. Department of Nursing, Ching Kuo Institute of Management and Health, Keelung City 20301, Taiwan; kellywang@tmu.edu.tw
7. Graduate Institute of Cancer Biology and Drug Discovery, College of Medical Science and Technology, Taipei Medical University, Taipei 11031, Taiwan; tsuichin@tmu.edu.tw
8. Department of Nutrition, I-Shou University, Kaohsiung City 82445, Taiwan; yonghan@isu.edu.tw
9. Graduate Institute of Metabolism and Obesity Sciences, College of Nutrition, Taipei Medical University, Taipei 11031, Taiwan
10. School of Food Safety, College of Nutrition, Taipei Medical University, Taipei 11031, Taiwan
11. Nutrition Research Center, Taipei Medical University Hospital, Taipei 11031, Taiwan
* Correspondence: bryanhsia@tmu.edu.tw; Tel.: +886-2-2736-1661 (ext. 6558)
† These authors equal contribution as the first author in this study.

Received: 22 January 2020; Accepted: 7 March 2020; Published: 10 March 2020

Abstract: Patients with triple-negative breast cancer have few therapeutic strategy options. In this study, we investigated the effect of isoliquiritigenin (ISL) on the proliferation of triple-negative breast cancer cells. We found that treatment with ISL inhibited triple-negative breast cancer cell line (MDA-MB-231) cell growth and increased cytotoxicity. ISL reduced cell cycle progression through the reduction of cyclin D1 protein expression and increased the sub-G1 phase population. The ISL-induced apoptotic cell population was observed by flow cytometry analysis. The expression of Bcl-2 protein was reduced by ISL treatment, whereas the Bax protein level increased; subsequently, the downstream signaling molecules caspase-3 and poly ADP-ribose polymerase (PARP) were activated. Moreover, ISL reduced the expression of total and phosphorylated mammalian target of rapamycin (mTOR), ULK1, and cathepsin B, whereas the expression of autophagic-associated proteins p62, Beclin1, and LC3 was increased. The decreased cathepsin B cause the p62 accumulation to induce caspase-8 mediated apoptosis. In vivo studies further showed that preventive treatment with ISL could inhibit breast cancer growth and induce apoptotic and autophagic-mediated apoptosis cell death. Taken together, ISL exerts an effect on the inhibition of triple-negative MDA-MB-231 breast cancer cell growth through autophagy-mediated apoptosis. Therefore, future studies of ISL as a supplement or alternative therapeutic agent for clinical trials against breast cancer are warranted.

Keywords: isoliquiritigenin (ISL); triple-negative breast cancer; apoptosis; autophagy

1. Introduction

Breast cancer is the most frequently diagnosed form of cancer and the leading cause of cancer-related deaths in women worldwide [1,2]. As a result of advanced and multidisciplinary therapeutic treatments, the five-year survival rate of breast cancer has greatly improved [3]. However, an increasing trend in the incidence rate of breast cancer has been observed, and the age of patients who are diagnosed with breast cancer has been reported to be trending younger [4].

Breast cancer is categorized into three major subtypes based on the presence or absence of molecular markers for estrogen receptors (ERs), progesterone receptors (PRs), and human epidermal growth factor 2 (ERBB2, formerly HER2/neu). The standard therapy for all patients with breast cancer is tumor eradication by surgical resection. Then, a systemic therapeutic strategy for breast cancer is determined by tumor stage and subtype. Patients with hormone-receptor-positive tumors receive endocrine therapy, such as treatment with hormone receptor antagonists, in combination with limited chemotherapy. For ERBB2-positive tumors, patients can receive ERBB2-targeted antibodies or a combination of small-molecule inhibitors and chemotherapy. Patients with triple-negative tumors (tumors without all three molecular markers) receive chemotherapy alone [5]. Previous studies have indicated that patients with triple-negative breast cancer have worse clinical outcomes compared with hormone-receptor-positive (ER+ and PR+) and ERBB2-positive (ERBB2+) breast cancer patients [6–8]. Systemic chemotherapy for patients with triple-negative breast cancer causes acute and chronic toxicities, including nausea, vomiting, neuropathy, sterility, and congestive heart failure [9,10]. Therefore, there is still an urgent need to optimize the current therapeutic strategy for patients with triple-negative breast cancer.

Numerous natural dietary phytochemicals have been observed to inhibit carcinogenesis and have attracted growing interest as complementary and alternative medicines as well as for cancer chemoprevention due to their wide availability, mildness, and low toxicity [11]. *Glycyrrhiza* species (licorice) are widely used as herbal medicine in Asia, as they exhibit highly effective antitussive, expectorant, and antipyretic activities [12,13]. Among the bioactive ingredients isolated from licorice, isoliquiritigenin (ISL; 2′,4,4′-trihydroxychalcone) has been reported to exert considerable biological activities. ISL has an anti-inflammatory effect on the inhibition of nucleotide-binding domain leucine-rich repeat (NLR) and pyrin domain containing receptor 3 (NLRP3) inflammasome-associated inflammatory diseases [14,15]; an antioxidative effect on the activation of the Nrf2-induced antioxidant system [16]; a hepatoprotective effect against CCl4-induced liver injury [17]; and a cardioprotective effect on the reduction of oxidative stress [18].

ISL has also been reported to exert potential antitumor activity on multistage carcinogenesis procession in cervical [19], ovarian [20], prostate [21], and lung cancers [22]. Previously, we found that ISL of 10 μM exhibits an inhibitory effect on Vascular endothelial growth factor (VEGF)-induced triple-negative breast cancer migration and invasion through downregulation of the PI3K-AKT-MAPK signaling pathway [23]. We also observed that a dose of ISL over 25 μM showed an antiproliferation effect on breast cancer cells [23]. Hence, in the present study, we used an in vitro culture system to explore the molecular mechanism underlying ISL-induced antiproliferation of triple-negative MDA-MB-231 breast cancer cells. Moreover, an MDA-MB-231 tumor xenograft mouse model was generated to examine the preventive effect of ISL on tumor growth in vivo.

2. Materials and Methods

2.1. Cell Line and Culture Condition

Human breast cancer MDA-MB-231 cell line was purchased from the Bioresource Collection and Research Center (BCRC, Hsinchu, Taiwan). MDA-MB-231 cells were maintained in Dulbecco's Modified Eagle Medium (DMEM)/F12 medium (Sigma-Aldrich, St. Louis, MO, USA) supplemented with 10% fetal bovine serum (FBS; CORNING, Manassas, VA, USA), 100 units/mL of penicillin, 100 μg/mL of streptomycin (CORNING), sodium bicarbonate (2.438 g/L; BioShop, Burlington, ON, Canada), and 4-(2-hydroxyethyl)piperazine-1-ethanesulfonic acid (HEPES; 5.986 g/L; BioShop) in a humidified incubator (37 °C, 5% CO_2). Dimethyl sulfoxide (DMSO) was used as a solvent of ISL, stock concentration was 100 mM, and the dosage was according to a previous study [24].

2.2. MTT Assay

MDA-MB-231 cells were seeded in 96-well plates (2.5×10^3 cells/well) and treated with ISL (Sigma-Aldrich, St. Louis, MO, USA) for 24, 48, and 72 h. The cell survival rate was analyzed using an MTT (3-(4,5-dimethyl-2-thiazolyl)-2,5-diphenyl-2H-tetrazolium bromide; Abcam, Cambridge, MA, USA) assay. At the end of incubation, serum-free DMEM medium with 0.5 mg/mL of MTT was substituted for conditional medium and incubated for an additional 3 h; subsequently, the media were removed. Crystal formazan was dissolved in 100 μL/well dimethyl sulfoxide (DMSO; ECHO Chemical Co. Ltd., Taipei, Taiwan). The optical density was measured by using a VERSA Max microplate reader (Molecular Devices, San Jose, CA, USA) at 570 and 630 nm as reference wavelengths.

2.3. Lactate Dehydrogenase (LDH) Assay

MDA-MB-231 cells were seeded in 96-well plates (2×10^4 cells/well) and treated with ISL for 24, 48, and 72 h. The medium LDH activity was detected using the LDH Cytotoxicity Assay Kit (Cayman Chemical Company, Ann Arbor, MI, USA). The procedure was performed according to the manufacturer's protocols. The absorbance was read at 450 nm with a VERSA Max microplate reader (Molecular Devices, San Jose, CA, USA).

2.4. Cell Counting

MDA-MB-231 cells (10^5 cells) were seeded in 6 cm culture dishes. After attaching overnight, cells were treated with ISL for 48 h. Cell morphologies were photographed using a light microscope (Olympus, Tokyo, Japan). Then, cells were detected by trypsinization and resuspended in the culture medium. Cell suspensions were mixed with 0.4% trypan blue solution (Gibco, Grand Island, NY, USA). The number of cells was counted using a hemocytometer under an inverted phase-contrast microscope at 200× magnification.

2.5. Flow Cytometry Analysis of Cell Cycle Distribution and Apoptosis

To analyze the cell cycle distribution, MDA-MB-231 (1×10^5) cells were seeded into 6 cm culture dishes. After attaching overnight, cells were treated with the vehicle or ISL at 25 and 50 μM in DMEM/F12 medium containing 10% FBS for 48 h. At the end of incubation, cells were detached by trypsinization. For the cell cycle distribution assay, the cells were fixed in 70% alcohol at −20 °C and stained with a propidium iodide (PI) solution (2 mg of DNAse-free RNAse A and 0.4 mL of 500 μg/mL PI was added to 10 mL of 0.1% Triton X-100 in phosphate buffered saline (PBS)) at room temperature for 30 min. For the apoptosis analysis, a commercial annexin V fluorescein isothiocyanate (FITC) apoptosis detection kit (BD Biosciences, San Jose, CA, USA) was used. The procedure was performed according to the manufacturer's protocols. The cell cycle distribution and apoptosis were analyzed using a BD FACSCanto flow cytometer (BD Biosciences, San Jose, CA, USA). A minimum of 10,000 cells per sample were collected and further analyzed using CellQuest software (BD Biosciences).

2.6. Protein Preparation and Western Blot Analysis

Cells were lysed in RIPA lysis buffer containing protease and phosphatase inhibitors (Roche, Mannheim, Baden-Württemberg, Germany). Protein was quantitated by the Bradford assay and then resolved using sodium dodecyl sulfate polyacrylamide gel electrophoresis (SDS-PAGE). After transfer to a polyvinylidene Fluoride (PVDF) membrane, 1.5% Bovine serum albumin (BSA) solution was used to block the empty space of the membrane. Subsequently, membranes were incubated with primary antibodies—CDK4 (1:1000; Cell Signaling, Boston, MA, USA), cyclin D1 (1:1000; Cell Signaling), Bcl-2 (1:1000; Cell Signaling), Bax (1:1000; Cell Signaling), caspase-3 (1:1000; Cell Signaling), PARP (1:1000; Cell Signaling), GAPDH (1:10,000; Proteintech, Rosemont, IL, USA), p62 (1:1000; Cell Signaling), LC3 (1:1000; Cell Signaling), Beclin1 (1:10,000; Novus Biologicals, Littleton, CO, USA), Caspase-8 (1:1000; Cell Signaling), and Cathepsin B (1:1000; Cell Signaling) and β-actin (1:1000; Santa Cruz Biotechnology, Santa Cruz, CA, USA)—at 4 °C overnight and then with a horseradish peroxidase (HRP)-conjugated secondary antibody (1:10,000) for 1 to 2 h. The visual signal was captured by an image analysis system (UVP BioChemi, Analytik Jena US, Upland, CA, USA). The band densities were determined as arbitrary absorption units using the Image-J software program Version 1.52t (NIH, Bethesda, MD, USA). The expression level of these target proteins was analyzed by three individual experiments. In the same quantitative protein sample, two different molecular weights of the target protein were determined by the same gel. The membrane was cut into two fractions and incubated with two different antibodies at the same time.

2.7. Cell Transfection

The siRNA of SQSTM1/p62 (SignalSilence® SQSTM1/p62 (Cell Signaling) were transfected into MDA-MB-231 cells with lipofectamine 3000 (Thermo Fisher Scientific, Waltham, MA, USA) reagent according to the manufacturer's instructions. Briefly, 100 nM sip62 was transfected into MDA-MB-231 and treated with 50 μM ISL for 48 h.

2.8. Tumor Xenograft Mouse Model

The animal studies were conducted according to the protocols approved by the Institutional Animal Care and Use Committee (IACUC) of China Medical University (permit number: 99-190-C). In this study, 5-week-old female Nude-Foxn1^{nu} mice were purchased from BioLASCO (Taipei, Taiwan). The mice were housed with 12 h of artificial illumination in a temperature-controlled room (22 ± 2 °C). Food and water were provided ad libitum. Mice were randomly divided into two groups: (1) mice treated with PBS (0.25 mL/mouse) by oral gavage once daily for 2 weeks as a control, and (2) mice pretreated with ISL (Sigma-Aldrich) at 2.5 or 5.0 mg/mL by oral gavage (0.25 mL/mouse) once daily for 2 weeks. After pretreatment for 2 weeks, mice were inoculated with 100 μL PBS/mouse only as the control or human breast MDA-MB-231 cancer cells (5×10^6 cells in 100 μL PBS/mouse) by a subcutaneous (s.c.) injection on the right flank, and they were constitutively treated with ISL for an additional 25 days. After implantation for 1 week, tumor volume was measured using calipers and the following formula was calculated: $0.5 \times \text{length} \times \text{width}^2$. At the end of the experiment, the mice were sacrificed by an intraperitoneal (i.p.) injection with an anesthetic mixture (30 μL/mouse; Zoletil:Rompun (v/v) = 1:1). The tumor tissues were weighed, photographed, and then fixed with 10% formalin for further immunohistochemistry (IHC) analysis.

2.9. IHC Staining

After deparaffinization, antigen retrieval, and blocking of peroxidase activity, the tissue sections were incubated with anti-Ki-67 antibody (1:200; Abcam, Cambridge, UK), anti-caspase-3 antibody (1:100; Cell Signaling), and anti-p62 antibody (1:100; Cell Signaling) at 4 °C overnight, separately. Slides were further incubated for 30 min with Super Enhance and polymer HRP. The bound antibody was elected by a species-specific secondary antibody and then developed with a 3-amino-9-ethylcarbazole (AEC) substrate. Images for IHC staining were captured with an EVOS® microscope (Thermo Fisher Scientific, Waltham, MA, USA).

2.10. Statistical Analysis

All quantitative results are expressed as mean ± standard error of the mean (SEM), which were analyzed using Prism version 6.0 software (GraphPad, San Diego, CA, USA). The statistically significant difference from the respective controls for each experiment was determined using one-way analysis of variance (ANOVA) for all groups. Student's unpaired t-test was used for comparison between two groups. Significance was accepted at * $p < 0.05$ and ** $p < 0.01$.

3. Results

3.1. ISL Suppressed Breast Cancer MDA-MB-231 Cell Growth

The MTT assay was performed to examine whether ISL can suppress breast cancer MDA-MB-231 cell growth. The results showed that cell viability was reduced by treatment of cells with ISL at 25 and 50 µM for 48 and 72 h (Figure 1a), whereas treatment of cells with ISL at 25 and 50 µM for 48 and 72 h was shown to increase cytotoxicity by LDH activity analysis (Figure 1b). The cell morphology and number were also monitored, and we found that the cell morphology switched from the spindle type to irregular by treatment of cells with ISL at 25 and 50 µM for 48 h (Figure 1c). Furthermore, a reduction of the cell number was observed by treatment of cells with 25 and 50 µM for 48 h (Figure 1d).

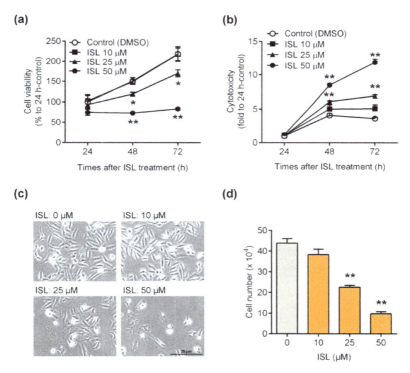

Figure 1. Inhibitory effects of isoliquiritigenin (ISL) on the proliferation of breast cancer MDA-MB-231 cells. (**a**) MDA-MB-231 cells were seeded in 96-well plates (3000 cells per well) with 100 µL per well culture medium. Cells were treated with ISL in various doses for 24, 48, and 72 h. At the end of incubation, cell viability was measured by MTT assay ($n = 3$). (**b**) MDA-MB-231 cells (2×10^4 cells per well) seeded in 96-well plates. Cells were treated with various doses for 24, 48, and 72 h. The cytotoxic effects of ISL was detected using the lactate dehydrogenase (LDH) assay kit ($n = 3$). (**c**,**d**) MDA-MB-231 cells (2×10^5 cells per well) were seeded in six-well plates. After treatment of cells with ISL for 48 h, the cell number was monitored ($n = 4$). Data are represented as mean ± SEM. * $p < 0.05$, ** $p < 0.01$ compared with the control group.

3.2. ISL Suppressed the Cell Cycle Progression of MDA-MB-231 Cells

Here, we studied whether ISL suppressed cell growth by inflecting the cell cycle progression. MDA-MB-231 cells were treated with ISL and then the cell cycle distribution was monitored using flow cytometry analysis. The results showed that treatment with ISL at 25 and 50 μM for 48 h reduced the cell population in the G1 phase in MDA-MB-231 cells (Figure 2a,b). In addition, the expression of G1/S gate-associated proteins CDK4 and cyclin D was detected. The results showed that treatment with ISL did not alter the expression of CDK4 (Figure 2c), but the expression of cyclin D1 was reduced by treatment with ISL at 25 and 50 μM for 48 h (Figure 2d).

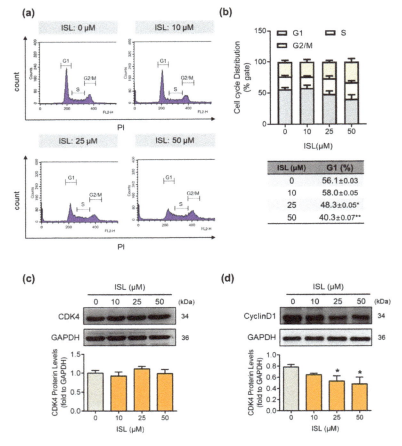

Figure 2. Cell cycle progression of MDA-MB-231 cells was suppressed by treatment with ISL. (**a**) MDA-MB-231 cells were treated with ISL for 48 h. Cells were stained with propidium iodide, and the cell cycle distributions were analyzed by the flow cytometry. (**b**) The quantitative results of cell distributions in MDA-MB-231 cells are shown. Cell-cycle-associated proteins (**c**) CDK4 and (**d**) CyclinD1 were analyzed using Western blot. Each target protein was normalized to GAPDH expression. Data are represented as mean ± SEM ($n = 3$). * $p < 0.05$, ** $p < 0.01$ compared with the control group.

3.3. ISL Induced Apoptotic Cell Death

To further examine whether ISL induced cell apoptosis in MDA-MB-231 cells, the cells were stained with annexin V FITC, and then the apoptotic cell population was analyzed by flow cytometry.

The results showed that the annexin-V-positive (early apoptotic phase) and annexin-plus-PI-positive (later apoptotic phase) cell populations were increased by ISL treatment at 25 and 50 µM for 48 h (Figure 3a,b). The expression of Bcl-2 protein decreased from ISL treatment at 50 µM for 48 h (Figure 3c,d), whereas treatment with ISL increased the expression of Bax protein (Figure 3c,e). In addition, the cleaved form of caspase-3 was induced by treatment with ISL at 25 and 50 µM for 48 h (Figure 3f,g), and the expression of cleaved PARP was also increased by treatment with ISL at 50 µM for 48 h (Figure 3f,h).

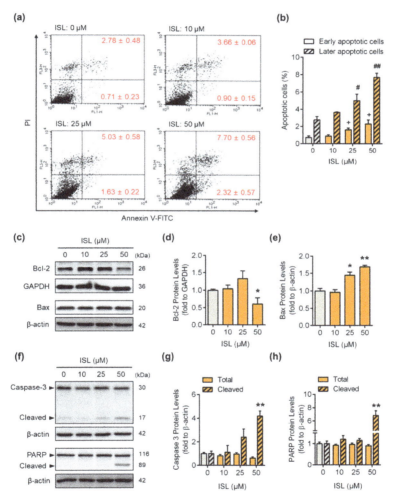

Figure 3. ISL induced cell apoptosis by upregulating apoptotic protein expression in MDA-MB-231 cells. (**a**) MDA-MB-231 cells were treated with ISL for 48 h. Cells were stained with propidium iodide and annexin V fluorescein isothiocyanate (FITC), and the apoptosis rates were analyzed by flow cytometry. (**b**) The quantitative data of apoptotic cell death in early and late phases are shown. (**c**) After treatment as indicated above, the anti- and proapoptotic proteins were monitored using Western blotting in MDA-MB-231 cells. (**d**,**e**) Each target protein was normalized to GAPDH or β-actin expression. (**f**) The expression of caspase-3 and its downstream molecule, PAPR, was monitored using Western blotting in MDA-MB-231 cells. (**g**,**h**) Each target protein was normalized to β-actin expression.

Data are represented as mean ± SEM (n = 3). * $p < 0.05$, ** $p < 0.01$ compared with the control group. + $p < 0.05$ compared with the early apoptotic phase of the control group. # $p < 0.05$, ## $p < 0.01$ compared with the later apoptotic phase of the control group.

3.4. ISL-Mediated p62 Accumulation Causes Autophagy-Mediated Apoptosis

The total and phosphorylated protein levels of mTOR decreased after treatment with ISL at 50 µM for 48 h (Figure 4a,b). The downstream molecule ULK1 was activated by mTOR regulation in the autophagy pathway [25]. We found that the total and phosphorylated protein levels of ULK1 decreased after ISL treatment at 10, 25, and 50 µM for 48 h (Figure 4a,c). The autophagosome-associated proteins p62, Beclin1, and LC3 were increased by treatment with ISL for 48 h (Figure 4d–f). In our results observing that p62 accumulation, with autophagy inhibitor bafilomycin (BAF), blocked p62 accumulation and caused cell toxicity (Supplementary Figure S1). In the autophagy process, lysosomal proteases are needed, such as Cathepsins degrades p62 [26]. Treatment with ISL at 50 µM for 48 h significant decreased cathepsin B protein expression (Figure 4g), which explains the p62 accumulation. Moreover, p62 accumulation was related to the increased apoptosis through the increase of caspase-8. Furthermore, we used siRNA to eliminate p62 function. After 48 h, ISL increased caspase-8 protein expression. However, ISL combined with siRNA treatment have a lower expression of caspase-8 than ISL treatment (Figure 4h).

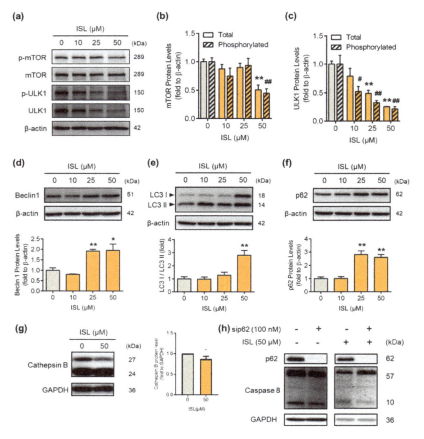

Figure 4. ISL treatment induced the expression of autophagy-associated proteins in MDA-MB-231 cells.

MDA-MB-231 cells were treated with ISL for 48 h. (**a**) The expression levels of mTOR and ULK1 protein were analyzed by Western blotting. The total and phosphorylated forms of (**b**) mTOR and (**c**) ULK1 were normalized to β-actin expression. The expression levels of (**d**) p62, (**e**) Beclin1, (**f**) LC3, and (**g**) Cathepsin B were also analyzed using Western blotting. MDA-MB-231 cells were used lipofectamine 3000 (Thermo Fisher Scientific) to transfect SignalSilence® SQSTM1/p62 (Cell Signaling), and treated with ISL for 48 h. The expression of (**h**) caspase-8 proteins were also analyzed using Western blotting. Data are represented as mean ± SEM ($n = 3$). * $p < 0.05$, ** $p < 0.01$ compared with the control group. # $p < 0.05$, ## $p < 0.01$ compared with the phosphorylated protein level of the control group.

3.5. Preventive Effects of ISL on Breast Cancer Cell Growth in a Xenograft Mouse Model

A breast cancer MDA-MB-231 cell xenograft mouse model was generated to evaluate whether ISL has a potential role in antitumor proliferation in vivo. After pretreatment with PBS or ISL at 2.5 and 5.0 mg/kg for 2 weeks, MDA-MB-231 cells were inoculated into mice, which were constitutively treated with ISL for an additional 25 days. The results showed that MDA-MB 231 tumors formed in all five mice by treatment with PBS. However, the number of MDA-MB 231 tumors was reduced by pretreatment with ISL at 2.5 mg/kg (three of five mice) and 5.0 mg/kg (two of five mice) (Figure 5a,c). Measuring the tumor volume elicited by MDAMB-231 cells showed that the tumor volume increased by treatment with PBS in a time-dependent manner, and ISL treatments significantly reduced them (Figure 5b). At the end of the experiment, tumor tissues were isolated and weighed. Treatments of mice with ISL at 2.5 and 5.0 mg/kg showed significantly reduced tumor weight compared with the vehicle group (Figure 5d). According to the IHC results, expression of the Ki-67 protein level was observed in tumor tissues for mice treated with ISL at 2.5 and 5.0 mg/kg, separately (Figure 6, top panel). Treatment with ISL at 5.0 mg/kg increased the caspase-3 expression level in tumor tissues (Figure 6, middle panel). Moreover, the expression of p62 was increased in tumor tissues by treatment with ISL at 2.5 and 5.0 mg/kg, separately (Figure 6, bottom panel).

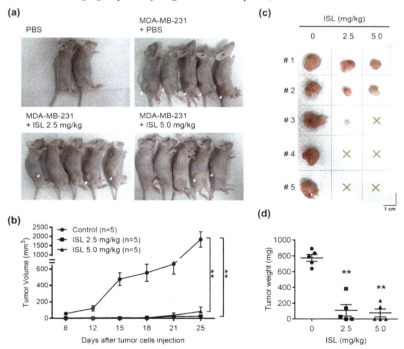

Figure 5. ISL suppressed tumor growth, as shown by the human MDA-MB-231 breast tumor xenograft

mouse model. Mice were pretreated with ISL at 2.5 and 5.0 mg/kg or PBS as control by oral gavage once daily for 2 weeks. Then, MDA-MB-231 tumor cells (5×10^6 cells per mouse) were implanted into mice by subcutaneous injection on the right flank and the mice received oral gavage of PBS and/or ISL for 25 days. At the end of the experiment, (**a**) the mice were photographed, (**b**) tumor size was monitored, (**c**) tumors were isolated and photographed, and (**d**) tumor weight was recorded. Data are represented as mean ± SEM (n = 5). ** $p < 0.01$ compared with the control group.

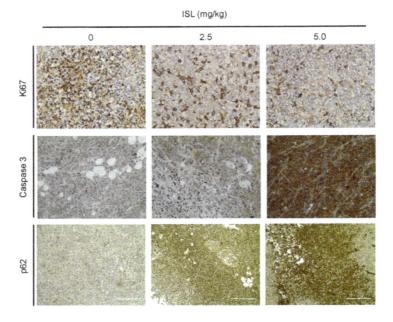

Figure 6. Representative results of IHC staining for Ki-67, caspase-3, and p62 protein expression levels in ISL-treated MDA-MB-231 breast tumor xenograft mouse model. MDA-MB-231 tumor cells (5×10^6 cells per mouse) were implanted and mice were treated with ISL for 25 days Then, tumor tissues were isolated and immunohistochemistry was performed to analyze Ki-67, caspase-3, and p62 protein levels. Images were photographed (200× magnification; scale bar = 200 µm).

3.6. ISL Suppressed VEGF Production and Capillary-Like Tube Formation of SVEC4-10 Cells

IHC analysis was performed to evaluate the expression of VEGF in tumor tissues by treatment with ISL. The results showed that the expression of VEGF was significantly reduced in treatment of mice with ISL at 2.5 and 5.0 mg/kg, separately (Supplementary Figure S2a). Measuring the serum VEGF level showed that treatment with ISL at 2.5 and 5.0 mg/kg reduced the serum VEGF level compared with the vehicle group (Supplementary Figure S2b). Additionally, to further investigate whether ISL has a potential role in antiangiogenesis, we performed a capillary-like tube formation assay. The results showed that the capability of capillary-like tube formation in SVEC4-10 cells was inhibited by treatment with ISL at 10 µM for 5 h (Supplementary Figure S3).

4. Discussion

Patients with triple-negative breast cancer are insensitive to endocrine therapy and ERBB2-targeted antibody treatment. Systemic chemotherapy for patients with triple-negative breast cancer is the predominant clinical treatment strategy [27,28]. Triple-negative breast cancer cells have been reported to exhibit genetic and phenotypical shifts in subpopulations, leading to resistance to the available standard drugs [29,30]. However, the functional differences made by genomically heterogeneous triple-negative breast cancer cells to chemoresistance remain poorly understood [31]. Natural ingredients from plants are widely explored to identify their availability for application in patients with breast cancer as an adjuvant approach [32]. For example, curcumin, a bioactive ingredient of the plant *Curcuma longa*, exerts an inhibitory effect on triple-negative MDA-MB-231 breast cancer cells through upregulation of p21 protein expression and enhancement of the Bax-to-Bcl-2 ratio [33,34]. A clinical phase I trial of curcumin has been reported to increase the sensitivity of breast cancer patients to docetaxel chemotherapy [35]. Similar effectiveness was observed for licorice roots (*Glycyrrhiza glabra*). The compounds derived from licorice have been widely investigated for their anticancer effects on several cancer cell types. Among the chalcone-type derivatives from licorice, ISL has been identified for cancer prevention in the last decade [12,36].

In the present study, we found that treatment of triple-negative MDA-MB-231 breast cancer cells with ISL inhibited cancer cell growth through blocking the cell cycle by reduction of cyclin D1 expression. The enhancement of the sub-G1 cell population by ISL treatment indicated that the apoptotic cell death pathway was active. Flow cytometry analysis was used to further confirm whether ISL induced MDA-MB-231 cell apoptosis. Our results showed that treatment with ISL increased the annexin-V- and PI-positive cell population. The expression of the anti-apoptotic protein Bcl-2 was reduced, whereas expression of the proapoptotic protein Bax was increased. Furthermore, the downstream molecules caspase-3 and PARP were active. In agreement with our findings, Peng et al. reported that treatment of MDA-MB-231 cells with ISL induced cell apoptosis, and they observed an enhancement of the ratio of Bax to Bcl-2 by ISL treatment [37]. Moreover, ISL treatment was also observed to induce cancer cell apoptosis in lung cancer [38], uterine leiomyoma [39], and endometrial cancer [40].

A clinical study reported that the expression of Beclin1 is lower in patients with breast and ovarian cancers [41]. Beclin1, a Bcl-2 homology 3 (BH3)-domain-only protein, is an important protein in initial autophagy. The anti-apoptotic Bcl-2 protein has been reported to interact with Beclin1 via its BH3 domain [42]. The Beclin1 and Bcl-2 complex suppresses the pre-autophagosomal structure combination, thereby inhibiting autophagy progression. In this study, the results showed that treatment with ISL reduced the expression of anti-apoptotic Bcl-2 and induced autophagy-induced- apoptosis. Our results found that ISL may have p62 accumulation. Additionally, p62 may activate caspase-8, which can induce apoptosis by autophagy inhibition. Caspase-8 activation promotes cell apoptosis [43]. In addition, we found that treatment with ISL induced autophagy-associated signaling proteins mTOR and ULK1 were downregulated by ISL treatment, while the expressions of p62, Beclin1, and LC3 autophagy-promoting proteins were all increased. However, during autophagy progression, p62 was enclosed by autolysosome and degraded by cathepsin B in lysosomes [44], although both p62 and LC3 II levels were increased by ISL. Therefore, ISL influences p62 accumulation by reduced cathepsin B to inhibit autophagy by blocked phagolysosome formation. Furthermore, ISL can activate apoptosis by increased caspase-8 protein expression.

Cathepsin B is a lysosomal protease and is related to autophagy progression and p62 degradation [45]. Cancer studies have shown that cathepsin B is related to cancer cell proliferation [46], and its ability to break down the cell base membrane can increase cancer cell invasion. Our results show that ISL can decrease cathepsin B protein expression to suppress p62 accumulation, increasing autophagy-mediated apoptosis and decreasing cancer cell proliferation-related Ki67 expression.

Cuendet et al. reported that administration of ISL for 1 week delays 7,12-dimethylbenz (*a*)anthracene (DMBA)-induced mammary carcinogenesis in female rats. However, this effect does not reduce the incidence of tumor formation [36,47]. In the present study, to examine whether ISL has preventive effectiveness for anti-proliferation of breast cancer cells, mice were administered with ISL by oral gavage for 2 weeks before implantation of MDA-MB-231 breast cancer cells. Our results showed that pretreatment of mice with ISL reduced the success rate of breast cancer cell implantation, thereby reducing tumor tissue formation. Despite tumor formation from the ISL-treated group, the expression of the Ki-67 protein, a hallmark of proliferation, was reduced, whereas the expressions of caspase-3 and p62 proteins were increased, suggesting that these tumors were undergoing apoptotic and autophagic cell death progression. In addition, we further evaluated the effects of ISL on anti-angiogenesis in breast cancer cells. Our results show that the protein expression and secretion of VEGF were reduced by treatment of mice with ISL. Consistent with our results, Wang et al. reported that treatment with ISL suppresses VEGF-induced proliferation of human umbilical vein endothelial cells (HUVECs) through interaction with the VEGF2R receptor and the reduction of VEGF-regulated downstream signaling [48]. Moreover, we also found that the capability of capillary-like tube formation of SVEC4-10 cells was abolished by treatment with ISL. ISL has a potential role in antiangiogenesis, which might contribute to the inhibitory effect of ISL on the migration and invasion ability of cancer cells.

The effectiveness of ISL in combination with traditional chemotherapeutic drugs has also been investigated. Wang et al. reported that ISL exerts synergistic effects with first-line chemotherapeutic drugs for breast cancer therapy [49]. Specifically, ISL chemosensitizes breast cancer stem cells by docking into the adenosine triphosphate (ATP) domain of GRP78 directly, which in turn suppresses β-catenin/ATP-binding cassette subfamily G2 (ABCG2) signaling [49]. The side-effects of systemic chemotherapeutic therapy were also investigated. ISL has been reported to enhance the toxicity-induced cell death of bladder cancer cells. Furthermore, ISL has been shown to attenuate cisplatin-induced normal renal proximal tubular cell death via activation of the nuclear factor (erythroid-derived 2)-like 2 (Nrf2)/heme oxygenase-1 (HO-1) regulatory pathway [50]. Similarly, the protective effect of ISL on the kidney and liver against cisplatin-induced colon cancer cell death has also been demonstrated [51]. These results suggest that the promotion of chemosensitivity and the lower toxicity of ISL should be further considered in clinical trials.

5. Conclusions

In conclusion, the results of this study showed that ISL effectively inhibited triple-negative MDA-MB-231 breast cancer cells through the activation of apoptotic cell death and cause p62 accumulation activated autophagy mediated apoptosis cell death progressions. In the mouse breast cancer tumor xenograft model, the preventive effect of ISL on antitumor growth was demonstrated (Figure 7). Therefore, supplementation with ISL may be an effective anticancer approach for use in clinical trials.

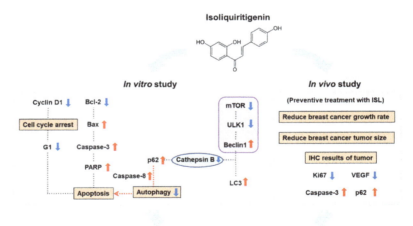

Figure 7. Schematic representation of the inhibitory effect of ISL on the proliferation of triple-negative breast cancer MDA-MB-231 cells through p62 accumulation and activation of apoptotic cell death programs. Blue arrow: decreased; Red arrow: increased.

Supplementary Materials: The following are available online at http://www.mdpi.com/2076-3921/9/3/228/s1, **Figure S1**: ISL and bafilomycin A1 (BAF) treatment induced the expression of autophagy-associated proteins and caused cell toxicity. MDA-MB-231 cells were seeded in 96-well plates (3000 cells per well). Cell were pretreated with BAF for 3 h, and combined with ISL for 48 h. At the end of incubation, (**a**) cell viability was measured by MTT assay. MDA-MB-231 cells were pretreated with BAF for 3 h, and combined with ISL for 48 h. The expression of (**b**) p62 and (**c**) Beclin1 protein were analyzed using western blotting. Data were represented as means ± SD (n = 3). a, $p < 0.05$, compared with control group. b, $p < 0.05$, compared with ISL group. **Figure S2**: Effects of ISL on the expression and secretion of VEGF. MDA-MB-231 tumor cells (5×106 cells per mouse) were implanted and mice were treated with ISL for 25 days. At the end of experiment, (**a**) tumor tissues were isolated and performed immunohistochemistry to analyze VEGF protein level. Images were photographed at 200× magnification. (**b**) Serum VEGF levels were measured using ELISA. Data represent as means ± SEM (n = 5 each group). ** $p < 0.01$ compared with the control group. **Figure S3**: Effects of ISL on the capillary-like tube formation. ISL inhibited capillary-like tube formation of mouse endothelial cells SVEC4-10 cells in matrigel. Images were photographed at 100× magnification.

Author Contributions: Conceptualization, P.-H.L., T.-M.S., H.-Y.C., T.-C.H., and S.-M.H.; experimentation, P.-H.L., T.-M.S., H.-Y.C., Y.-F.C., T.-H.W., K.-L.W., and S.-M.H.; data analysis and figure preparation, P.-H.L., T.-M.S., H.-Y.C., and S.-M.H.; methodology and resources Y.-F.C., C.-K.S., S.-C.L., and Y.-H.H.; writing—original draft preparation, P.-H.L., Y.-F.C., and S.-M.H.; writing—review and editing, P.-H.L. and S.-M.H.; editing and approval of the final version of the manuscript, S.-M.H. All authors have read and agreed to the published version of the manuscript.

Funding: This research was funded by the Ministry of Science and Technology (MOST), Taiwan, Republic of China, under grant numbers MOST103-2313-B-038-003-MY3, MOST106-2320-B-038-064-MY3, MOST106-2811-B-038-033-, MOST 108-2314-B-039-009-MY3 and MOST 108-2314-B-468-001- and the Council of Agriculture, Taiwan, China, under grant numbers 106AS-16.4.1-ST-a4 and 107AS-13.4.1-ST-a6 and China Medical University, under grant numbers CMU108-MF-25.

Conflicts of Interest: The authors declare no conflict of interest.

References

1. Bray, F.; Ferlay, J.; Soerjomataram, I.; Siegel, R.L.; Torre, L.A.; Jemal, A. Global cancer statistics 2018: GLOBOCAN estimates of incidence and mortality worldwide for 36 cancers in 185 countries. *CA A Cancer J. Clin.* **2018**, *68*, 394–424. [CrossRef] [PubMed]
2. Ferlay, J.; Colombet, M.; Soerjomataram, I.; Mathers, C.; Parkin, N.M.; Piñeros, M.; Znaor, A.; Bray, F. Estimating the global cancer incidence and mortality in 2018: GLOBOCAN sources and methods. *Int. J. Cancer* **2018**, *144*, 1941–1953. [CrossRef] [PubMed]
3. Chen, L.; Linden, H.M.; Anderson, B.O.; Li, C.I. Trends in 5-year survival rates among breast cancer patients by hormone receptor status and stage. *Breast Cancer Res. Treat.* **2014**, *147*, 609–616. [CrossRef] [PubMed]
4. Global Burden of Disease Cancer Collaboration. The Global Burden of Cancer 2013. *JAMA Oncol.* **2015**, *1*, 505–527. [CrossRef]
5. Waks, A.G.; Winer, E.P. Breast Cancer Treatment: A Review Breast Cancer Treatment in 2019. *JAMA* **2019**, *321*, 288–300. [CrossRef]
6. Reddy, S.M.; Barcenas, C.H.; Sinha, A.K.; Hsu, L.; Moulder, S.L.; Tripathy, D.; Hortobagyi, G.N.; Valero, V. Long-term survival outcomes of triple-receptor negative breast cancer survivors who are disease free at 5 years and relationship with low hormone receptor positivity. *Br. J. Cancer* **2017**, *118*, 17–23. [CrossRef]
7. Dent, R.; Trudeau, M.E.; Pritchard, K.I.; Hanna, W.M.; Kahn, H.K.; Sawka, C.A.; Lickley, L.A.; Rawlinson, E.; Sun, P.; Narod, S.A. Triple-Negative Breast Cancer: Clinical Features and Patterns of Recurrence. *Clin. Cancer Res.* **2007**, *13*, 4429–4434. [CrossRef]
8. Lin, N.U.; Vanderplas, A.; Hughes, M.E.; Theriault, R.L.; Edge, S.B.; Wong, Y.-N.; Blayney, U.W.; Niland, J.C.; Winer, E.P.; Weeks, J.C. Clinicopathologic features, patterns of recurrence, and survival among women with triple-negative breast cancer in the National Comprehensive Cancer Network. *Cancer* **2012**, *118*, 5463–5472. [CrossRef]
9. Al-Mahmood, S.; Sapiezynski, J.; Garbuzenko, O.B.; Minko, T. Metastatic and triple-negative breast cancer: Challenges and treatment options. *Drug Deliv. Transl. Res.* **2018**, *8*, 1483–1507. [CrossRef]
10. Rock, E.; DeMichele, A. Nutritional Approaches to Late Toxicities of Adjuvant Chemotherapy in Breast Cancer Survivors. *J. Nutr.* **2003**, *133* (Suppl. 1), 3785S–3793S. [CrossRef]
11. Landis-Piwowar, K.; Iyer, N.R. Cancer Chemoprevention: Current State of the Art. *Cancer Growth Metastasis* **2014**, *7*, 19–25. [CrossRef] [PubMed]

12. Kao, T.-C.; Wu, C.-H.; Yen, G.-C. Bioactivity and Potential Health Benefits of Licorice. *J. Agric. Food Chem.* **2014**, *62*, 542–553. [CrossRef] [PubMed]
13. Ramalingam, M.; Kim, H.; Lee, Y.; Lee, Y.-I. Phytochemical and Pharmacological Role of Liquiritigenin and Isoliquiritigenin From Radix Glycyrrhizae in Human Health and Disease Models. *Front. Aging Neurosci.* **2018**, *10*, 348. [CrossRef] [PubMed]
14. Honda, H.; Nagai, Y.; Matsunaga, T.; Okamoto, N.; Watanabe, Y.; Tsuneyama, K.; Hayashi, H.; Fujii, I.; Ikutani, M.; Hirai, Y.; et al. Isoliquiritigenin is a potent inhibitor of NLRP3 inflammasome activation and diet-induced adipose tissue inflammation. *J. Leukoc. Boil.* **2014**, *96*, 1087–1100. [CrossRef] [PubMed]
15. Watanabe, Y.; Nagai, Y.; Honda, H.; Okamoto, N.; Yamamoto, S.; Hamashima, T.; Ishii, Y.; Tanaka, M.; Suganami, T.; Sasahara, M.; et al. Isoliquiritigenin Attenuates Adipose Tissue Inflammation in vitro and Adipose Tissue Fibrosis through Inhibition of Innate Immune Responses in Mice. *Sci. Rep.* **2016**, *6*, 23097. [CrossRef] [PubMed]
16. Zeng, J.; Chen, Y.; Ding, R.; Feng, L.; Fu, Z.; Yang, S.; Deng, X.; Xie, Z.; Zheng, S. Isoliquiritigenin alleviates early brain injury after experimental intracerebral hemorrhage via suppressing ROS- and/or NF-κB-mediated NLRP3 inflammasome activation by promoting Nrf2 antioxidant pathway. *J. Neuroinflamm.* **2017**, *14*, 119. [CrossRef] [PubMed]
17. Kuang, Y.; Lin, Y.; Li, K.; Song, W.; Ji, S.; Qiao, X.; Zhang, Q.-Y.; Ye, M. Screening of hepatoprotective compounds from licorice against carbon tetrachloride and acetaminophen induced HepG2 cells injury. *Phytomedicine* **2017**, *34*, 59–66. [CrossRef]
18. Zhang, X.; Zhu, P.; Zhang, X.; Ma, Y.; Li, W.; Chen, J.-M.; Guo, H.-M.; Bucala, R.; Zhuang, J.; Li, J. Natural Antioxidant-Isoliquiritigenin Ameliorates Contractile Dysfunction of Hypoxic Cardiomyocytes via AMPK Signaling Pathway. *Mediat. Inflamm.* **2013**, *2013*, 390890. [CrossRef]
19. Yuan, X.; Zhang, B.; Gan, L.; Wang, Z.H.; Yu, B.C.; Liu, L.L.; Zheng, Q.; Wang, Z.P. Involvement of the mitochondrion-dependent and the endoplasmic reticulum stress-signaling pathways in isoliquiritigenin-induced apoptosis of HeLa cell. *Biomed. Environ. Sci.* **2013**, *26*, 268–276.
20. Yuan, X.; Yu, B.; Wang, Y.; Jiang, J.; Liu, L.; Zhao, H.; Qi, W.; Zheng, Q. Involvement of endoplasmic reticulum stress in isoliquiritigenin-induced SKOV-3 cell apoptosis. *Recent Patents Anti-Cancer Drug Discov.* **2013**, *8*, 191–199. [CrossRef]
21. Kanazawa, M.; Satomi, Y.; Mizutani, Y.; Ukimura, O.; Kawauchi, A.; Sakai, T.; Baba, M.; Okuyama, T.; Nishino, H.; Miki, T. Isoliquiritigenin inhibits the growth of prostate cancer. *Eur. Urol.* **2003**, *43*, 580–586. [CrossRef]
22. Chen, C.; Shenoy, A.K.; Padia, R.; Fang, D.-D.; Jing, Q.; Yang, P.; Su, S.-B.; Huang, S. Suppression of lung cancer progression by isoliquiritigenin through its metabolite 2, 4, 2', 4'-Tetrahydroxychalcone. *J. Exp. Clin. Cancer Res.* **2018**, *37*, 243. [CrossRef]
23. Wang, K.-L.; Hsia, S.-M.; Chan, C.-J.; Chang, F.-Y.; Huang, C.; Bau, D.-T.; Wang, P.S. Inhibitory effects of isoliquiritigenin on the migration and invasion of human breast cancer cells. *Expert Opin. Ther. Targets* **2013**, *17*, 337–349. [CrossRef] [PubMed]
24. Chen, H.-Y.; Huang, T.-C.; Shieh, T.-M.; Wu, C.-H.; Lin, L.-C.; Hsia, S.-M. Isoliquiritigenin Induces Autophagy and Inhibits Ovarian Cancer Cell Growth. *Int. J. Mol. Sci.* **2017**, *18*, 2025. [CrossRef] [PubMed]
25. Kim, J.; Kundu, M.; Viollet, B.; Guan, K.-L. AMPK and mTOR regulate autophagy through direct phosphorylation of Ulk1. *Nat. Cell Biol.* **2011**, *13*, 132–141. [CrossRef] [PubMed]
26. Puissant, A.; Fenouille, N.; Auberger, P. When autophagy meets cancer through p62/SQSTM1. *Am. J. Cancer Res.* **2012**, *2*, 397–413.
27. Denkert, C.; Liedtke, C.; Tutt, A.; Von Minckwitz, G. Molecular alterations in triple-negative breast cancer—the road to new treatment strategies. *Lancet* **2017**, *389*, 2430–2442. [CrossRef]
28. Harbeck, N.; Gnant, M. Breast cancer. *Lancet* **2017**, *389*, 1134–1150. [CrossRef]
29. Volm, M.; Efferth, T. Prediction of Cancer Drug Resistance and Implications for Personalized Medicine. *Front. Oncol.* **2015**, *5*, 282. [CrossRef]
30. Kim, C.; Gao, R.; Sei, E.; Brandt, R.; Hartman, J.; Hatschek, T.; Crosetto, N.; Foukakis, T.; Navin, N. Chemoresistance Evolution in Triple-Negative Breast Cancer Delineated by Single-Cell Sequencing. *Cell* **2018**, *173*, 879–893. [CrossRef]

31. Echeverria, G.V.; Ge, Z.; Seth, S.; Zhang, X.; Jeter-Jones, S.; Zhou, X.; Cai, S.; Tu, Y.; McCoy, A.; Peoples, M.D.; et al. Resistance to neoadjuvant chemotherapy in triple-negative breast cancer mediated by a reversible drug-tolerant state. *Sci. Transl. Med.* **2019**, *11*, eaav0936. [CrossRef] [PubMed]

32. Lopes, C.; Dourado, A.; Oliveira, R.C.S. Phytotherapy and Nutritional Supplements on Breast Cancer. *BioMed Res. Int.* **2017**, *2017*, 7207983. [CrossRef] [PubMed]

33. Yoon, M.J.; Kim, E.; Lim, J.H.; Kwon, T.K.; Choi, K. Superoxide anion and proteasomal dysfunction contribute to curcumin-induced paraptosis of malignant breast cancer cells. *Free. Radic. Boil. Med.* **2010**, *48*, 713–726. [CrossRef] [PubMed]

34. Liu, D.; Chen, Z. The Effect of Curcumin on Breast Cancer Cells. *J. Breast Cancer* **2013**, *16*, 133–137. [CrossRef] [PubMed]

35. Bayet-Robert, M.; Kwiatowski, F.; Leheurteur, M.; Gachon, F.; Planchat, E.; Abrial, C.; Mourret-Reynier, M.-A.; Durando, X.; Barthomeuf, C.; Chollet, P. Phase I dose escalation trial of docetaxel plus curcumin in patients with advanced and metastatic breast cancer. *Cancer Boil. Ther.* **2010**, *9*, 8–14. [CrossRef] [PubMed]

36. Bode, A.M.; Dong, Z. Chemopreventive Effects of Licorice and Its Components. *Curr. Pharmacol. Rep.* **2015**, *1*, 60–71. [CrossRef]

37. Peng, F.; Tang, H.; Liu, P.; Shen, J.; Guan, X.-Y.; Xie, X.; Gao, J.; Xiong, L.; Jia, L.; Chen, J.; et al. Isoliquiritigenin modulates miR-374a/PTEN/Akt axis to suppress breast cancer tumorigenesis and metastasis. *Sci. Rep.* **2017**, *7*, 9022. [CrossRef]

38. Tian, T.; Sun, J.; Wang, J.; Liu, Y.; Liu, H. Isoliquiritigenin inhibits cell proliferation and migration through the PI3K/AKT signaling pathway in A549 lung cancer cells. *Oncol. Lett.* **2018**, *16*, 6133–6139. [CrossRef]

39. Lin, P.-H.; Kung, H.-L.; Chen, H.-Y.; Huang, K.-C.; Hsia, S.-M. Isoliquiritigenin Suppresses E2-Induced Uterine Leiomyoma Growth through the Modulation of Cell Death Program and the Repression of ECM Accumulation. *Cancers* **2019**, *11*, 1131. [CrossRef]

40. Wu, C.-H.; Chen, H.-Y.; Wang, C.-W.; Shieh, T.-M.; Huang, T.-C.; Lin, L.-C.; Wang, K.-L.; Hsia, S.-M. Isoliquiritigenin induces apoptosis and autophagy and inhibits endometrial cancer growth in mice. *Oncotarget* **2016**, *7*, 73432–73447. [CrossRef]

41. Liang, X.H.; Jackson, S.; Seaman, M.; Brown, K.; Kempkes, B.; Hibshoosh, H.; Levine, B. Induction of autophagy and inhibition of tumorigenesis by beclin 1. *Nature* **1999**, *402*, 672–676. [CrossRef] [PubMed]

42. Liang, X.H.; Kleeman, L.K.; Jiang, H.H.; Gordon, G.; Goldman, J.E.; Berry, G.; Herman, B.; Levine, B. Protection against Fatal Sindbis Virus Encephalitis by Beclin, a Novel Bcl-2-Interacting Protein. *J. Virol.* **1998**, *72*, 8586–8596. [CrossRef]

43. Islam, A.; Sooro, M.A.; Zhang, P. Autophagic Regulation of p62 is Critical for Cancer Therapy. *Int. J. Mol. Sci.* **2018**, *19*, 1405. [CrossRef] [PubMed]

44. Vallecillo-Hernández, J.; Barrachina, M.D.; Ortiz-Masia, M.D.; Coll, S.; Esplugues, J.V.; Calatayud, S.; Hernández, C. Indomethacin Disrupts Autophagic Flux by Inducing Lysosomal Dysfunction in Gastric Cancer Cells and Increases Their Sensitivity to Cytotoxic Drugs. *Sci. Rep.* **2018**, *8*, 3593. [CrossRef] [PubMed]

45. Choi, Y.K.; Cho, S.-G.; Choi, Y.-J.; Yun, Y.J.; Lee, K.M.; Lee, K.; Yoo, H.-H.; Shin, Y.C.; Ko, S.-G. SH003 suppresses breast cancer growth by accumulating p62 in autolysosomes. *Oncotarget* **2016**, *8*, 88386–88400. [CrossRef] [PubMed]

46. Bao, W.; Fan, Q.; Luo, X.; Cheng, W.-W.; Wang, Y.-D.; Li, Z.-N.; Chen, X.-L.; Wu, D. Silencing of Cathepsin B suppresses the proliferation and invasion of endometrial cancer. *Oncol. Rep.* **2013**, *30*, 723–730. [CrossRef] [PubMed]

47. Cuendet, M.; Guo, J.; Luo, Y.; Chen, S.-N.; Oteham, C.P.; Moon, R.C.; Van Breemen, R.B.; Marler, L.E.; Pezzuto, J.M. Cancer chemopreventive activity and metabolism of isoliquiritigenin, a compound found in licorice. *Cancer Prev. Res.* **2010**, *3*, 221–232. [CrossRef]

48. Wang, Z.; Wang, N.; Han, S.; Wang, N.; Mo, S.; Yu, L.; Huang, H.; Tsui, K.; Shen, J.; Chen, J. Dietary Compound Isoliquiritigenin Inhibits Breast Cancer Neoangiogenesis via VEGF/VEGFR-2 Signaling Pathway. *PLoS ONE* **2013**, *8*, e68566. [CrossRef]

49. Wang, N.; Wang, Z.; Peng, C.; You, J.; Shen, J.; Han, S.; Chen, J. Dietary compound isoliquiritigenin targets GRP78 to chemosensitize breast cancer stem cells via β-catenin/ABCG2 signaling. *Carcinogenesis* **2014**, *35*, 2544–2554. [CrossRef]

50. Moreno-Londoño, Á.P.; Bello-Alvarez, C.; Pedraza-Chaverri, J. Isoliquiritigenin pretreatment attenuates cisplatin induced proximal tubular cells (LLC-PK1) death and enhances the toxicity induced by this drug in bladder cancer T24 cell line. *Food Chem. Toxicol.* **2017**, *109 Pt 1*, 143–154.
51. Lee, C.K.; Son, S.H.; Park, K.K.; Park, J.H.Y.; Lim, S.S.; Chung, W.-Y. Isoliquiritigenin inhibits tumor growth and protects the kidney and liver against chemotherapy-induced toxicity in a mouse xenograft model of colon carcinoma. *J. Pharmacol. Sci.* **2008**, *106*, 444–451. [CrossRef] [PubMed]

© 2020 by the authors. Licensee MDPI, Basel, Switzerland. This article is an open access article distributed under the terms and conditions of the Creative Commons Attribution (CC BY) license (http://creativecommons.org/licenses/by/4.0/).

Article

Phenol-Rich *Feijoa sellowiana* (Pineapple Guava) Extracts Protect Human Red Blood Cells from Mercury-Induced Cellular Toxicity

Fabiana Tortora [1], Rosaria Notariale [1], Viviana Maresca [2], Katrina Vanessa Good [3], Sergio Sorbo [4], Adriana Basile [2], Marina Piscopo [2,*] and Caterina Manna [1,*]

[1] Department of Precision Medicine, School of Medicine, University of Campania "Luigi Vanvitelli", via Luigi de Crecchio, 80138 Naples, Italy
[2] Department of Biology, University of Naples Federico II, via Cupa Nuova Cinthia, 80126 Naples, Italy
[3] Department of Biochemistry & Microbiology, University of Victoria, Victoria, BC V8W 3R4, Canada
[4] Ce.S.M.A, Microscopy Section, University of Naples Federico II, via Cupa Nuova Cinthia, 80126 Naples, Italy
* Correspondence: marina.piscopo@unina.it (M.P.); caterina.manna@unicampania.it (C.M.)

Received: 6 June 2019; Accepted: 8 July 2019; Published: 11 July 2019

Abstract: Plant polyphenols, with broadly known antioxidant properties, represent very effective agents against environmental oxidative stressors, including mercury. This heavy metal irreversibly binds thiol groups, sequestering endogenous antioxidants, such as glutathione. Increased incidence of food-derived mercury is cause for concern, given the many severe downstream effects, ranging from kidney to cardiovascular diseases. Therefore, the possible beneficial properties of *Feijoa sellowiana* against mercury toxicity were tested using intact human red blood cells (RBC) incubated in the presence of $HgCl_2$. Here, we show that phenol-rich (10–200 μg/mL) extracts from the *Feijoa sellowiana* fruit potently protect against mercury-induced toxicity and oxidative stress. Peel and pulp extracts are both able to counteract the oxidative stress and thiol decrease induced in RBC by mercury treatment. Nonetheless, the peel extract had a greater protective effect compared to the pulp, although to a different extent for the different markers analyzed, which is at least partially due to the greater proportion and diversity of polyphenols in the peel. Furthermore, *Feijoa sellowiana* extracts also prevent mercury-induced morphological changes, which are known to enhance the pro-coagulant activity of these cells. These novel findings provide biochemical bases for the pharmacological use of *Feijoa sellowiana*-based functional foods in preventing and combating mercury-related illnesses.

Keywords: feijoa extracts; mercury; red blood cells; oxidative stress; glutathione; thiol groups; functional food

---

1. Introduction

Feijoa sellowiana (Feijoa), commonly known as pineapple guava, is an evergreen shrub in the Mytraceae family that is native to South America. It is commonly cultivated in tropical and subtropical countries, such as Brazil, Uruguay, Paraguay, and Argentina, but its cultivation has been extended to other countries, including Italy. *Feijoa* fruit, an intensely fragrant dark green oval berry, is commonly eaten fresh or as a variety of commercially available processed foods, such as jam, ice cream, and yoghurt [1,2].

The advances in chemical composition and biological activities of different botanical parts of *Feijoa* have recently been summarized in a mini-review by Fan Zhu [3]. The fruit is the most utilized botanical part of *Feijoa*, and its nutritional value is generally defined by the presence of dietary fiber, essential amino acids, potassium, and vitamins, including vitamin C. Recent studies have added to *Feijoa* nutritional properties, including high folic acid content, and particularly high iodine (3 mg/100 g wet weight) [4]. This fruit also contains pharmacologically relevant bioactive phytochemicals in the pulp and peel, including polyphenols, the major contributors of its bioactivities [3]. Plant polyphenols are widely known for their antioxidant activity [5], and *Feijoa* contains a variety of these compounds, including epigallocatechin, procyanidin B2, epigallocatechin gallate, myricetin-3-O-galactoside, epicatechin gallate, quercetin-3-galactoside, and quercetin-3-rhamnoside [6]. *Feijoa* extract has demonstrated antioxidant activity in a variety of human blood cells, as well as in vivo in rats [7,8]. Potential mechanisms of this activity have been studied, first by Rossi et al., wherein *Feijoa* extract suppressed nitric oxide production in murine macrophages via either attenuation of the nuclear factor κB (NF-κB) or activation of mitogen-activated protein kinase (MAPK) [9]. Additional activities include anti-microbial, anti-inflammatory, anti-cancer, and anti-diabetic capacities [10–15]. In human intestinal epithelial cells, *Feijoa* fruit improved lactase and sucrose-isomaltase activity and reduced proliferation, but was not cytotoxic, and also prevented lipid peroxidation [16].

The claimed *Feijoa* biological activities suggest that its bioactive components have great potential to be developed into functional food or natural supplements [17]. In addition to diet, *Feijoa* fruit extract is used for industrial purposes, including cosmetic preparation, due to its emollient, elasticizing, and dermo-purifying properties. Furthermore, *Feijoa* extracts have recently been considered as a potential sunscreen resource, due to their proven high absorption in the UV region [18].

Heavy metals are common environmental pollutants and come from natural and anthropogenic sources [19–21]. In recent decades, their contamination has increased dramatically [22–24]. In this respect, attention has focused on how bioactive phytochemicals may counteract heavy metal-induced body burden [25]. Of particular concern is the heavy metal mercury (Hg), which exists in several redox forms, each endowed with different bioavailabilities and toxicities [19,20]. It reacts adversely within organisms, but also remains and bioaccumulates within aquatic animal tissue, which is the major Hg source to humans. Hg poisoning in humans manifests as neurological, kidney, immune, and respiratory disorders, and is especially toxic in pregnant women [26–30]. Evidence from human and animal studies suggests that mercury also affects reproductive function [31], as do other heavy metals, such as copper [32–34] and cadmium [35]. Moreover, high levels of Hg may induce or exacerbate cardiovascular diseases [36–38]. Hg travels through the bloodstream and binds with high affinity to sulphydryl groups (-SH) [39]. This effectively increases the concentration of reactive oxygen species (ROS) in blood and tissue cells by inactivating or sequestering antioxidants, such as glutathione (GSH) [40]. Recent data from our group showed that hydroxytyrosol, an antioxidant phenol present in high concentrations in virgin olive oil, has the potential to modulate Hg toxicity in red blood cells (RBC), confirming the viability of designing nutritional strategies to counteract the adverse effects of Hg exposure in humans [41–43].

The aim of this study was to extend the relatively new research on the beneficial properties of *Feijoa* by testing the role of its components against Hg toxicity using intact human RBC. These anucleated cells, free of cellular organelles, represent a particularly important cellular model to study heavy metal toxicity [44,45]. First, we evaluate the ability of *Feijoa* peel and pulp acetonic extracts to protect human RBC from Hg-induced hemolysis. Then, the antioxidant properties of *Feijoa* extracts are measured in RBC by their ability to counteract the Hg-induced increase in ROS cellular generation. Protection against GSH, as well as membrane thiol depletion, are also measured. Moreover, confocal laser scanning microscopy was used to determine the protective effect of *Feijoa* extracts on Hg-treated RBC morphological alterations, as well as microvesicle (MV) formation. Exposure, even at low Hg concentrations, induces morphological changes that cause an increase in the procoagulant activity of these cells [46].

2. Materials and Methods

2.1. Chemicals

DCFH-DA (2,7-Dichlorodihydrofluorescein diacetate), mercuric chloride ($HgCl_2$), and DTNB (5,5-dithiobis (2-nitrobenzoic acid), or Ellman's reagent) were from Sigma Chemical Co. (St. Louis, Missouri, USA). Anti-Glicophorin A antibody (FITC) was purchased from antibodies-online.com (ABIN6253946, antibodies-online GmbH, Aachen, Germany).

2.2. Pulp and Peel Acetonic Extraction of Feijoa Fruits

A total of 60 g of fruit, subdivided into pulp and peel, either fresh or after storage at −5 °C, were treated with 0.8% water solution of Triton X-100 to remove epiphytic hosts normally found on the surface. They were then extensively washed in tap and distilled water and dried on filter paper. Peel and pulp separately underwent extraction with acetone for 15 min in a liquefier blender until homogenized. The resulting homogenate was centrifuged at $1800 \times g$ for 10 min; the supernatant was filtered, and the solvent was evaporated under reduced pressure at 45 °C for about 24 h. The dry extracts were solubilized with pure dimethyl sulfoxide (DMSO).

2.3. Preparation of Red Blood Cells and Treatment with $HgCl_2$

Whole blood was obtained with informed consent from healthy volunteers at Campania University "Luigi Vanvitelli" in Naples, Italy. It was deprived of leucocytes and platelets by filtration in a nylon net and washed twice with isotonic saline solution (0.9% NaCl); the resulting intact RBC were resuspended in buffer A (5 mM Tris-HCl pH 7.4, 0.9% NaCl, 1 mM $MgCl_2$, and 2.8 mM glucose) to obtain a 10% hematocrit, as previously described [41]. Intact RBC were incubated at 37 °C with 40 μM $HgCl_2$ for 4 h or 24 h. For experiments with *Feijoa* pulp and peel extracts, stock solutions, prepared in DMSO as above described, were diluted in buffer A to a final DMSO concentration of about 0.02%, in order to avoid DMSO cytotoxicity. As a control, the effect of this volume of DMSO on RBC was evaluated and found to be negligible (data not shown). RBC from each donor were used for a single assay in triplicate. Each experiment was repeated on RBC obtained from three different donors.

2.4. Hemolysis Assay

RBC hemolysis extent was determined spectrophotometrically, according to Tagliafierro et al. [41]. After simultaneous treatment with $HgCl_2$ and *Feijoa* extracts for 24 h, the reaction mixture was centrifuged at $1100 \times g$ for 5 min, and the released hemoglobin (Hb) in the supernatant was evaluated by measuring the absorption at 540 nm (A). As a positive control, packed RBC were used hemolyzed with ice-cold distilled water at 40:1 v/v, and by measuring the A540 of the supernatant obtained centrifuging the suspension at $1500 \times g$ for 10 min (B). The percentage of hemolysis was calculated as the ratio of the readings (A/B) × 100%.

2.5. Determination of Reactive Oxygen Species

ROS generation was determined using the dichlofluorescein (DCF) assay, according to Tagliafierro et al. [41]. Using this method, 250 μL of intact RBC (hematocrit 10%) were incubated with the non-polar, non-fluorescent 2′,7′-dichlorodihydrofluorescin diacetate (DCFH-DA) at a final concentration of 10 μM for 15 min at 37 °C. After centrifuging at room temperature at $1200 \times g$ for 5 min, the supernatant was removed, and the hematocrit was re-adjusted to 10% with buffer A. RBC were then treated concurrently with $HgCl_2$ and *Feijoa* extracts in the dark for 4 h. After the incubation, 20 μL of RBC were diluted in 2 mL of water, and the fluorescence intensity of the oxidized derivative DCF was recorded ($\lambda_{exc} 502$; $\lambda_{em} 520$). The results were expressed as fluorescence intensity/mg of Hb.

2.6. Quantification of Intracellular Glutathione

Intracellular GSH content was determined spectrophotometrically by reacting with DTNB reagent, according to Van den Berg et al. [47]. After co-incubation with $HgCl_2$ and *Feijoa* extracts for 4 h, the samples (0.25 mL) were centrifuged, and the cells were lysed by the addition of 0.6 mL of ice-cold water. Proteins were precipitated with 0.6 mL ice-cold metaphosphoric acid solution (1.67 g metaphosphoric acid, 0.2 g EDTA, and 30 g NaCl in 100 mL of water). After incubation at 4 °C for 5 min, the protein precipitate was removed by centrifugation at $18,000\times g$ for 10 min, and 0.45 mL of the supernatant was mixed with an equal volume of 0.3 M Na_2HPO_4. Then, 100 µL of DTNB solution (20 mg DTNB plus 1% of sodium citrate in 100 mL of water) was then added to the sample, and after a 10 min incubation at room temperature, the absorbance of the sample was read against the blank at 412 nm.

2.7. Estimation of Free Sulfhydryl Groups in Isolated Red Blood Cell Membranes

Free sulfhydryl groups in membrane proteins (2.5 µg/µL for each sample estimated by Bradford assay) were assayed, according to the method of Ellman [48]. To do this, 650 µL of $HgCl_2$-treated RBC were washed three times in 40 volumes of 5 mM sodium phosphate buffer pH 8.8, centrifuged at $10,000\times g$ for 20 min at 4 °C, then washed several times with the same buffer, for the complete removal of Hb. Then, 1 mL of 0.1 M Tris–HCl pH 7.5 was added to 50 µL membrane protein. The colorimetric reaction was started by adding 50 µL of 10 mM DTNB in methanol. After 15 min of incubation at room temperature, the absorbance was read against the blank at 412 nm. Blanks were run for each sample in which DTNB was not added to methanol.

2.8. Morphological Analysis of Red Blood Cells

To investigate the possible protective effect of *Feijoa* extracts on alterations in the Hg-induced erythrocytes' shape, we treated the cells both with 40 µM $HgCl_2$ as well as 20 or 80 µg/mL of *Feijoa* peel and pulp extracts for 4h at 37 °C. After incubation, erythrocytes were washed twice with phosphate-buffered saline pH 7.4 (PBS), and counted in a Burker chamber. The confocal laser scanning microscope analyses were performed according to Nguyen [49], with few modifications. In brief, the cells were then fixed with 2% formaldehyde for 1 h at 4°C, then washed several times and incubated with anti-human anti glycophorin A FITC antibody for 30 min at 4°C in the dark. Afterwards, the samples were placed on glass slides and air-dried for 1 h. The slides were dipped quickly, and gently washed stepwise with ethanol from 50% to 75%, 90%, and then 100% for dehydration. Finally, cells were fixed in 2% formaldehyde and washed three times with PBS. For confocal laser scanning microscope imaging, several randomly selected frames from each sample were captured for morphological observation and statistical strength. Excitation and emission filters were set at 488 nm and 550–600 nm, respectively.

2.9. Statistical Analysis

Data were expressed as mean ± standard error of the mean (SEM). The significance of differences was determined by one-way ANOVA followed by a post hoc Tukey's multiple comparisons test. GraphPad Prism 5 was utilized for statistical analysis.

3. Results

3.1. Feijoa Peel and Pulp Extracts Protect Against Hg-Induced Hemolysis.

The rescue of Hg-induced hemolysis by *Feijoa* extracts was assayed separately for peel and pulp, and the results are shown in Figures 1 and 2, respectively. It can be seen that 24 h treatment of RBC with 40 µM $HgCl_2$ resulted in approximately 13–17% hemolysis, compared to 1–2% in negative controls, as expected based on our previous work [41]. *Feijoa* peel extract potently reduced Hg-induced hemolysis compared to that of the pulp, with a significant 3% drop in hemolysis at 10 µg/mL, and a steady reduction of about 1% with each doubling of peel extract (Figure 1). *Feijoa* pulp treatment reduced cellular lysis in similar proportions, but the required protective extract concentration to do so was almost eight-fold greater than that of the peel extract (Figure 2). No cytotoxic effect was found by either *Feijoa* extracts up to the maximum concentration utilized in this study (data not shown).

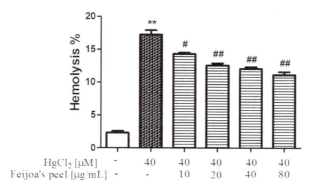

Figure 1. *Feijoa* peel acetonic extract reduces Hg-induced hemolysis. Cells were treated with 40 μM HgCl$_2$ for 24 h with increasing acetonic extract concentrations. Data are the means ± standard errors of the mean (SEM) (n = 9). Statistical significance was calculated by one-way ANOVA followed by Tukey's test. ** ($p < 0.05$) indicates a significant difference from cells lacking HgCl$_2$ treatment. # ($p < 0.05$) and ## ($p < 0.01$) indicate significant differences from cells lacking *Feijoa* extract treatment.

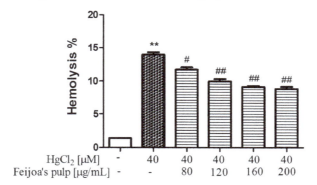

Figure 2. *Feijoa* pulp extract reduces Hg-induced hemolysis. Cells were treated with 40 μM HgCl$_2$ for 24 h in the presence of increasing concentrations of extract. Data are the means ± SEM (n = 9). Statistical significance was calculated with one-way ANOVA followed by Tukey's test. ** ($p < 0.05$) indicates a significant difference from cells lacking HgCl$_2$ treatment. # ($p < 0.05$) and ## ($p < 0.01$) indicate significant differences from cells lacking *Feijoa* extract treatment.

3.2. Feijoa Peel and Pulp Extracts Reduce Reactive Oxygen Species Production in Red Blood Cells

The fluorescence probe DCF assay elucidated the protective role of *Feijoa* extracts against oxidative stress in RBC, as reported in Figures 3 and 4. ROS production increased nearly two-fold in Hg-treated RBC compared to the negative control. In contrast, co-incubation with 10, 20, 40, or 80 μg/mL of both peel and pulp acetonic extracts incrementally reduced ROS production in RBC. Similar to hemolysis, *Feijoa* peel extract prevented ROS production more potently than the pulp, and remarkably reduced fluorescence levels by approximately 50% at 10 μg/mL, to near control values at the highest extract concentration. At the same concentrations, *Feijoa* pulp extract also significantly reduced ROS production compared to the non-*Feijoa* protected RBC, reaching a maximum of about 50% at 80 μg/mL.

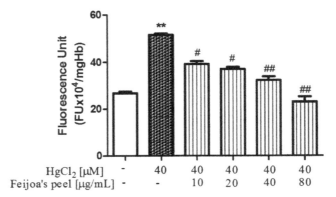

Figure 3. *Feijoa* peel extract protects against Hg-induced reactive oxygen species (ROS) production in red blood cells (RBC). Cells were treated with 40 μM HgCl$_2$ for 4 h in the presence of increasing concentrations of extract. ROS production was determined by fluorescence unit means of the dichlofluorescein (DCF) probe. Data are the means ± SEM ($n = 9$). Statistical significance was calculated with one-way ANOVA followed by Tukey's test. ** ($p < 0.05$) indicates a significant difference from cells lacking HgCl$_2$ treatment. # ($p < 0.05$) and ## ($p < 0.01$) indicate significant differences from cells lacking *Feijoa* extract treatment.

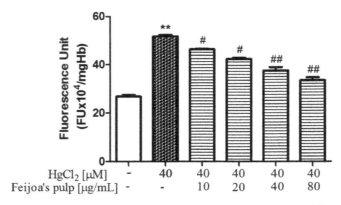

Figure 4. *Feijoa* pulp extract protects against Hg-induced ROS production in RBC. Cells were treated with 40 μM HgCl$_2$ for 4 h in the presence of increasing concentrations of extract. ROS production was determined by fluorescence unit means of the DCF probe. Data are the means ± SEM ($n = 9$). Statistical significance was calculated with one-way ANOVA followed by Tukey's test. ** ($p < 0.05$) indicates a significant difference from cells lacking HgCl$_2$ treatment. # ($p < 0.05$) and ## ($p < 0.01$) indicate significant differences from cells lacking *Feijoa* extract treatment.

3.3. Peel and Pulp Extracts Prevent Hg-Induced Glutathione and Membrane Thiol Depletion in Red Blood Cells

GSH depletion is a key mechanism of Hg toxicity due to the weakening of the antioxidant defense system. We therefore evaluated the possible protective effect of *Feijoa* peel and pulp extracts on this specific, Hg-induced metabolic condition. As shown in Figure 5, 4 h treatments of RBC with 40 μM HgCl$_2$ reduce GSH levels by about 40%. Co-incubation with 20, 80, or 100 μg/mL of *Feijoa* peel extract prevented GSH depletion by about 20% with each concentration, such that GSH levels were unchanged from healthy control levels at the latter two peel extract concentrations. For the pulp extract, data indicate that 20 μg/mL had no effect on GSH levels, while significant protection was observed at 80 and 100 μg/mL, to a maximum of about 90% GSH levels compared to controls.

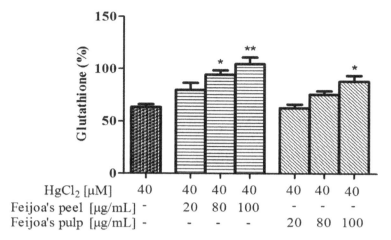

Figure 5. *Feijoa* peel and pulp extracts protect against Hg-induced glutathione (GSH) decrease in RBC. Cells were treated with 40 μM HgCl$_2$ for 4 h in the presence of increasing concentrations of *Feijoa* peel or pulp. Data are the means ± SEM (n = 9). Statistical significance was calculated with one-way ANOVA followed by Tukey's test. * ($p < 0.05$) and ** ($p < 0.01$) indicate significant differences from cells lacking *Feijoa* extract treatment.

Based on these results, it seemed appropriate to evaluate the efficacy of peel and pulp extracts in reducing the Hg-induced depletion of membrane thiols, using membranes obtained from intact RBC after incubation with HgCl$_2$ (Figure 6). Exposure to 40 μM HgCl$_2$ reduced the level of membrane thiols by about 40%. This depletion was significantly counteracted by about 45% and 75% given co-incubation with 40 and 80 μg/mL of peel extract, respectively. Again, the pulp extract was less protective than the peel at the same concentrations, such that membrane thiol depletion was counteracted only at 80 μg/mL, by about 50%.

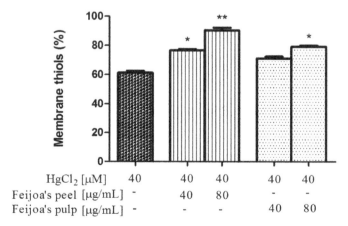

Figure 6. *Feijoa* peel and pulp extracts protect against Hg-induced membrane thiol depletion in RBC. Cells were treated with 40 μM HgCl$_2$ for 4 h in the presence of increasing concentrations of *Feijoa* peel or pulp. Data are the means ± SEM (n = 9). Statistical significance was calculated with one-way ANOVA followed by Tukey's test. * ($p < 0.05$) and ** ($p < 0.01$) indicate significant differences from cells lacking *Feijoa* extract treatment.

3.4. Peel and Pulp Extracts of Feijoa Reduce Microvesicles Released from Red Blood Cells

To investigate the protective role of *Feijoa* extracts on erythrocyte morphological changes and MV formation known to be induced by Hg treatment [41,46], cells treated with HgCl$_2$ and peel or pulp extracts, as described in the Materials and Methods section, were analyzed with confocal microscopy. Hg treatment was associated with loss of the typical erythrocyte biconcave shape, as well as the formation of MV clearly discernible on cell membranes (not observable in the control) (Figure 7, Panel A). Cell treatment with *Feijoa* extracts completely restored the typical biconcave shape at 20 µg/mL and 80 µg/mL for peel and pulp, respectively (Figure 7C–F).

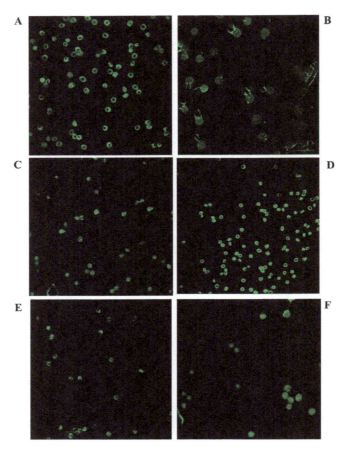

Figure 7. Peel and pulp extracts of *Feijoa* reduce microvesicles (MV) released from RBC. Untreated cells are shown in (**A**). Cells were treated with 40 µM HgCl$_2$ for 4 h (**B**) and concurrently treated with 20 or 80 µg/mL of *Feijoa* peel (**C** and **D**, respectively) and pulp (**E** and **F**, respectively) extracts. RBC were stained with Annexin V-FITC.

4. Discussion

Mercury is not only highly toxic, but is an increasingly pervasive dietary heavy metal. As a matter of fact, concerns on the effect of Hg exposure on human health are not only limited to occupationally exposed workers, but also to the general population, mainly via contaminated food ingestion. Although in some European populations the overall Hg daily intake is below the tolerable amount [50,51], appreciable proportions of large fish populations are reported to contain levels of this heavy metal exceeding this amount, up to 2.22 mg/kg wet weight, including anglerfish (*Lophius piscatorius*) and black-bellied angler (*Lophius budegassa*) [52]. Discovering analogous means to simultaneously combat and protect against diet-based Hg toxicity is therefore crucial to public health. In this respect, phytochemicals able to counteract structural and metabolic alterations associated with heavy metal exposure are attractive for the reduction of their toxicity [53–56]. Data from our group indicate that hydroxytyrosol, an olive oil-derived phenolic antioxidant, has the potential to modulate the toxic effects exerted by Hg in human RBC [41–43].

To expand data on the potential role of nutrition in heavy metal toxicity, intact human RBC were exposed to 40 µM $HgCl_2$, in line with our previous studies. Several markers of cellular toxicity were then evaluated to test the protective effect of *Feijoa* fruit extracts. According to data reported in similar experimental conditions, RBC treatment with 40 µM $HgCl_2$ for 4 h results in a doubling of ROS production, as indicated by DCF fluorescence [41]. Hg-induced ROS generation follows a significant decrease of GSH, which builds up a pro-oxidative microenvironment and renders cells more susceptible to ROS-mediated oxidative damage. A significant decrease in membrane thiols is also detectable in Hg-treated cells. The resulting hemolysis is significantly increased and measurable later, at 24 h.

Here we show the first evidence of *Feijoa* fruit extract protection against $HgCl_2$-induced toxic effects in human RBC. The acetonic extracts of both the pulp and peel were able to counteract oxidative stress and cellular thiol decrease in Hg-treated RBC. The peel extract had a greater protective effect compared to the pulp, although to varying extents for the different markers analyzed, which is at least partially due to the greater proportion and diversity of polyphenols in the peel [3,57,58]. Interestingly, the protective effect of the peel from ROS production is only two-fold, compared to an eight-fold effect against overall cytotoxicity indicated by hemolysis. Whereas Hg sequesters and inactivates GSH by binding to sulfhydryl groups, polyphenols act on the resulting ROS by virtue of their hydrogen and electron transfer abilities. The presence of additional bioactive compounds with different activities in the peel (i.e., chelating properties) can also be hypothesized. Remarkably, as little as 10 µg/mL of *Feijoa* peel extract significantly affects all the tested markers, and 80 µg/mL completely prevents Hg-induced ROS production.

The data presented in this paper, although obtained from in vitro studies on human cells, also offer significant experimental evidence that *Feijoa* extracts prevent Hg-induced RBC shape alteration in RBC, which could be taken into account for future clinical investigations. In fact, although a particularly high GSH concentration may partially protect RBC from Hg's toxic effects, chronic exposure could affect RBC viability and induce morphological changes, also affecting cardiovascular disease. As mentioned before, Hg exposure enhances pro-coagulant activity of these cells, resulting in a contributing factor for Hg-related thrombotic disease [46]. In our previous studies, we raised the fascinating hypothesis that metabolic and shape modification of RBC may be regarded as a clinical biomarker, indicating increased cardiovascular risk in Hg-exposed individuals [41,42].

Our findings, in agreement with the literature data, strengthen the nutritional relevance of *Feijoa* bioactive compounds to the claimed health-promoting effects of this fruit. There is growing interest in utilizing *Feijoa* fruit for human consumption, due to its appetizing quality and its claimed health benefits. *Feijoa* fruit is an excellent source of vitamins and nonessential nutrients, as well as a variety of bioactive compounds endowed with significant antioxidant, antibacterial, and anti-inflammatory activities [10–13]. In this respect, there is a general agreement that the health-promoting effects of fruit and vegetable intake result from the combined properties and synergistic action of all bioactive constituents, including polyphenols [59,60]. These compounds can improve health due to their strong antioxidant activity, counteracting oxidative stress-induced cellular dysfunctions and modulating key mechanisms implicated in the development of oxidative stress-related human pathologies. Polyphenols are very useful in combating the deleterious effects of heavy metals. For example, Sobeh et al. [61] isolated and identified two compounds from the leaves of *Syzygium samarangense* (myricitrin and 3,5-di-O-methyl gossypetin), both showing antioxidant activities [62,63] and strongly reducing intracellular ROS accumulation and carbonyl content, while also protecting the intercellular GSH levels in keratinocytes (HaCaT) after exposure to sodium arsenite, one of the more toxic environmental heavy metals [61].

Feijoa has been proposed as an ideal candidate for nutraceutical strategies in the development of functional foods [64]. The data reported in this paper expands upon the known beneficial effects of *Feijoa* fruit, particularly related to chronic human exposure to heavy metal. In this respect, an interesting observation is that the very low active concentrations utilized in our study could be approached in vivo upon daily intake of *Feijoa* fruit. In this respect, some studies indicate that *Feijoa* fruit extracts are well tolerated in animal models. Karami et al. [65] demonstrated the hepatoprotective activity of methanolic extract of *Feijoa* fruit in a concentration range of 10–100 mg/kg, using the isolated rat liver perfusion system. The same group also investigated nephroprotective effects of leaf extracts (10–40 mg/kg) on renal injury induced by acute doses of ecstasy (MDMA) in mice [66]. Moreover, in a recent study, *Feijoa* leaf extract was shown to be devoid of toxicity in rats up to 2 g/Kg, [11]. Finally, we have confirmed by MTT test, on human leucocytes as well (data not shown), that treatment for 24 h with 5, 50, and 500 μg/mL of acetonic extracts of *F. sellowiana* did not induce significant cytotoxic effects, as already demonstrated on either the Caco-2 or HT-29 cell lines [16].

The food industry is increasingly interested in the utilization of non-edible parts of fruits. Phytochemicals are proposed for designing foods with added functional value, aiming to beneficially affect target functions in the body and reduce the risk of diseases. These compounds are present in large quantities in waste products from the agri-food supply chain, especially peels and seeds. Our data, showing a greater protective effect from the *Feijoa* peel on Hg cytotoxicity than from the pulp, corroborates this rationale. Recovering and using such a waste product, normally destined to magnify industrial waste production, would give new life to the less noble part of the fruit. This is further in line with recent studies that propose the potential utilization of *Feijoa* fruit peel for added processing and functional value. As demonstrated by Sun-Waterhouse et al. [64], the extracts produced from *Feijoa* waste, such as the peel, retain high pectin content, which is advantageous for food applications. Moreover, the possibility to utilize *Feijoa* peel-containing food packaging film for the inhibition of foodborne bacteria was recently demonstrated [67].

In conclusion, the novel beneficial properties of *Feijoa* reported in this paper, regarding its efficacy to reduce heavy metal toxicity in human RBC, provide biochemical bases for the use of *Feijoa*-based functional foods or pharmacological preparations in preventing and combating mercury-related illnesses.

Author Contributions: Conceptualization, C.M. and M.P.; methodology, F.T., R.N., S.S., and A.B.; investigation, F.T., R.N., and V.M.; data curation, F.T. and R.N.; formal analysis, F.T., R.N., and A.B.; writing—original draft preparation, C.M., M.P., and K.V.G.; writing—review and editing, C.M., M.P., K.V.G., and V.M.; supervision, C.M. and M.P.; project administration, C.M.

Funding: This research received no external funding.

Conflicts of Interest: The authors declare no conflict of interest.

References

1. Sharpe, R.H.; Sherman, W.B.; Miller, E.P. Feijoa history and improvement. *Proc. Fla. State Hort. Soc.* **1996**, *106*, 134–139.
2. Weston, R.J. Bioactive products from fruit of the feijoa (Feijoa sellowiana, Myrtaceae): A review. *Food Chem.* **2010**, *121*, 923–926. [CrossRef]
3. Zhu, F. Chemical and biological properties of feijoa (Acca sellowiana). *Trends Food Sci. Technol.* **2018**, *81*, 121–131. [CrossRef]
4. Ferrara, L.; Montesano, D. Nutritional characteristics of Feijoa sellowiana fruit. The iodine content. *Riv. Di Sci. Dell Aliment. (Italy)* **2001**, *30*, 353–356.

5. Koch, W. Dietary Polyphenols-Important Non-Nutrients in the Prevention of Chronic Noncommunicable Diseases. A Systematic Review. *Nutrients* **2019**, *11*, 1039. [CrossRef] [PubMed]
6. Beyhan, Ö.; Elmastas, M.; Gedikli, F. Total phenolic compounds and antioxidant capacity of leaf, dry fruit and fresh fruit of feijoa (Acca sellowiana, Myrtaceae). *J. Med. Plants Res.* **2010**, *4*, 1065–1072.
7. Ielpo, M.T.L.; Moscatiello, V.; Satriano, S.M.R.; Vuotto, M.L.; Basile, A. Inhibitory activity of feijoa sellowiana fruits of human pmn oxidative metabolism. *Pharmacol. Res.* **1997**, *35*, 69.
8. Keles, H.; Ince, S.; Küçükkurt, I.; Tatli, I.I.; Akkol, E.K.; Kahraman, C.; Demirel, H.H. The effects of Feijoa sellowiana fruits on the antioxidant defense system, lipid peroxidation, and tissue morphology in rats. *Pharm. Biol.* **2012**, *50*, 318–325. [CrossRef]
9. Rossi, A.; Rigano, D.; Pergola, C.; Formisano, C.; Basile, A.; Bramanti, P.; Senatore, F.; Sautebin, L. Inhibition of Inducible Nitric Oxide Synthase Expression by an Acetonic Extract fromFeijoa sellowiana Berg. Fruits. *J. Agric. Food Chem.* **2007**, *55*, 5053–5061. [CrossRef]
10. Vuotto, M.L.; Basile, A.; Moscatiello, V.; De Sole, P.; Castaldo-Cobianchi, R.; Laghi, E.; Ielpo, M.T. Antimicrobial and antioxidant activities of Feijoa sellowiana fruit. *Int. J. Antimicrob. Agents* **2000**, *13*, 197–201. [CrossRef]
11. Mosbah, H.; Louati, H.; Boujbiha, M.A.; Chahdoura, H.; Snoussi, M.; Flamini, G.; Ascrizzi, R.; Bouslema, A.; Achour, L.; Selmi, B. Phytochemical characterization, antioxidant, antimicrobial and pharmacological activities of Feijoa sellowiana leaves growing in Tunisia. *Ind. Crop. Prod.* **2018**, *112*, 521–531. [CrossRef]
12. Monforte, M.T.; Fimiani, V.; Lanuzza, F.; Naccari, C.; Restuccia, S.; Galati, E.M. Feijoa sellowiana Berg Fruit Juice: Anti-Inflammatory Effect and Activity on Superoxide Anion Generation. *J. Med. Food* **2014**, *17*, 455–461. [CrossRef] [PubMed]
13. Peng, Y.; Bishop, K.S.; Ferguson, L.R.; Quek, S.Y. Screening of Cytotoxicity and Anti-Inflammatory Properties of Feijoa Extracts Using Genetically Modified Cell Models Targeting TLR2, TLR4 and NOD2 Pathways, and the Implication for Inflammatory Bowel Disease. *Nutrients* **2018**, *10*, 1188. [CrossRef] [PubMed]
14. Piscopo, M.; Tenore, G.C.; Notariale, R.; Maresca, V.; Maisto, M.; De Ruberto, F.; Heydari, M.; Sorbo, S.; Basile, A. Antimicrobial and antioxidant activity of proteins from Feijoa sellowiana Berg. fruit before and after in vitro gastrointestinal digestion. *Nat. Prod. Res.* **2019**, 1–5. [CrossRef] [PubMed]
15. Motohashi, N.; Kawase, M.; Shirataki, Y.; Tani, S.; Saito, S.; Sakagami, H.; Kurihara, T.; Nakashima, H.; Wolfard, K.; Mucsi, I.; et al. Biological activity of feijoa peel extracts. *Anticancer Res.* **2000**, *20*, 4323–4329.
16. Turco, F.; Palumbo, I.; Andreozzi, P.; Sarnelli, G.; De Ruberto, F.; Esposito, G.; Basile, A.; Cuomo, R. Acetonic Extract from the Feijoa sellowiana Berg. Fruit Exerts Antioxidant Properties and Modulates Disaccharidases Activities in Human Intestinal Epithelial Cells. *Phytother. Res.* **2016**, *30*, 1308–1315. [CrossRef] [PubMed]
17. Sun-Waterhouse, D.; Sun-Waterhouse, D. The development of fruit-based functional foods targeting the health and wellness market: A review. *Int. J. Food Sci. Technol.* **2011**, *46*, 899–920. [CrossRef]
18. Ebrahimzadeh, M.A.; Enayatifard, R.; Khalili, M.; Ghaffarloo, M.; Saeedi, M.; Charati, J.Y. Correlation between Sun Protection Factor and Antioxidant Activity, Phenol and Flavonoid Contents of some Medicinal Plants. *Iran. J. Pharm. Res. IJPR* **2014**, *13*, 1041–1047.
19. Driscoll, C.T.; Mason, R.P.; Chan, H.M.; Jacob, D.J.; Pirrone, N. Mercury as a Global Pollutant: Sources, Pathways, and Effects. *Environ. Sci. Technol.* **2013**, *47*, 4967–4983. [CrossRef]
20. Spiegel, S.J. New mercury pollution threats: A global health caution. *Lancet* **2017**, *390*, 226–227. [CrossRef]
21. Sundseth, K.; Pacyna, J.M.; Pacyna, E.G.; Pirrone, N.; Thorne, R.J. Global Sources and Pathways of Mercury in the Context of Human Health. *Int. J. Environ. Res. Public Health* **2017**, *14*, 105. [CrossRef] [PubMed]
22. Basile, A.; Loppi, S.; Piscopo, M.; Paoli, L.; Vannini, A.; Monaci, F.; Sorbo, S.; Lentini, M.; Esposito, S. The biological response chain to pollution: A case study from the "Italian Triangle of Death" assessed with the liverwort Lunularia cruciata. *Environ. Sci. Pollut. Res.* **2017**, *24*, 26185–26193. [CrossRef] [PubMed]
23. Maresca, V.; Fusaro, L.; Sorbo, S.; Siciliano, A.; Loppi, S.; Paoli, L.; Monaci, F.; Karam, E.A.; Piscopo, M.; Guida, M.; et al. Functional and structural biomarkers to monitor heavy metal pollution of one of the most contaminated freshwater sites in Southern Europe. *Ecotoxicol. Environ. Saf.* **2018**, *163*, 665–673. [CrossRef] [PubMed]
24. Piscopo, M.; Ricciardiello, M.; Palumbo, G.; Troisi, J. Selectivity of metal bioaccumulation and its relationship with glutathione S-transferase levels in gonadal and gill tissues of Mytilus galloprovincialis exposed to Ni (II), Cu (II) and Cd (II). *Rend. Lincei* **2016**, *27*, 737–748. [CrossRef]
25. Officioso, A.; Tortora, F.; Manna, C. Nutritional Aspects of Food Toxicology: Mercury Toxicity and Protective Effects of Olive Oil Hydroxytyrosol. *J. Nutr. Food Sci.* **2016**, *6*, 6.

26. Andrew, A.S.; Chen, C.Y.; Caller, T.A.; Tandan, R.; Henegan, P.L.; Jackson, B.P.; Hall, B.P.; Bradley, W.G.; Stommel, E.W. Toenail mercury Levels are associated with amyotrophic lateral sclerosis risk. *Muscle Nerve* **2018**, *58*, 36–41. [CrossRef]
27. Carocci, A.; Rovito, N.; Sinicropi, M.S.; Genchi, G. Mercury toxicity and neurodegenerative effects. In *Reviews of Environmental Contamination and Toxicology*; Springer: Berlin/Heidelberg, Germany, 2014; Volume 229, pp. 1–18.
28. Miller, S.; Pallan, S.; Gangji, A.S.; Lukic, D.; Clase, C.M. Mercury-associated nephrotic syndrome: A case report and systematic review of the literature. *Am. J. Kidney Dis.* **2013**, *62*, 135–138. [CrossRef]
29. Silbergeld, E.K.; Silva, I.A.; Nyland, J.F. Mercury and autoimmunity: Implications for occupational and environmental health. *Toxicol. Appl. Pharmacol.* **2005**, *207*, 282–292. [CrossRef]
30. Tinkov, A.A.; Ajsuvakova, O.P.; Skalnaya, M.G.; Popova, E.V.; Sinitskii, A.I.; Nemereshina, O.N.; Gatiatulina, E.R.; Nikonorov, A.A.; Skalny, A.V. Mercury and metabolic syndrome: A review of experimental and clinical observations. *Biometals* **2015**, *28*, 231–254. [CrossRef]
31. Henriques, M.C.; Loureiro, S.; Fardilha, M.; Herdeiro, M.T. Exposure to mercury and human reproductive health: A systematic review. *Reprod. Toxicol.* **2019**, *85*, 93–103. [CrossRef]
32. Lettieri, G.; Maione, M.; Ranauda, M.A.; Mele, E.; Piscopo, M. Molecular effects on spermatozoa of Mytilus galloprovincialis exposed to hyposaline conditions. *Mol. Reprod. Dev.* **2019**, *86*, 650–660. [CrossRef] [PubMed]
33. Piscopo, M.; Trifuoggi, M.; Scarano, C.; Gori, C.; Giarra, A.; Febbraio, F. Relevance of arginine residues in Cu(II)-induced DNA breakage and Proteinase K resistance of H1 histones. *Sci. Rep.* **2018**, *8*, 7414. [CrossRef] [PubMed]
34. Vassalli, Q.A.; Caccavale, F.; Avagnano, S.; Murolo, A.; Guerriero, G.; Fucci, L.; Ausió, J.; Piscopo, M. New insights into protamine-like component organization in Mytilus galloprovincialis' sperm chromatin. *DNA Cell Biol.* **2015**, *34*, 162–169. [CrossRef] [PubMed]
35. De Guglielmo, V.; Puoti, R.; Notariale, R.; Maresca, V.; Ausió, J.; Troisi, J.; Verrillo, M.; Basile, A.; Febbraio, F.; Piscopo, M. Alterations in the properties of sperm protamine-like II protein after exposure of Mytilus galloprovincialis (Lamarck 1819) to sub-toxic doses of cadmium. *Ecotoxicol. Environ. Saf.* **2019**, *169*, 600–606. [CrossRef] [PubMed]
36. Houston, M.C. The Role of Mercury in Cardiovascular Disease. *J. Cardiovasc. Dis. Diagn.* **2014**, *2*, 170. [CrossRef]
37. Genchi, G.; Sinicropi, M.S.; Carocci, A.; Lauria, G.; Catalano, A. Mercury Exposure and Heart Diseases. *Int. J. Environ. Res. Public Health* **2017**, *14*, 74. [CrossRef] [PubMed]
38. Hu, X.F.; Singh, K.; Chan, H.M. Mercury Exposure, Blood Pressure, and Hypertension: A Systematic Review and Dose–response Meta-analysis. *Environ. Heal. Perspect.* **2018**, *126*, 076002. [CrossRef]
39. Rooney, J.P.K. The role of thiols, dithiols, nutritional factors and interacting ligands in the toxicology of mercury. *Toxicology* **2007**, *234*, 145–156. [CrossRef]
40. Oram, P.D.; Fang, X.; Fernando, Q.; Letkeman, P.; Letkeman, D. The formation of constants of mercury(II)–glutathione complexes. *Chem. Res. Toxicol.* **1996**, *9*, 709–712. [CrossRef]
41. Tagliafierro, L.; Officioso, A.; Sorbo, S.; Basile, A.; Manna, C. The protective role of olive oil hydroxytyrosol against oxidative alterations induced by mercury in human erythrocytes. *Food Chem. Toxicol.* **2015**, *82*, 59–63. [CrossRef]
42. Officioso, A.; Alzoubi, K.; Lang, F.; Manna, C. Hydroxytyrosol inhibits phosphatidylserine exposure and suicidal death induced by mercury in human erythrocytes: Possible involvement of the glutathione pathway. *Food Chem. Toxicol.* **2016**, *89*, 47–53. [CrossRef] [PubMed]
43. Officioso, A.; Panzella, L.; Tortora, F.; Alfieri, M.L.; Napolitano, A.; Manna, C. Comparative Analysis of the Effects of Olive Oil Hydroxytyrosol and Its 5-S-Lipoyl Conjugate in Protecting Human Erythrocytes from Mercury Toxicity. *Oxid. Med. Cell Longev.* **2018**, *2018*, 9042192. [CrossRef] [PubMed]
44. Harisa, G.I.; Mariee, A.D.; Abo-Salem, O.M.; Attiaa, S.M. Erythrocyte nitric oxide synthase as a surrogate marker for mercury-induced vascular damage: The modulatory effects of naringin. *Environ. Toxicol.* **2014**, *29*, 1314–1322. [CrossRef] [PubMed]
45. Ahmad, S.; Mahmood, R. Mercury chloride toxicity in human erythrocytes: Enhanced generation of ROS and RNS, hemoglobin oxidation, impaired antioxidant power, and inhibition of plasma membrane redox system. *Environ. Sci. Pollut. Res. Int.* **2019**, *26*, 5645–5657. [CrossRef] [PubMed]

46. Lim, K.-M.; Kim, S.; Noh, J.-Y.; Kim, K.; Jang, W.-H.; Bae, O.-N.; Chung, S.-M.; Chung, J.-H. Low-level mercury can enhance procoagulant activity of erythrocytes: A new contributing factor for mercury-related thrombotic disease. *Environ. Health Perspect.* **2010**, *118*, 928–935. [CrossRef] [PubMed]

47. Van den Berg, J.J.; Op den Kamp, J.A.; Lubin, B.H.; Roelofsen, B.; Kuypers, F.A. Kinetics and site specificity of hydroperoxide-induced oxidative damage in red blood cells. *Free Radic. Biol. Med.* **1992**, *12*, 487–498. [CrossRef]

48. Ellman, G.L. Tissue sulfhydryl groups. *Arch. Biochem. Biophys.* **1959**, *82*, 70–77. [CrossRef]

49. Nguyen, D.B.; Wagner-Britz, L.; Maia, S.; Steffen, P.; Wagner, C.; Kaestner, L.; Bernhardt, I. Regulation of phosphatidylserine exposure in red blood cells. *Cell Physiol Biochem.* **2011**, *28*, 847–856. [CrossRef]

50. Arnich, N.; Sirot, V.; Rivière, G.; Jean, J.; Noël, L.; Guérin, T.; Leblanc, J.-C. Dietary exposure to trace elements and health risk assessment in the 2nd French Total Diet Study. *Food Chem. Toxicol.* **2012**, *50*, 2432–2449. [CrossRef]

51. Koch, W.; Karim, M.R.; Marzec, Z.; Miyataka, H.; Himeno, S.; Asakawa, Y. Dietary intake of metals by the young adult population of Eastern Poland: Results from a market basket study. *J. Trace. Elem. Med. Biol.* **2016**, *35*, 36–42. [CrossRef]

52. Okpala, C.O.R.; Sardo, G.; Vitale, S.; Bono, G.; Arukwe, A. Hazardous properties and toxicological update of mercury: From fish food to human health safety perspective. *Crit. Rev. Food Sci. Nutr.* **2018**, *58*, 1986–2001. [CrossRef]

53. Kunwar, A.; Jayakumar, S.; Bhilwade, H.N.; Bag, P.P.; Bhatt, H.; Chaubey, R.C.; Priyadarsini, K.I. Protective effects of selenocystine against γ-radiation-induced genotoxicity in Swiss albino mice. *Radiat. Environ. Biophys.* **2011**, *50*, 271–280. [CrossRef] [PubMed]

54. Kalender, S.; Uzun, F.G.; Demir, F.; Uzunhisarcıklı, M.; Aslanturk, A. Mercuric chloride-induced testicular toxicity in rats and the protective role of sodium selenite and vitamin E. *Food Chem. Toxicol.* **2013**, *55*, 456–462. [CrossRef] [PubMed]

55. Karunasagar, D.; Krishna, M.V.B.; Rao, S.V.; Arunachalam, J. Removal and preconcentration of inorganic and methyl mercury from aqueous media using a sorbent prepared from the plant Coriandrum sativum. *J. Hazard. Mater.* **2005**, *118*, 133–139. [CrossRef] [PubMed]

56. Manini, P.; Panzella, L.; Eidenberger, T.; Giarra, A.; Cerruti, P.; Trifuoggi, M.; Napolitano, A. Efficient Binding of Heavy Metals by Black Sesame Pigment: Toward Innovative Dietary Strategies to Prevent Bioaccumulation. *J. Agric. Food Chem.* **2016**, *64*, 890–897. [CrossRef]

57. Peng, Y.; Bishop, K.S.; Quek, S.Y. Extraction Optimization, Antioxidant Capacity and Phenolic Profiling of Extracts from Flesh, Peel and Whole Fruit of New Zealand Grown Feijoa Cultivars. *Antioxidants* **2019**, *8*, 141. [CrossRef] [PubMed]

58. Vieira, F.G.K.; Borges, G.D.S.C.; Copetti, C.; Di Pietro, P.F.; Nunes, E.D.C.; Fett, R. Phenolic compounds and antioxidant activity of the apple flesh and peel of eleven cultivars grown in Brazil. *Sci. Hortic.* **2011**, *128*, 261–266. [CrossRef]

59. Wang, S.; Zhu, F. Dietary antioxidant synergy in chemical and biological systems. *Crit. Rev. Food Sci. Nutr.* **2017**, *57*, 2343–2357. [CrossRef]

60. Sánchez-Riaño, A.M.; Solanilla-Duque, J.F.; Méndez-Arteaga, J.J.; Váquiro-Herrera, H.A. Bioactive potential of colombian feijoa in physiological ripening stage. *J. Saudi Soc. Agric. Sci.* **2019**. [CrossRef]

61. Sobeh, M.; Petruk, G.; Osman, S.; El Raey, M.A.; Imbimbo, P.; Monti, D.M.; Wink, M. Isolation of Myricitrin and 3, 5-di-O-Methyl Gossypetin from Syzygium samarangense and Evaluation of their Involvement in Protecting Keratinocytes against Oxidative Stress via Activation of the Nrf-2 Pathway. *Molecules* **2019**, *24*, 1839. [CrossRef]

62. Sobeh, M.; Youssef, F.S.; Esmat, A.; Petruk, G.; El-Khatib, A.H.; Monti, D.M.; Ashour, M.L.; Wink, M. High resolution UPLC-MS/MS profiling of polyphenolics in the methanol extract of Syzygium samarangense leaves and its hepatoprotective activity in rats with CCl4-induced hepatic damage. *Food Chem. Toxicol.* **2018**, *113*, 145–153. [CrossRef] [PubMed]

63. Sobeh, M.; Mahmoud, M.F.; Petruk, G.; Rezq, S.; Ashour, M.L.; Youssef, F.S.; El-Shazly, A.M.; Monti, D.M.; Abdel-Naim, A.B.; Wink, M. Syzygium aqueum: A Polyphenol- Rich Leaf Extract Exhibits Antioxidant, Hepatoprotective, Pain-Killing and Anti-inflammatory Activities in Animal Models. *Front. Pharm.* **2018**, *9*, 566. [CrossRef] [PubMed]

64. Sun-Waterhouse, D.; Wang, W.; Waterhouse, G.I.N.; Wadhwa, S.S. Utilisation Potential of Feijoa Fruit Wastes as Ingredients for Functional Foods. *Food Bioprocess Technol.* **2013**, *6*, 3441–3455. [CrossRef]
65. Karami, M.; Saeidnia, S.; Nosrati, A. Study of the Hepatoprotective Activity of Methanolic Extract of Feijoa sellowiana Fruits Against MDMA using the Isolated Rat Liver Perfusion System. *Iran. J. Pharm. Res.* **2013**, *12*, 85–91. [PubMed]
66. Karami, M.; Karimian Nokabadi, F.; Ebrahimzadeh, M.A.; Naghshvar, F. Nephroprotective effects of Feijoa Sellowiana leaves extract on renal injury induced by acute dose of ecstasy (MDMA) in mice. *Iran. J. Basic Med. Sci.* **2014**, *17*, 69–72. [PubMed]
67. Widsten, P.; Mesic, B.B.; Cruz, C.D.; Fletcher, G.C.; Chycka, M.A. Inhibition of foodborne bacteria by antibacterial coatings printed onto food packaging films. *J. Food Sci. Technol.* **2017**, *54*, 2379–2386. [CrossRef] [PubMed]

© 2019 by the authors. Licensee MDPI, Basel, Switzerland. This article is an open access article distributed under the terms and conditions of the Creative Commons Attribution (CC BY) license (http://creativecommons.org/licenses/by/4.0/).

Article

Evaluation of the Influence of Process Parameters on the Properties of Resveratrol-Loaded NLC Using 2^2 Full Factorial Design

Andréa A. M. Shimojo [1,2,*], Ana Rita V. Fernandes [2], Nuno R. E. Ferreira [3], Elena Sanchez-Lopez [2,4,5], Maria H. A. Santana [1] and Eliana B. Souto [2,6,*]

1. Department of Engineering of Materials and Bioprocesses-School of Chemical Engineering, University of Campinas, Campinas 13083-970, Brazil
2. Department of Pharmaceutical Technology, Faculty of Pharmacy, University of Coimbra, 3000-548 Coimbra, Portugal
3. CQ Pharma, (FFUC), Pólo das Ciências da Saúde, Azinhaga de Santa Comba, 3000-548 Coimbra, Portugal
4. Department of Pharmacy, Pharmaceutical Technology and Physical Chemistry, Faculty of Pharmacy, and Institute of Nanoscience and Nanotechnology (IN2UB), University of Barcelona, 08007 Barcelona, Spain
5. Networking Research Centre of Neurodegenerative Disease (CIBERNED), Instituto de Salud Juan Carlos III, 28049 Madrid, Spain
6. CEB-Centre of Biological Engineering, University of Minho, Campus de Gualtar, 4710-057 Braga, Portugal
* Correspondence: lshimojo51@gmail.com (A.A.M.S.); ebsouto@ebsouto.pt (E.B.S.); Tel.: +55-19-995-380140 (A.A.M.S.); +351-239-488-400 (E.B.S.)

Received: 17 July 2019; Accepted: 30 July 2019; Published: 3 August 2019

Abstract: Resveratrol (RSV) is a natural antioxidant commonly found in grapes, berries, and nuts that has shown promising results in the treatment of a variety of degenerative and age-related diseases. Despite the proven beneficial results on reduction of reactive oxidant species (ROS) and on inflammatory process, RSV shows various limitations including low long-term stability, aqueous solubility, and bioavailability, restricting its applications in the medical-pharmaceutical area. To overcome these limitations, it has been applied in pharmaceutical formulations as nanostructured lipid carriers (NLC). Thus, the present study focuses on the optimization of the production process of NLC. NLC was produced by high shear homogenization (HSH) and ultrasound method (US) using Compritol® ATO C888 as solid lipid and Miglyol 812® as liquid lipid. In order to obtain an optimized formulation, we used a 2^2 full factorial design with triplicate of central point investigating the effects of the production process parameters; shear intensity and homogenization time, on the mean particle size (PS) and polydispersity index (PDI). Instability index, encapsulation efficiency, and production yield were also evaluated. As the PS and PDI values obtained with 6 min of shear at 19,000 rpm and 10 min of shear and 24,000 rpm were similar, the instability index (<0.1) was also used to select the optimal parameters. Based on the results of the experimental design and instability index, it was concluded that the shear rate of 19,000 rpm and the shear time of 6 min are the optimal parameters for RSV-loaded NLC production. Factorial design contributed therefore to optimize the variables of the NLC production process from a small number of experiments.

Keywords: resveratrol; nanostructured lipid carriers (NLC); factorial design; high shear homogenization; ultrasound method

1. Introduction

Reactive oxygen species (ROS) are directly associated with a variety of degenerative and age-related diseases, and other pathologies, including different types of cancers. ROS are generated as by-products of cellular metabolism and their excessive production can damage lipids, proteins, and DNA, different cells, and tissues [1].

Resveratrol (RSV) (trans-3,4,5-trihydroxystilbene) is a natural antioxidant commonly found in grapes, berries, and nuts that has shown promising results in the treatment of a variety of degenerative and age-related diseases. It blocks the activation of nuclear factor κB (NF-κB), reducing the generation of ROS and pro-inflammatory cytokines (interleukin IL-1 β and IL-6), which results in the inhibition of chondrocyte apoptosis, inflammation, and the progression of several diseases [2,3]. It is also a direct inhibitor of cyclooxygenase 2 (COX-2) which produces pro-inflammatory lipid mediators (leukotrienes and prostaglandins) responsible for pain sensation [4].

In recent years, RSV has also shown beneficial results in modulation of tissue regeneration, microcirculation, the function of peripheral nerves, production of anti-inflammatory cytokines, and insulin [5–9]. However, RSV shows low long-term stability, rapid metabolism, and release; low aqueous solubility (0.05 mg/mL) and bioavailability. RSV is also unstable under the influence of light, certain pH levels, and temperature, which causes isomerization or degradation of RSV, making it difficult to apply in the medical–pharmaceutical area [10–13]. To overcome these limitations, RSV has been employed in pharmaceutical formulations in different drug delivery systems (DDS), including microparticulate system [14,15], micro/nanocapsules [16], cyclodextrin complexes [17,18], solid lipid nanoparticles (SLN) [19–22], nanosuspensions [23], vesicular systems liposomes [24], niosomes [25], nanosponges [18,26], microspheres [27], transfersomes and ethosomes [14,15,28], and nanostructured lipid carriers (NLC) [19,20]. The DDS are employed to improve the physicochemical stability of loaded drugs, provide a sustained-release profile, increase plasma half-life, decrease the risk of immunogenicity, improve the drug solubility and thereby its bioavailability and therapeutic activity, enhance antioxidant activity, and improve the permeation and targeted delivery [14,15].

NLC consists of a mixture of solid and liquid lipids, which creates an imperfect crystalline structure, providing more space between the lipid chains and the matrix [29,30]. The main advantages of the use of NLC are high encapsulation efficiency and storage stability, the possibility of controlling the release of several drugs, low toxicity due to the absence of solvents in the production process, and the possibility of production in an industrial scale [31,32]. Moreover, Gokce et al. [14] observed that RSV-loaded NLC penetrated deeper into the skin [19]. Jose et al. showed a negligible release of resveratrol over several hours, corroborating the high stability of RSV-loaded NLC [20].

Experimental designs have been the most used tools to simultaneously analyze the influence of different variables on the properties of NLC, aiming to ensure the high product quality, the economy of production, and reduction of production time, allowing the scale-up of the process [33].

To evaluate the optimum experimental conditions for NLC production, several authors have assessed the influence of different factors which can affect the final properties of formulations, including type, ratio, and concentration of lipids and stabilizers, cycle numbers, time and intensity of homogenization, and pressure [34–36]. Thus, this work reports the effects of the production process parameters, shear intensity, and homogenization time of RSV-loaded NLC by means of a 2^2 factorial design with triplicate of the central point, measuring the mean particle size (PS) and polydispersity index (PDI) as the dependent variables.

2. Materials and Methods

2.1. Materials

Hyaluronic acid sodium salt (HA, MW = 1.01–1.8 MDa) was purchased from Lifecore Biomedical Co. (Chaska, MN, USA). Resveratrol (RSV) was obtained from Sigma-Aldrich Co. (St. Louis, MO, USA). Tween® 80 (Polysorbate 80) was purchased from Uniqema (Everberg, Belgium). Compritol® ATO C-888 was purchased from Gattefossé (Nanterre, France) and Miglyol 812® from Sasol Chemical Industries Ltd. (Homburg, Germany). Poloxamer® 188 (Pluronic F68, Kolliphor® P188) was obtained from BASF (Ludwigshafen, Germany). All other reagents and solvents were analytical grade and used exactly as received. Ultra-purified water was obtained from Milli-Q® Plus system, home supplied.

2.2. Preparation of RSV-Loaded NLC

RSV-loaded NLC were prepared through the high shear homogenization method associated with sonication [19,20]. The lipid phase was composed of 255 mg Compritol® 888 ATO (C-888), 45 mg Miglyol 812®, and 10 mg RSV, and the aqueous phase was composed of 150 mg of Poloxamer®, 188 (P188), and 75 mg of Tween® 80 (Tw 80) in 12.5 mL bidistilled water were heated at 85 °C, separately. The aqueous phase was poured into the lipid phase and stirred with Ultra Turrax homogenizer Ystral GmbH X10/25 (Dottingen, Germany), followed by sonication for 15 min at 70% intensity (14 watts) using a Sonics and Materials Vibra-Cell™ CV18 (Newtown, CT, USA), according to Table 1. The particles were dispersed in 12.5 mL of bidistilled water and kept at −20 °C for 10 min. Blank (i.e. non-loaded NLC) was prepared in a similar way, without the RSV. The formulations were stored at 4 °C.

Table 1. Preparation of resveratrol-loaded nanostructured lipid carriers (NLC). Factorial design, providing the lower (−1), upper (+1), and (0) central point level values for each variable.

Factor	Coded Levels		
	−1	0	+1
Shear intensity (rpm)	13,000	19,000	24,000
Shear time (minutos)	2	6	10

2.3. 2^2 Factorial Design

The influence of the shear intensity and homogenization time on the NLC properties was evaluated using a 2^2 factorial design with triplicate runs of the central point to estimate the experimental error, composed of 2 variables, which were set at 2-levels each (Table 2). The mean particle size and polydispersity index (PDI) were the dependent variables. The design required a total of 7 experiments. Each factor, the lower and higher values of the lower and upper levels, was represented by a (−1) and a (+1), and the central point was represented by (0), as summarized in Table 1. The data were analyzed using STATISTICA 7.0.

Table 2. A 2^2 full factorial experimental design layout. The formulation codes of NLC with resveratrol are named as NLC-RSV(number of the experiment) and the one without resveratrol as NLC(number of the experiment).

Formulation Code	Coded Factor Level	
	Factor 1	Factor 2
NLC$_1$ or NLC-RSV$_1$	−1	−1
NLC$_2$ or NLC-RSV$_2$	+1	−1
NLC$_3$ or NLC-RSV$_3$	−1	+1
NLC$_4$ or NLC-RSV$_4$	+1	+1
NLC$_5$ or NLC-RSV$_5$	0	0
NLC$_6$ or NLC-RSV$_6$	0	0
NLC$_7$ or NLC-RSV$_7$	0	0

2.4. Characterization of RSV-Loaded NLC

2.4.1. Particle Size and Polydispersity Index

The particle size (PS) and polydispersity index (PDI) of NLC were measured at 25 °C using photon correlation spectroscopy (PCS) (dynamic light scattering, DLS, Zetasizer Nano NS, Malvern Instruments, Malvern, Worcs, UK). The measurements were carried out using a He–Ne laser at 633 nm and 4.0 mW power, with a back-scattering detection angle of 173° after dilution of formulations with ultra-purified water. The average hydrodynamic diameter was recorded based on the observed diameters weighted by the number size distribution. The polydispersity index (PDI) was also calculated from cumulative analysis of the measured DLS intensity autocorrelation function (a dimensionless number that ranges from 0 to 1). PS and PDI of NLC were determined in triplicate. For each measurement, the NLC was diluted in Milli-Q® water to an appropriate concentration to avoid multiple scattering.

2.4.2. Zeta Potential

The zeta potential was determined by applying an electric field across the samples, and the value of the zeta potential was obtained by measuring the velocity of the electrophoretic mobility of the particles using the laser Doppler anemometry technique. The measurements were performed in triplicate for each sample at 25 °C using a Malvern Zetasizer Nano ZS (Malvern Instruments, Worcs, UK). Milli-Q® water was used to dilute the NLC to a proper concentration. The zeta potential was calculated using the Helmholtz-Smoluchowsky equation included in the software of the system. Values are presented as the mean of triplicate runs per sample.

2.4.3. Stability Index

Analytic centrifuge LUMiSizer (LUM GmbH, Dias de Sousa, Portugal), which accelerates the destabilization phenomena, was used to evaluate the simulated long-term physical stability of NLC. Briefly, the samples without prior dilution were placed in rectangular test tubes (optical path of 2 mm) and exposed to centrifugal force at 10,000 rpm, measuring 300 profiles in intervals of 10 s at 25 °C. These experiments allowed differentiation between various instability mechanisms at an accelerated rate. Extrapolated results were used to estimate dispersion shelf life in minutes. To simply assess the physical stability of NLC, the instability index was calculated by the delivered software (SepView 6.0; LUM, Berlin, Germany). The index was quantified by the clarification at a given separation time, divided by the maximum clarification, according to Hoffmann and Schrader [37].

2.4.4. Entrapment Efficiency

The drug entrapment efficiency (EE) was determined by UV-visible spectrometry at 303 nm, using the Synergy™ HTX Multi-Mode Microplate Reader (Biotek Instruments, Winooski, VT, USA). Briefly, NLC was placed in the dialysis bag (cutoff 14 kDa) (Sigma-Aldrich, St. Louis, MO, USA). Then, the bags were placed in centrifuge tubes, covered with a mixture of ethanol and water 1:1 (v/v), and centrifuged for 1.5 h at 5000 rpm in centrifuge Laborzentrifugen 3K15 (Sigma, Osterode am Herz, Germany). The mixture of ethanol and water was analyzed for RSV content through the standard curve allowing the quantity of free drug to be determined. The encapsulated amount of RSV was calculated by subtracting the free amount of RSV from the total amount present in the dispersion. The measurements were performed in triplicate. The EE percentage was calculated by the Equation (1).

$$EE = \frac{(amount\ of\ initial\ RSV - amount\ of\ free\ RSV)}{amount\ of\ initial\ RSV} \times 100 \tag{1}$$

2.4.5. Production Yield

The NLC formulations were frozen for 12 h at −80 °C and lyophilized at a pressure of 0.1 mbar for 24 h at −40 °C using a Telstar LyoQuest Freeze Dryer (Barcelona, Spain). The production yield (Y_{NLC}) was calculated by Equation (2).

$$Y_{NLC}\ (\%) = \frac{(initial\ amount - final\ amount)}{initial\ amount} \times 100 \tag{2}$$

2.4.6. Morphology

The morphology was performed with a transmission electron microscope Tecnai G2 Spirit Biotwin (FEI Company, Eindhoven, The Netherlands). The samples were stained with 2% (w/v) phosphotungstic acid and placed on copper grids for viewing by transmission electron microscopy (TEM).

2.5. Statistical Analysis

Statistical differences were determined using analysis of variance (ANOVA), followed by Tukey's test for comparisons between groups. The significance level was taken as 95% ($p < 0.05$). Factorial design data were analyzed using STATISTICA 7.0.

3. Results and Discussion

The lipid composition and concentration used in the preparation of NLC were chosen based on work reported by Gokce et al. [19] in which the optimal liquid lipid concentration (Miglyol 812®) was 15% of the whole lipid phase (Compritol® 888 ATO and Miglyol 812®) [19]. Poloxamer® 188 and Tween® 80 were used as stabilizers of the formulations. Compritol® 888 ATO is a solid lipid composed of glycerol tribehenate (28–32%), glycerol dibehenate (52–54%), and glycerol monobehenate (12–18%). The main fatty acid is behenic acid (C_{22}) (>85%), but other fatty acids (C_{16}–C_{20}) are also present [38]. Miglyol 812® is a medium chain triglyceride composed mainly of caprylic (C8:0; 50–80%) and capric (C10:0; 20–50%) fatty acids with a minor level of caproic (C6:0; ≤2%), lauric (C12:0; ≤3%), and myristic (C14:0; ≤1%) fatty acids [39].

Mohammadi et al. [39] and Kovasevic et al. [40] demonstrated that Miglyol 812® can be used in the range of 10–60% of total lipid without affecting the mean particle size and the distribution of NLC.

The choice of stabilizers is also a very important step in the preparation of NLC formulations because they control the particle size and the stability, preventing their aggregation during storage [39,41]. Currently, the non-ionic surfactants Poloxamer 188® and Tween® 80 are the most used for the preparation of these formulations [42]. According to Tamjidi et al. [42], the steric repulsion is the major colloidal interaction among NLC stabilized with non-ionic surfactants, yielding good stability to the variations of concentration and pH of electrolytes, and to the freeze–thaw stages. Moreover, non-ionic surfactants have lower toxicity and irritation potential than ionic ones [43].

However, NLC prepared with non-ionic surfactants may undergo a weak flocculation, as well as requiring large amounts of surfactants to cover the particle surface compared to those stabilized by electrostatic repulsion [44].

In Poloxamer® 188, the hydrophobic polypropylene oxide chains are adsorbed onto the particle surface as the "anchor chain", while the hydrophilic polyethylene oxide chains are pulled out from the surface to the aqueous medium, thereby creating a stabilizer layer [45]. In addition, Poloxamer® 188 exhibits low toxicity, can control release and targeted delivery applications, and is stable at high temperatures [46]. Tween® 80 is a polyethoxylated sorbitan and oleic acid derivative that has high surface activity and low toxicity [39]. The coating with Tween® 80 improves the stability of the lipid present in NLC by hydration in the surface layer [47,48].

In this work, we used a combination of these surfactants because they produce a layer at the interface, generating high coverage as well as adequate viscosity to improve the stability and synergism in the particle size reduction [39]. Aiming to obtain particles in the nanometer range, NLC was produced by an association of high shear homogenization (HSH) [19] and ultrasound method (US) [20]. HSH produced particles in the micrometer range (pre-emulsion) and the ultrasound method reduced the microparticles to the nanometer range.

The effects of the formulation variables (independent variables)-shear intensity and homogenization time on the response parameters (dependent variables)-mean particle size (PS) and polydispersity index (PDI), were evaluated using full factorial design 2^2 with triplicate of the central point. For the factorial design study, a total of seven experiments were required. Zeta potential, encapsulation efficiency (EE), production yield (Y), and instability index were also measured.

Table 3 shows the influence of shear intensity and homogenization time on NLC production (RSV-loaded NLC and NLC without RSV = placebo).

The combination of HSH and US methods produced placebos (NLC) with sizes ranging between 100 nm and 260 nm, and RSV-loaded NLC (NLC-RSV) with sizes ranging between 125 nm and 190 nm.

Particle size of less than 200 nm was attributed to the efficiency of the emulsion step. Gokce et al. [49] observed that the Compritol® 888 ATO tends to return to solid form during mixing because this lipid is a mixture of mono-, di-, and triglycerides. It is known that the longer the fatty acid triglyceride, the higher the temperature needed to convert it from the solid state to liquid (melt) state. However, the presence of Miglyol® 812 helps to distribute the heat energy more homogeneously due to the high concentration of unsaturated fatty acids reducing the melting point of the system. This results in a more efficient emulsification, which in turn has an effect on the size of the particles formed. After cooling, the pre-emulsion shows smaller particles, which may result in even smaller nanoparticles [50,51]. Thus, the stability is related to the lipid composition, since NLC presents a disordered lipid matrix conferred by the presence of liquid lipid and to polysorbate surfactant (Tween® 80) used in its preparation [52]. All NLC formulations showed a PDI of above 0.2 and negative ZP around −12 mV.

The PDI has an important effect on the physical stability and uniformity (distribution) of NLC. The values should be as low as possible to ensure the long-term stability. PDI values of 0.1–0.25 show a narrow size distribution, while PDI values greater than 0.5 indicate a very broad distribution [53]. The PDI values obtained from placebo and RSV-loaded NLC above 0.2 indicated a non-monodisperse distribution with the presence of aggregated suggesting lower long-term stability. This type of distribution is usual in NLC produced using the HSH and US method, where it is very difficult to achieve a unimodal distribution of sizes [20].

ZP is also an indirect measurement of the long-term physical stability of NLC. It relates to the trend of particles to aggregate. According to Lakshimi and Kumar (2010), in electrostatically stabilized NLC, a good stability is achieved in ZP above ±30 mV, whereas in a combination of electrostatic and steric stabilization, a minimum of ZP of ±20 mV is desirable [53,54]. In addition, ZP of ±0–5 mV produces a maximum flocculation [32,55,56]. As shown in Table 3, all NLC had a negative ZP around −12 mV, indicating moderate stability regardless of RSV incorporation, suggesting that RSV did not significantly alter the ZP of the formulations ($p > 0.05$).

Besides the ZP, the long-term stability was also assessed by the instability index. The instability index is a dimensionless number between 0 (more stable) and 1 (more unstable), calculated based on the clarification at a given separation time, divided by the maximum clarification. For that, we used the LUMiSizer® equipment, which allows the measurement of the transmitted light intensity during centrifugation, as a function of time and position, over the entire sample length [57,58].

In spite of the ZP values, with the exception of the NLC-RSV$_2$ and NLC-RSV$_4$ formulations, the dispersion analysis indicated a good simulated physical stability of the NLC containing RSV, expressed as instability index (<0.05). This observation suggests that these particles will remain stable and have a good dispersion quality in long-term storage.

The results of encapsulation efficiency showed that a large amount of RSV (EE > 92%) was incorporated in all RSV-loaded NLC formulations, suggesting its preferential partition into lipid matrix of the nanoparticles [15]. Gokce et al. [14] also obtained the EE of 91% using the same formulation. In addition, the production yield of both placebo and RSV-loaded NLC was found to be satisfactory, with an average above 60%.

Figure 1 shows the micrographs obtained by TEM of NLC and NLC-RSV. TEM analysis confirmed the colloidal sizes of particles. NLC was almost spherical and uniform in shape with smooth surfaces, while NLC-RSV showed more amorphous shapes. No crystallization of RSV was observed on the surface of NLC-RSV. Thus, our study suggests that the lipid matrix used solidified upon cooling, but it remained in the amorphous state, helping with the accommodation of RSV in a lipid matrix [38].

Table 3. Influence of shear intensity and homogenization time on the production of NLC.

	Shear Time (minutes)	Shear Intensity (rpm)	PS ± SD (nm)	PDI ± SD	Zeta ± SD (mV)	Entrapment Efficiency (%)	Production Yield (%)	Instability Index *
NLC_1	2	13,000	263.6 ± 3.9	0.465 ± 0.020	−12.97 ± 0.50	-	97 ± 2	0.022
NLC_2	10	13,000	190.3 ± 1.8	0.429 ± 0.025	−10.02 ± 0.24	-	78 ± 4	0.044
NLC_3	2	24,000	111.6 ± 1.1	0.236 ± 0.001	−13.93 ± 0.47	-	73.5 ± 0.6	0.040
NLC_4	10	24,000	178.3 ± 1.6	0.478 ± 0.022	−12.03 ± 0.51	-	66 ± 1	0.034
NLC_5	6	19,000	109.1 ± 0.7	0.265 ± 0.008	−10.03 ± 0.23	-	71 ± 3	0.043
NLC_6	6	19,000	103.4 ± 1.5	0.279 ± 0.036	−11.00 ± 0.46	-	68 ± 13	0.03
NLC_7	6	19,000	106.4 ± 0.1	0.271 ± 0.006	−12.83 ± 0.35	-	65.8 ± 0.9	0.032
$NLC\text{-}RSV_1$	2	13,000	164.3 ± 3.2	0.619 ± 0.010	−13.23 ± 0.21	94.6 ± 0.4	76.1 ± 0.9	0.073
$NLC\text{-}RSV_2$	10	13,000	187.7 ± 0.9	0.464 ± 0.003	−8.43 ± 0.15	94.4 ± 0.1	78 ± 2	0.493
$NLC\text{-}RSV_3$	2	24,000	153.1 ± 1.0	0.366 ± 0.005	−12.07 ± 0.40	96.3 ± 0.1	72 ± 3	0.041
$NLC\text{-}RSV_4$	10	24,000	142.9 ± 2.5	0.353 ± 0.016	−11.67 ± 0.70	94.2 ± 0.1	69 ± 1	0.244
$NLC\text{-}RSV_5$	6	19,000	125.3 ± 0.5	0.357 ± 0.009	−12.87 ± 0.38	92.9 ± 0.4	67.9 ± 0.6	0.024
$NLC\text{-}RSV_6$	6	19,000	139.4 ± 0.7	0.339 ± 0.006	−12.90 ± 0.27	93.15 ± 0.04	62 ± 4	0.025
$NLC\text{-}RSV_7$	6	19,000	146.6 ± 3.2	0.444 ± 0.061	−13.00 ± 0.53	92.0 ± 0.3	63 ± 11	0.011

SD = standard deviation; PDI = polydispersity index. * Instability index measured at t = 200 s.

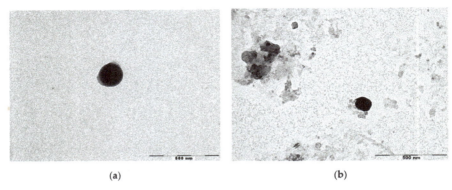

Figure 1. Micrographs obtained by TEM of (**a**) NLC$_4$ and (**b**) NLC-RSV$_4$. Scale bar = 500 nm.

Figure 2 shows the Pareto chart of the standardized effects and Figure 3 shows the surface response charts of experimental design for the production of placebos. As shown in Figure 2a,b, the PS and their PDI were not significantly influenced by tested parameters; neither was the interaction between variables.

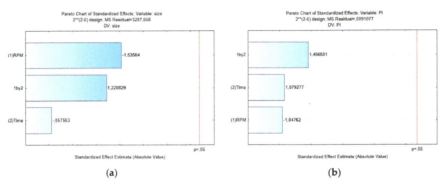

Figure 2. Pareto charts of the standardized effects for the placebo obtained for (**a**) particle size (Z-average) and (**b**) polydispersity index (PDI).

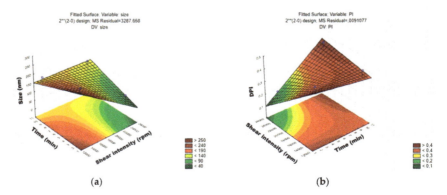

Figure 3. Surface response charts of experimental design of the placebo obtained for (**a**) particle size (Z-average) and (**b**) polydispersity index (PDI).

For the mean particle size, the p-value obtained by shear intensity was −1.53564, homogenization time was −0.57553, and the interaction was 1.220829, while for the PDI, the *p*-value obtained by shear intensity was −1.04762, homogenization time was 1.079277, and the interaction was 1.456501. These parameters and their interaction were reported not to be statistically significant. However, the response surface charts of experimental design (Figure 3a,b), show that increasing the shear intensity decreases the average size and the PDI. Moreover, in Figure 3, we observed that the average PS is slightly affected by the homogenization time, while PDI is not affected.

Comparing NLC_1 with NLC_2 and NLC_3 with NLC_4, we observed two trends where the PS goes down in NLC_1/NLC_2 and where PS goes up in NLC_3/NLC_4 by increasing the homogenization time.

Thus, although neither variable is statistically significant when the placebos are subjected to a lower homogenization time and shear intensity, they tend to be larger, i.e., approximately 263 nm, and the PDI is >0.40. The placebo produced with shear intensity of 19,000 rpm and homogenization time of 6 min showed a smaller PS, around 105 nm.

The influence of each independent variable and their interactions on RSV-loaded NLC were also evaluated by Pareto charts (Figure 4) and surface response (Figure 5). As shown in Figure 4a,b, the PS and their PDI were not significantly influenced by tested parameters; neither was the interaction between variables. For the mean particle size, the p-value obtained by shear intensity was −1.50165, homogenization time was 0.3316068, and the interaction was −0.84409, while for the PDI, the *p*-value obtained by shear intensity was −2.85191, homogenization time was −127996, and the interaction was 1.081873. These parameters and their interaction were reported not to be statistically significant. However, the response surface charts of experimental design (Figure 5a,b), shows that increasing the shear intensity decreases the average size and the PDI. Moreover, in Figure 5, we observed that both particle size and PDI are slightly affected by the homogenization time. Thus, although neither variable is statistically significant when the RSV-loaded NLC are subjected to a smaller homogenization time and intensity shear, the PDI is >0.54. We observed that smaller particles are obtained by increasing shear intensity. However, comparing $NLC\text{-}RSV_1$ with $NLC\text{-}RSV_2$ and $NLC\text{-}RSV_3$ with $NLC\text{-}RSV_4$, we observed two trends where the PS goes up in $NLC\text{-}RSV_1/NLC\text{-}RSV_2$ and where PS goes down in $NLC\text{-}RSV_3/NLC\text{-}RSV_4$ by increasing the homogenization time.

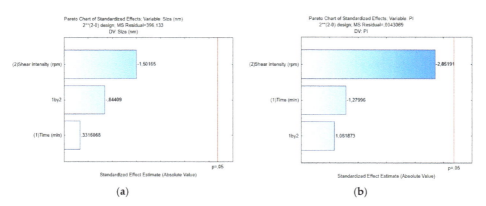

Figure 4. Pareto charts of the standardized effects for RSV-loaded NLC obtained for (**a**) particle size (Z-average) and (**b**) polydispersity index (PDI).

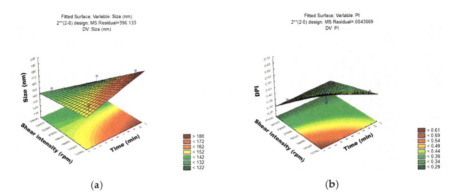

Figure 5. Surface response charts of experimental design of RSV-loaded NLC obtained for (**a**) particle size (Z-average) and (**b**) polydispersity index (PDI).

The RSV-loaded NLC produced at the central point with the shear intensity of 19,000 rpm and homogenization time of 6 min showed a smaller PS, around 135 nm.

As the experimental results of PS and PDI of NLC-RSV$_4$ were similar to the results obtained for NLC-RSV$_5$, NLC-RSV$_6$, and NLC-RSV$_7$, we also used the instability index to select as optimal parameter.

Thus, based on the results of the experimental design and instability index, it was concluded that the shear rate of 19,000 rpm and the shear time of 6 min are the optimal parameters for RSV-loaded NLC production.

4. Conclusions

This study attempted to design and optimize RSV-loaded NLC prepared by a combination of high shear homogenization and ultrasound method. After selecting the critical process variables affecting particle size (PS) and polydispersity index (PDI), a 2^2 factorial design with triplicate of the central point was employed to plan and perform the experiments. Zeta potential, morphology, drug entrapment efficiency, production yield, and stability index were also measured. RSV-loaded NLC and NLC without RSV (placebo) were prepared. Optimized NLC formulation was prepared based on the predicted optimum levels of the independent variables, shear intensity, and homogenization time of the factorial design using Pareto charts and surface response charts, and on instability index. Thus, optimal parameters for NLC were obtained using shear intensity of 19,000 rpm and shear time of 6 min, producing NLC with PS around 135 nm and DPI around 0.4. Moreover, these production process parameters produced particles with high entrapment efficiency (~93%) and production yield (~65%).

Author Contributions: A.A.M.S., M.H.A.S., A.R.V.F., and E.M.B.S. contributed for the conceptualization, methodology, validation, formal analysis, and investigation. A.A.M.S., A.R.V.F., E.S.-L., and N.R.E.F. contributed for the Writing—Original draft preparation. A.A.M.S., M.H.A.S., and E.M.B.S. contributed for supervision, Writing—Review and editing, project administration, resources, and funding acquisition. All authors have made a substantial contribution to the work.

Funding: The work has received financial support from FAPESP (Fundação de Amparo à Pesquisa do Estado de São Paulo), Project's numbers 2017/034968 and 2014/27200-2 and from the Portuguese Science and Technology Foundation, Ministry of Science and Education (FCT/MEC) through national funds, and co-financed by FEDER, under the Partnership Agreement PT2020 for the project M-ERA-NET/0004/2015-PAIRED.

Conflicts of Interest: The authors declare no conflict of interest.

References

1. Montesano-Gesualdi, N.; Chirico, G.; Catanese, M.T.; Pirozzi, G.; Esposito, F. AROS-29 is involved in adaptive response to oxidative stress. *Free Radic. Res.* **2006**, *40*, 467–476. [CrossRef] [PubMed]

2. Elmali, N.; Esenkaya, I.; Harma, A.; Ertem, K.; Turkoz, Y.; Mizrak, B. Effect of resveratrol in experimental osteoarthritis in rabbits. *Inflamm. Res.* **2010**, *54*, 158–162. [CrossRef]

3. Li, W.; Cai, L.; Zhang, Y.; Cui, L.; Shen, G. Intra-Articular Resveratrol Injection Prevents Osteoarthritis Progression in a Mouse Model by Activating SIRT1 and Thereby Silencing HIF-2a. *J. Orthop. Res.* **2015**, *33*, 1061–1070. [CrossRef]

4. Limagne, E.; Lançon, A.; Delmas, D.; Cherkaoui-Malki, M.; Latruffe, N. Resveratrol Interferes with IL1-β-Induced Pro-Inflammatory Paracrine Interaction between Primary Chondrocytes and Macrophages. *Nutrients* **2016**, *8*, 280. [CrossRef] [PubMed]

5. van Ginkel, P.R.; Sareen, D.; Subramanian, L.; Walker, Q.; Darjatmoko, S.R.; Lindstrom, M.J.; Kulkarni, A.; Albert, D.M.; Polans, A.S. Resveratrol inhibits tumor growth of human neuroblastoma and mediates apoptosis by directly targeting mitochondria. *Clin. Cancer Res.* **2007**, *13*, 5162–5169. [CrossRef] [PubMed]

6. Csaki, C.; Keshishzadeh, N.; Fischer, K.; Shakibaei, M. Regulation of inflammation signalling by resveratrol in human chondrocytes in vitro. *Biochem. Pharmacol.* **2008**, *75*, 677–687. [CrossRef]

7. Bashmakov, Y.K.; Assaad-Khalil, S.; Petyaev, I.M. Resveratrol may be beneficial in treatment of diabetic foot syndrome. *Med. Hypotheses* **2011**, *77*, 364–367. [CrossRef]

8. Cooley, J.; Broderick, T.L.; Al-Nakkash, L.; Plochocki, J.H. Effects of resveratrol treatment on bone and cartilage in obese diabetic mice. *J. Diabetes Metab. Disord.* **2015**, *14*, 1–7. [CrossRef]

9. Gokce, E.H.; Tanrıverdi, S.T.; Eroglu, I.; Tsapis, N.; Gokce, G.; Tekmen, I.; Fattal, E.; Ozer, O. Wound healing effects of collagen-laminin dermal matrix impregnated with resveratrol loaded hyaluronic acid-DPPC microparticles in diabetic rats. *Eur. J. Pharm. Biopharm.* **2017**, *119*, 17–27. [CrossRef]

10. Francioso, A.; Mastromarino, P.; Restignoli, R.; Boffi, A.; d'Erme, M.; Mosca, L. Improved Stability of trans-Resveratrol in Aqueous Solutions by Carboxymethylated (1,3/1,6)-β-D-Glucan. *J. Agric. Food Chem.* **2014**, *62*, 1520–1525. [CrossRef]

11. Robinson, K.; Mock, C.; Liang, D. Pre-formulation studies of resveratrol. *Drug Dev. Ind. Pharm.* **2015**, *41*, 1464–1469. [CrossRef] [PubMed]

12. Cottart, C.-H.; Nivet-Antoine, V.; Laguillier-Morizot, C.; Beaudeux, J.L. Resveratrol bioavailability and toxicity in humans. *Mol. Nutr. Food Res.* **2010**, *54*, 7–16. [CrossRef] [PubMed]

13. Zupančič, S.; Lavrič, Z.; Kristl, J. Stability and solubility of trans-resveratrol are strongly influenced by pH and temperature. *Eur. J. Pharm. Biopharm.* **2015**, *93*, 196–204. [CrossRef]

14. Devi, P.; Sharma, P.; Rathore, C.; Negi, P. Novel Drug Delivery Systems of Resveratrol to Bioavailability and Therapeutic Effects. In *Resveratrol-Adding Life to Years, Not Adding Years to Life*; IntechOpen Limited: London, UK, 2019; Chapter 2. [CrossRef]

15. Shindikar, A.; Singh, A.; Nobre, M.; Kirolikar, S. Curcumin and Resveratrol as Promising Natural Remedies with Nanomedicine Approach for the Effective Treatment of Triple Negative Breast Cancer. *J. Oncol.* **2016**, *2016*, 1–13. [CrossRef] [PubMed]

16. Figueiro, F.; Bernardi, A.; Frozza, R.L.; Terroso, T.; Zanotto-Filho, A.; Jandrey, E.H.; Moreira, J.C.; Salbego, C.G.; Edelweiss, M.I.; Pohlmann, A.R.; et al. Resveratrol-loaded lipid-core nanocapsules treatment reduces in vitro and in vivo glioma growth. *J. Biomed. Nanotechnol.* **2013**, *9*, 516–526. [CrossRef] [PubMed]

17. Lu, Z.; Cheng, B.; Hu, Y.; Zhang, Y.; Zou, G. Complexation of resveratrol with cyclodextrins: Solubility and antioxidant activity. *Food Chem.* **2009**, *113*, 17–20. [CrossRef]

18. Trotta, F.; Zanetti, M.; Cavalli, R. Cyclodextrin-based nanosponges as drug carriers. *Beilstein J. Org. Chem.* **2012**, *8*, 2091–2099. [CrossRef] [PubMed]

19. Gokce, E.H.; Korkmaz, E.; Dellera, E.; Sandri, G.; Bonferoni, M.C.; Ozer, O. Resveratrol-loaded solid lipid nanoparticles versus nanostructured lipid carriers: Evaluation of antioxidant potential for dermal applications. *Int. J. Nanomed.* **2012**, *7*, 1841–1850. [CrossRef] [PubMed]

20. Jose, S.; Anju, S.S.; Cinu, T.A.; Aleykutty, N.A.; Thomas, S.; Souto, E.B. In vivo pharmacokinetics and biodistribution of resveratrol-loaded solid lipid nanoparticles for brain delivery. *Int. J. Pharm.* **2014**, *474*, 6–13. [CrossRef] [PubMed]

21. Teskac, K.; Kristl, J. The evidence for solid lipid nanoparticles mediated cell uptake of resveratrol. *Int. J. Pharm.* **2010**, *390*, 61–69. [CrossRef] [PubMed]

22. Serini, S.; Cassano, R.; Corsetto, P.A.; Rizzo, A.; Calviello, G.; Trombino, S. Omega-3 PUFA Loaded in Resveratrol-Based Solid Lipid Nanoparticles: Physicochemical Properties and Antineoplastic Activities in Human Colorectal Cancer Cells In Vitro. *Int. J. Mol. Sci.* **2018**, *19*, 586. [CrossRef] [PubMed]

23. Kobierski, S.; Kwakye, K.O.; Muller, R.H.; Keck, C.M. Resveratrol nanosuspensions for dermal application-production, characterization, and physical stability. *Die Pharm.* **2009**, *64*, 741–747.
24. Bonechi, C.; Martini, S.; Ciani, L.; Lamponi, S.; Rebmann, H.; Rossi, C.; Ristori, S. Using liposomes as carriers for polyphenolic compounds: The case of trans-resveratrol. *PLoS ONE* **2012**, *7*, 1–11. [CrossRef] [PubMed]
25. Pando, D.; Gutierrez, G.; Coca, J.; Pazos, C. Preparation and characterization of niosomes containing resveratrol. *J. Food Eng.* **2013**, *117*, 227–234. [CrossRef]
26. Patel, E.K.; Oswal, R.J. Nanosponge and microsponges: A Novel Drug Delivery System. *Int. J. Res. Pharm. Chem.* **2012**, *2*, 2281–2781.
27. Nam, J.B.; Ryu, J.H.; Kim, J.W.; Chang, I.S.; Suh, K.D. Stabilization of resveratrol immobilized in monodisoerse cyano-functionalized porous polymeric microspheres. *Polymer (Guildf)*. **2005**, *46*, 8956–8963. [CrossRef]
28. Scognamiglio, I.; Stefano, D.D.; Campani, V.; Mayol, L.; Carnuccio, R.; Fabbrocini, G.; Ayala, F.; La Rotonda, M.I.; De Rosa, G. Nanocarriers for topical administration of resveratrol: A comparative study. *Int. J. Pharm.* **2013**, *440*, 179–187. [CrossRef] [PubMed]
29. Radtke, M.; Souto, E.B.; Müller, R.H. Nanostructured Lipid Carriers: A Novel Generation of Solid Lipid Drug Carriers. *Pharm. Technol. Eur.* **2005**, *17*, 45–50.
30. Montenegro, L.; Lai, F.; Offerta, A.; Sarpietro, M.G.; Micicche, L.; Maccioni, A.M.; Valenti, D.; Fadda, A.M. From nanoemulsions to nanostructured lipid carriers: A relevantdevelopment in dermal delivery of drugs and cosmetics. *J. Drug Deliv. Sci. Technol.* **2016**, *32*, 100–112. [CrossRef]
31. Souto, E.B.; Muller, R.H. Lipid nanoparticles: Effect on bioavailability and pharmacokinetic changes. *Handb. Exp. Pharmacol.* **2011**, *197*, 115–141.
32. Müller, R.H.; Mäder, K.; Gohla, S. Solid lipid nanoparticles (SLN) for controlled drug delivery—A review of the state of the art. *Eur. J. Pharm. Biopharm.* **2000**, *50*, 161–177. [CrossRef]
33. Araujo, J.; Gonzalez-Mira, E.; Egea, M.A.; Garcia, M.L.; Souto, E.B. Optimization and physicochemical characterization of a triamcinolone acetonide-loaded NLC for ocular antiangiogenic applications. *Int. J. Pharm.* **2010**, *393*, 167–175. [CrossRef] [PubMed]
34. Severino, P.; Santana, M.H.A.; Souto, E.B. Optimizing SLN and NLC by 22 full factorial design: Effect of homogenization technique. *Mater. Sci. Eng. C Mater. Biol. Appl.* **2012**, *32*, 1375–1379. [CrossRef] [PubMed]
35. Shah, N.V.; Seth, A.K.; Balaraman, R.; Aundhia, C.J.; Maheshwari, R.A.; Parmar, G.R. Nanostructured lipid carriers for oral bioavailability enhancement of raloxifene: Design and in vivo study. *J. Adv. Res.* **2016**, *7*, 423–434. [CrossRef] [PubMed]
36. Sütő, B.; Weber, S.; Zimmer, A.; Farkas, G.; Kelemen, A.; Budai-Szűcs, M.; Berkó, S.; Szabó-Révész, P.; Csányi, E. Optimization and design of an ibuprofen-loaded nanostructured lipid carrier with a 2^3 full factorial design. *Chem. Eng. Res. Des.* **2015**, *104*, 488–496. [CrossRef]
37. Hoffmann, W.; Schrader, K. Dispersion analysis of spreadable processed cheese with low content of emulsifying salts by photocentrifugation. *Int. J. Food Sci. Technol.* **2015**, *50*, 950–957. [CrossRef]
38. Souto, E.B.; Mehnert, W.; Müller, R.H. Polymorphic behaviour of Compritol®888 ATO as bulk lipid and as SLN and NLC. *J. Microencapsul.* **2006**, *23*, 417–433. [CrossRef] [PubMed]
39. Mohammadi, M.; Pezeshk, A.; Abbasi, M.M.; Ghanbarzadeh, B.; Hamishehkar, H. Vitamin D3-Loaded Nanostructured Lipid Carriers as a Potential Approach for Fortifying Food Beverages; in Vitro and in Vivo Evaluation. *Adv. Pharm. Bull.* **2017**, *7*, 61–71. [CrossRef] [PubMed]
40. Kovacevic, A.; Savic, S.; Vuleta, G.; Müller, R.H.; Keck, C.M. Polyhydroxy surfactants for the formulation of lipid nanoparticles (SLN and NLC): Effects on size, physical stability and particle matrix structure. *Int. J. Pharm.* **2011**, *406*, 163–172. [CrossRef]
41. Trotta, M.; Debernardi, F.; Caputo, O. Preparation of solid lipid nanoparticles by a solvent emulsification-diffusion technique. *Int. J. Pharm.* **2003**, *257*, 153–160. [CrossRef]
42. Tamjidi, F.; Shahedi, M.; Varshosaz, J.; Nasirpour, A. Nanostructured lipid carriers (NLC): A potential delivery system for bioactive food molecules. *Innov. Food Sci. Emerg. Technol.* **2013**, *19*, 29–43. [CrossRef]
43. McClements, D.J.; Rao, J. Food-grade nanoemulsions: Formulation, fabrication, properties, performance, biological fate, and potential toxicity. *Crit. Rev. Food Sci. Nutr.* **2011**, *51*, 285–330. [CrossRef] [PubMed]
44. Hunter, R.J. *Foundations of Colloid Science*; Oxford University Press: Oxford, UK, 1986.
45. Ghosh, I.; Bose, S.; Vippagunta, R.; Harmon, F. Nanosuspension for improving the bioavailability of a poorly soluble drug and screening of stabilizing agents to inhibit crystal growth. *Int. J. Pharm.* **2011**, *409*, 260–268. [CrossRef] [PubMed]

46. Trujillo, C.C.; Wright, A.J. Properties and stability of solid lipid particle dispersions based on canola stearin and Poloxamer 188. *J. Am. Oil Chem. Soc.* **2010**, *87*, 715–730. [CrossRef]
47. Teeranachaideekul, V.; Souto, E.B.; Junyaprasert, V.B.; Müller, R.H. Cetyl palmitate-based NLC for topical delivery of Coenzyme Q10—Development, physicochemical characterization and in vitro release studies. *Eur. J. Pharm. Biopharm.* **2007**, *67*, 141–148. [CrossRef] [PubMed]
48. Lim, S.J.; Kim, C.K. Formulation parameters determining the physicochemical characteristics of solid lipid nanoparticles loaded with all-trans retinoic acid. *Int. J. Pharm.* **2002**, *243*, 135–146. [CrossRef]
49. Gokce, E.H.; Sandri, G.; Bonferoni, M.C.; Rossi, S.; Ferrari, F.; Güneri, T.; Caramella, C. Cyclosporine A loaded SLNs: Evaluation of cellular uptake and corneal cytotoxicity. *Int. J. Pharm.* **2008**, *364*, 76–86. [CrossRef] [PubMed]
50. Souto, E.B.; Wissing, S.A.; Barbosa, C.M.; Müller, R.H. Evaluation of the physical stability of SLN and NLC before and after incorporation into hydrogel formulations. *Eur. J. Pharm. Biopharm.* **2004**, *58*, 83–90. [CrossRef]
51. Timms, R.E. Fractional crystallization—The fat modification process for the 21st century. *Eur. J. Lipid Sci. Technol.* **2005**, *107*, 48–57. [CrossRef]
52. Saupe, A.; Wissing, S.A.; Lenk, A.; Schmidt, C.; Müller, R.H. Solid lipid nanoparticles (SLN) and nanostructured lipid carriers (NLC)—Structural investigations on two different carrier systems. *Biomed. Mater. Eng.* **2005**, *15*, 393–402.
53. Lakshmi, P.; Kumar, G.A. Nanosuspension technology: A review. *Int. J. Pharm. Pharm Sci.* **2010**, *2*, 35–40.
54. Mitri, K.; Shegokar, R.; Gohla, S.; Anselmi, C.; Müller, R.H. Lipid nanocarriers for dermal delivery of lutein: Preparation, characterization, stability and performance. *Int. J. Pharm.* **2011**, *414*, 267–275. [CrossRef]
55. Müller, R.H.; Jacobs, C.; Kayser, O. Nanosuspensions as particulate drug formulations in therapy. Rationale for development and what we can expect for the future. *Adv. Drug Deliv. Rev.* **2001**, *47*, 3–19. [CrossRef]
56. Schwarz, C.; Mehnert, W.; Lucks, J.S.; Müller, R.H. Solid lipid nanoparticles (SLN) for controlled drug delivery. I. Production, characterization and sterilization. *J. Control. Release* **1994**, *30*, 83–96. [CrossRef]
57. Hou, Z.; Gao, Y.; Yuan, F.; Liu, Y.; Li, C.; Xu, D. Investigation into the physicochemical stability and rheological properties of b-carotene emulsion stabilized by soybean soluble polysaccharides and chitosan. *J. Agric. Food Chem.* **2010**, *58*, 8604–8611. [CrossRef]
58. Caddeo, C.; Manconi, M.; Fadda, A.M.; Lai, F.; Lampis, S.; Diez-Sales, O.; Sinico, C. Nanocarriers for antioxidant resveratrol: Formulation approach, vesicle self-assembly and stability evaluation. *Colloids Surf. B Biointerfaces* **2013**, *111*, 327–332. [CrossRef]

© 2019 by the authors. Licensee MDPI, Basel, Switzerland. This article is an open access article distributed under the terms and conditions of the Creative Commons Attribution (CC BY) license (http://creativecommons.org/licenses/by/4.0/).

Article

Preparation and Evaluation of Resveratrol-Loaded Composite Nanoparticles Using a Supercritical Fluid Technology for Enhanced Oral and Skin Delivery

Eun-Sol Ha [1], Woo-Yong Sim [1], Seon-Kwang Lee [1], Ji-Su Jeong [1], Jeong-Soo Kim [2], In-hwan Baek [3], Du Hyung Choi [4], Heejun Park [5], Sung-Joo Hwang [6] and Min-Soo Kim [1],*

[1] College of Pharmacy, Pusan National University, 63 Busandaehak-ro, Geumjeong-gu, Busan 46241, Korea; edel@pusan.ac.kr (E.-S.H.); popo923@pusan.ac.kr (W.-Y.S.); lsk7079@pusan.ac.kr (S.-K.L.); sui15@pusan.ac.kr (J.-S.J.)
[2] Dong-A ST Co. Ltd., Giheung-gu, Yongin, Gyeonggi 446-905, Korea; ttung2nd@naver.com
[3] College of Pharmacy, Kyungsung University, 309, Suyeong-ro, Nam-gu, Busan 48434, Korea; baek@ks.ac.kr
[4] Department of Pharmaceutical Engineering, Inje University, Gyeongnam 621-749, Korea; choidh@inje.ac.kr
[5] Department of Industrial and Physical Pharmacy, College of Pharmacy, Purdue University, 575 Stadium Mall Drive, West Lafayette, IN 47907, USA; pharmacy4336@gmail.com
[6] College of Pharmacy and Yonsei Institute of Pharmaceutical Sciences, Yonsei University, 85 Songdogwahak-ro, Yeonsu-gu, Incheon 21983, Korea; sjh11@yonsei.ac.kr
* Correspondence: minsookim@pusan.ac.kr; Tel.: +82-51-510-2813

Received: 2 November 2019; Accepted: 13 November 2019; Published: 14 November 2019

Abstract: We created composite nanoparticles containing hydrophilic additives using a supercritical antisolvent (SAS) process to increase the solubility and dissolution properties of *trans*-resveratrol for application in oral and skin delivery. Physicochemical properties of *trans*-resveratrol-loaded composite nanoparticles were characterized. In addition, an in vitro dissolution–permeation study, an in vivo pharmacokinetic study in rats, and an ex vivo skin permeation study in rats were performed. The mean particle size of all the composite nanoparticles produced was less than 300 nm. Compared to micronized *trans*-resveratrol, the *trans*-resveratrol/hydroxylpropylmethyl cellulose (HPMC)/poloxamer 407 (1:4:1) nanoparticles with the highest flux (0.792 μg/min/cm^2) exhibited rapid absorption and showed significantly higher exposure 4 h after oral administration. Good correlations were observed between in vitro flux and in vivo pharmacokinetic data. The increased solubility and flux of *trans*-resveratrol generated by the HPMC/surfactant nanoparticles increased the driving force on the gastrointestinal epithelial membrane and rat skin, resulting in enhanced oral and skin delivery of *trans*-resveratrol. HPMC/surfactant nanoparticles produced by an SAS process are, thus, a promising formulation method for *trans*-resveratrol for healthcare products (owing to their enhanced absorption via oral administration) and for skin application with cosmetic products.

Keywords: resveratrol; solubility; nanoparticle; correlation; supercritical fluid; bioavailability

1. Introduction

Trans-resveratrol is abundant in various foods, such as grapes, peanuts, and berries, and is usually taken as a dietary supplement. Chemically, *trans*-resveratrol is known as 3,5,4′-trihydroxystilbene, a non-flavonoid polyphenolic compound produced by plants in response to injury or attack by bacteria and fungi [1]. When exposed to UV light, the typically low transformation of *trans*-resveratrol to *cis*-resveratrol is accelerated [2]. *Trans*-resveratrol has been shown to have several beneficial properties, including anti-aging, anticancer, antidiabetic, anti-inflammatory, antioxidant, cardioprotective, and neuroprotective activities [3–6]. Unfortunately, due to its poor water solubility, instability, short plasma half-life, and extensive metabolism in the intestine and liver, clinical uses of *trans*-resveratrol are limited to oral administration [1,7–9]. *Trans*-resveratrol is a Biopharmaceutical Classification System (BCS) class II compound with an insolubility in aqueous solutions of pH 1.0 to pH 7.5 and high permeability [1]. Due to these properties, *trans*-resveratrol is quickly metabolized, therefore skin application may be an alternative to oral administration [10–12]. Various formulation strategies, including the use of liposomes, solid dispersions, cyclodextrin complexes, solid lipid nanoparticles, emulsions, polymeric micelles, polymeric nanoparticles, and nanocrystals, have been evaluated to attempt to overcome current limitations of *trans*-resveratrol [13–21]. In particular, a pharmacokinetic study of twelve healthy volunteers given an oral administration of a capsule formulation of resveratrol via solubilization with micelles consisting of polysorbate 80, polysorbate 20, and medium chain triacylglycerol yielded increases in the area under the plasma concentration versus time curve (AUC) and maximum plasma concentration (C_{max}) of resveratrol of 5.0-fold and 10.6-fold, respectively, compared to volunteers ingesting a resveratrol powder [19]. In a study with rabbits, a solid dispersion of resveratrol produced an AUC value that was 3-fold higher compared to rabbits who ingested a resveratrol/magnesium dihydroxide solid dispersion [20]. However, in a study of resveratrol and piperine cocrystals, concentration of resveratrol in saturated solution at various conditions was decreased compared to pure resveratrol, resulting in decreased oral bioavailability [21]. It is, thus, critical to enhance the in vitro solubility and dissolution properties of resveratrol to improve its absorption by the body, and thus increase its biological performance [22,23].

We hypothesize that the rapid dissolution rate and high degree of supersaturation (solubility) of *trans*-resveratrol formulated with amorphous composite nanoparticles is directly related to increases in *trans*-resveratrol absorption. In this study, composite nanoparticles containing hydrophilic additives were produced using the supercritical antisolvent (SAS) process to increase the solubility and dissolution properties of *trans*-resveratrol for application by oral and skin delivery. Supercritical carbon dioxide (SC-CO_2) was used as an antisolvent, and it has significant safety advantages. The airborne concentration at 25 °C, considering a high threshold limit value (TLV) of 5000 ppm, is a safe environment that a workforce may be exposed to daily without adverse effects. In addition, SC-CO_2 is highly dense, permeable, has a high solvent power and diffusion rate, and is usually miscible with organic solvents [24]. These properties can cause a higher supersaturation during the SAS process, hence reducing the critical energy barrier for nucleation, which leads to faster nucleation, and therefore the precipitation of more and smaller particles [25]. Physicochemical characterization of composite nanoparticles was carried out using particle size and specific surface measurements, scanning electron microscopy, powder X-ray diffraction, differential scanning calorimetry, and kinetic solubility analysis. In addition, an in vitro dissolution–permeation study was performed to compare the *trans*-resveratrol flux of different composite nanoparticles prepared by an SAS process. An in vivo pharmacokinetic study in rats was also performed. We also investigated the correlation between in vitro flux data and in vivo pharmacokinetic data on *trans*-resveratrol. Finally, we performed ex vivo skin permeation studies using rats to investigate the use of *trans*-resveratrol-loaded composite nanoparticles for skin delivery.

2. Materials and Methods

2.1. Materials

Trans-resveratrol was supplied by Ningbo Liwah Pharmaceutical Co., Ltd. (Zhejiang, China), and micronized to a mean particle size of 2.6 μm and purity of 99.1% by using an air jet mill. Polyvinylpyrrolidone K 12/25/30/90 (PVP K12/K25/K30/K90), polyvinylpyrrolidone vinyl acetate 64 (PVP VA 64), polyvinyl caprolactampolyvinyl acetate-polyethylene glycol graft copolymer (Soluplus®), macrogol (15)-hydroxystearate (Kolliphor™ HS 15), D-α-Tocopherol polyethylene glycol 1000 succinate (TPGS), and poly (ethylene glycol)-block-poly (propylene glycol)-block-poly(ethylene glycol) (Poloxamer 188/407) were obtained from BASF (Ludwigshafen, Germany). Hydroxylpropylmethyl cellulose (HPMC 3 cp/4.5 cp/6 cp) and low viscosity hydroxylpropyl cellulose (HPC-SSL) were provided by Shin-Etsu chemical Co., Ltd. (Tokyo, Japan), and Nippon

Soda Co., Ltd. (Japan), respectively. Polyethylene glycol 6000, sodium carboxymethylcellulose (CMC), and sodium lauryl sulfate (SLS) were purchased from Sigma-Aldrich Co., Ltd. (St. Louis, MO, USA). Sucrose laurate (Ryoto™ Ester L-1695, Mitsubishi-Kagaku Foods Co., Tokyo, Japan) was gifted by Namyung commercial Co., Ltd. (Seoul, Korea). Sorbitan monolaurate (Span® 20), sorbitan monostearate (Span® 60), sorbitan monooleate (Span® 80), and polysorbate 20/60/80 (Tween® 20/60/80) were purchased from Daejung Chemicals and Metals Co., Ltd. (Siheung-si, Korea). Propylene glycol dicaprolate/dicaprate (Labrafac™ PG, Gattefossè, Saint-Priest, France), medium chain triglycerides (Labrafac™ Lipophile WL1349, Gattefossè, Saint-Priest, France), oleoyl polyoxyl-6 glycerides (Labrafil® M 1944CS, Gattefossè, Saint-Priest, France), linoleoyl polyoxyl-6 glycerides (Labrafil® M 2125CS, Gattefossè, Saint-Priest, France), lauroyl polyoxyl-6 glycerides (Labrafil® M 2130CS, Gattefossè, Saint-Priest, France), caprylocaproyl polyoxyl-8 glycerides (Labrasol®, Gattefossè, Saint-Priest, France), propylene glycol monolaurate type II (Lauroglycol™ 90, Gattefossè, Saint-Priest, France), and polyglyceryl-3 dioleate (Plurol® Oleique CC 497, Gattefossè, Saint-Priest, France) were kindly provided by Masung Chemicals (Seoul, Korea). All organic solvents and reagents used were either of high-performance liquid chromatography (HPLC) grade or analytical grade and were purchased from Honeywell Burdick and Jackson (Muskegon, MI, USA) or Daejung Chemicals and Metals Co., Ltd. (Siheung-si, Korea), respectively.

2.2. Solubility Studies of Trans-Resveratrol in Aqueous Solutions Containing Various Additives

The solubility of *trans*-resveratrol in aqueous solutions was measured to test different polymers and surfactants. First, aqueous solutions containing 1% polymer or 1% surfactant (w/v) were prepared and excess resveratrol was added to an amber vial containing 10 mL of each solution. Samples were sonicated for 1 h then incubated in a shaking water bath at 37 °C for 72 h to reach equilibrium. Next, samples were centrifuged at 12,000× g for 15 min and filtered through a 0.2 μm glass fiber syringe filter. Then, 1 mL of filtrate was taken into an amber volumetric flask and diluted with methanol. The concentration of resveratrol was then determined using a Shimadzu HPLC system (Shimadzu, Tokyo, Japan) consisting of an SPD-20A ultraviolet–visible (UV/VIS) detector, CBM-20A communications bus module, SIL-20AC autosampler, LC-20AT liquid chromatograph, DGU-20A 5R degassing unit, and a C18 analytical column (4.6 × 150 mm, 5 μm, Shiseido, Tokyo, Japan). HPLC analysis conditions were as follows: a 40% acetonitrile in water mobile phase, 0.8 mL/min flow rate, 30 °C column temperature, 10 μL injection volume, and UV detector wavelength of 303 nm.

2.3. Nanoparticle Preparation Using an SAS Process

For preparation of *trans*-resveratrol-loaded composite nanoparticles, an SAS process was applied using the equipment as previously described [26,27]. CO_2 gas was liquefied using a cooler, raised to the required temperature using a heat exchanger, and was pumped through an ISCO™ pump (Model 260D, Teledyne Technologies Inc., Thousand Oaks, CA, USA) into a high-pressure precipitation vessel. Alongside an SAS process, drug solutions were prepared by dissolving resveratrol/polymer (HPMC 6 cp)/surfactant (gelucire 44/14, TPGS, or poloxamer 407)) in a mixture of methanol and dichloromethane (1:1, w/w) in an amber vial. When the system reached a certain temperature and pressure (40 °C and 12 MPa), the drug solution was injected through the nozzle into a high-pressure vessel at a constant rate (1 g/min), along with supercritical carbon dioxide (40 g/min). The temperature of the precipitation vessel was maintained by circulating water with a temperature-controlled bath circulator. After injection of the drug solution was complete, additional carbon dioxide was applied to remove residual solvent dissolved in supercritical carbon dioxide. The pressure in the precipitation vessel was then slowly reduced to atmospheric pressure using the back-pressure regulator and the precipitated particles inside the vessel were collected.

2.4. Trans-Resveratrol Content Analysis

Levels of *trans*-resveratrol within the composite particles were determined using HPLC analysis of sample solutions. Composite particles were dissolved in a mixture of methanol and dichloromethane (1:1, w/w). After dilution with methanol, the concentration of *trans*-resveratrol was determined using a Shimadzu Prominence HPLC system (Shimadzu, Tokyo, Japan). The encapsulation efficiency (%) was calculated by dividing the measured concentration by the theoretical concentration and then multiplying this by 100.

2.5. Scanning Electron Microscopy (SEM)

The morphology of the samples was examined using a scanning electron microscope (SUPRA 25 or 40, Zeiss, Oberkochen, Germany) operating at a voltage of 5 kV. Before observation, samples were fixed to aluminum stubs with double-sided adhesive carbon tape and then gold coated at a pressure of 8–10 Pa for 1 min to increase electrical conductivity of the sample.

2.6. Particle Size Measurements

The average particle size of the samples was determined with dynamic light scattering (ELSZ-1000, Otsuka Electronics, Tokyo, Japan). Samples were sufficiently dispersed in mineral oil and sonicated for 10 min, then size measurements performed at least four times.

2.7. Specific Surface Area Measurements

The specific surface area of composite nanoparticles was measured with a Micromeritics TriStar II 3020 instrument (Micromeritics, Norcross, GA, USA), using the adsorption of nitrogen at the temperature of liquid nitrogen.

2.8. Differential Scanning Calorimetry (DSC)

DSC analysis was performed using a thermal analyzer (DSC25, TA instruments, Inc., New Castle, DE, USA). Prior to each analysis, temperature and heat capacity calibration were performed using high purity indium and aluminum oxide sapphire, respectively, with a temperature range of 0 to 390 °C, a modulation rate of 0.6 °C/min every 40 s, and a scan speed of 5 °C/min. Next, 2–4 mg of sample was weighed and placed in a pre-weighed aluminum hermetic pan, then sealed with an aluminum cover. Closed, empty aluminum pans were used as reference samples. Analysis was carried out by heating samples from 0 to 350 °C at a heating rate of 10 °C/min under a nitrogen purge of 400 mL/min.

2.9. Powder X-ray Diffraction (PXRD)

Powder X-ray diffraction analysis for samples was performed from 5° to 60° using an X-ray Diffractometer (Xpert 3, Panalytical, Almelo, Netherlands) with Ni-filtered Cu-Kα radiation. Data were collected at a scanning speed of 3°/min and a step size of 0.01.

2.10. Kinetic Solubility Study

A sample of 100 mg *trans*-resveratrol was placed in a water-jacked beaker containing 50 mL of distilled water maintained at 37 °C with magnetic stirring at 300 rpm. At predetermined time intervals, 3 mL of sample was withdrawn from the medium and centrifuged at 12,000× *g* for 15 min, then filtered using a 0.2 μm glass fiber syringe filter to remove insoluble material. After dilution with methanol, the concentration of *trans*-resveratrol was quantified using HPLC analysis. All sample measurements were repeated six times.

2.11. Flux Measurements via In Vitro Dissolution and Permeation Studies

To compare *trans*-resveratrol flux between different composite nanoparticles prepared by an SAS process and to evaluate the correlation between in vitro flux data and in vivo pharmacokinetic data for *trans*-resveratrol, a flux measurement study of *trans*-resveratrol was carried out using a miniaturized dissolution–permeation apparatus (μFLUX^{TM} apparatus, Pion Inc., Billerica, MA, USA) [28]. This apparatus contains a horizontal diffusion cell composed of a donor cell, membrane, and receiver cell. The membrane (diffusion area of 1.54 cm^2) used to separate the donor and receiver cells was prepared by impregnating support material (polyvinylidenfluoride, 0.45 μm pore size, 70% porous, 120 μm thickness) with 50 μL of GIT^{TM} lipid solution consisting of 20% phospholipid in a dodecane lipid solution (Pion Inc., Billerica, MA, USA). The donor cell was filled with 16 mL of simulated intestinal fluid (pH 6.8, with pancreatin), while the receiver cell was filled with 16 mL of acceptor sink buffer (ASB, Pion Inc., Billerica, MA, USA), consisting of a hydroxyethyl piperazine ethane sulfonicacid (HEPES)-based pH 7.4 buffer containing surfactant micelles to ensure the sink condition of *trans*-resveratrol. The temperature of the diffusion cell was maintained at 37 °C by circulating water through the heating block with a temperature bath circulator. A 32 mg sample of *trans*-resveratrol was placed in the donor cell with magnetic stirring at 150 rpm. The concentration of drug in both the donor cell and receiver cell was quantified using UV fiber optic probes (2 mm path length for the donor cell and 20 mm path length for the receiver cell) connected with Pion Rainbow spectrometers, using a wavelength of 306 nm. Data were collected every 1 min for the first 30 min, then every 5 min for the next 240 min. The calibration curve for *trans*-resveratrol levels in simulated intestinal fluid (pH 6.8, with pancreatin) in the donor cell was generated by diluting a stock solution of *trans*-resveratrol in a mixture of ethanol and simulated intestinal fluid at pH 6.8 (with pancreatin). The calibration curve for *trans*-resveratrol levels in the receiver cell was generated using serial dilutions of a stock solution of *trans*-resveratrol in acceptor sink buffer. All sample tests were repeated four times. Flux (*J*), the mass transfer through the membrane, is calculated using Equation (1):

$$J = \frac{dm}{Sdt} = \frac{V}{S} \cdot \frac{dc}{dt} \tag{1}$$

where dc/dt is the slope of the concentration of *trans*-resveratrol vs. the time profile in the range with linear slope (except for lag time), V is the volume (mL) of medium in the donor cell, and S is the permeation area (cm^2).

2.12. Pharmacokinetic Study of Oral Delivery in Rats

The animal study protocol is in compliance with institutional guidelines for the care and use of laboratory animals and was approved by the ethics committee of Kyungsung University (No. 17-004A). To investigate the oral bioavailability of *trans*-resveratrol composite nanoparticles, the in vivo pharmacokinetics of *trans*-resveratrol in male Sprague–Dawley (SD) rats were evaluated. Thirty-six male SD rats (200 ± 10 g; Hyochang Science, Daegu, Korea) were divided into six treatment groups of six rats each. The six experimental groups received either micronized *trans*-resveratrol, *trans*-resveratrol/HPMC composite nanoparticles (1:4 or 1:5), or *trans*-resveratrol/HPMC/surfactant composite nanoparticles (1:4:1) at *trans*-resveratrol doses of 20 mg/kg by oral administration. Samples were dispersed in 1 mL of water immediately prior to oral dosing. Blood samples (approximately 0.25 mL each) were collected in heparinized tubes from the jugular vein of the treated rats at 0.25, 0.5, 0.75, 1, 1.5, 2, 4, 6, 8, and 12 h after dosing. Blood samples were centrifuged at 12,000 × g for 10 min at 4 °C. The amount of *trans*-resveratrol in plasma was determined using HPLC, following a previously reported analytical method [29]. Data were then used to determine the maximum plasma concentration of *trans*-resveratrol (C_{max}), and the time required to reach C_{max} (T_{max}) and the area under the plasma concentration versus time curve ($AUC_{0 \to 12\,h}$) were calculated using the linear trapezoidal method.

2.13. Ex Vivo Skin Permeation Study of Skin Delivery

For application of *trans*-resveratrol-loaded composite nanoparticles in skin delivery, ex vivo skin permeation studies were carried out for 24 h at 32 °C with vertical static-type Franz diffusion cells, using the skin of SD rats as the diffusion membrane. Firstly, after anesthesia, rat abdominal skin was shaved using electric and hand razors, then removed surgically. Skin samples were then cleaned of adherent subcutaneous fat and immersed in cold normal saline solution (pH 7.4) for 2 h. Powdered samples of *trans*-resveratrol-loaded composite nanoparticles were placed on the skin membrane surfaces, with effective diffusion areas of 1.86 cm^2. Skin membrane surfaces and sampling ports were later covered with parafilm and aluminum foil to minimize the influx of the external compounds and the degradation of *trans*-resveratrol from light. The loading dose of *trans*-resveratrol was 0.5 mg/cm^2 [30] The receptor cell was filled with 11.5 mL of a 60:40 mixture of phosphate-buffered saline (pH 7.4) and ethanol and stirred with a magnetic bar at 150 rpm to ensure uniform mixing. At predetermined time intervals, 0.2 mL of receptor medium was taken from the receptor compartment and replaced by an equal volume of fresh medium (32 °C). Samples were then centrifuged for 15 min at 12,000× g, and a 10 μL aliquot of the supernatant was injected into an HPLC system, as described above. The amount of *trans*-resveratrol permeated per unit area of the skin was then calculated. All sample tests were repeated six times.

2.14. Data Analysis

Data are expressed as mean ± standard deviation (n = 4 or 6). To evaluate the statistical significance of differences between groups, one-way ANOVA was carried out, followed by least significant difference (LSD) and Student-Newman-Keuls (SNK) tests using SPSS 25.0 software (IBM SPSS Statistics, IBM Corporation, Armonk, NY, USA).

3. Results and Discussions

Trans-resveratrol-loaded composite nanoparticles were prepared using an SAS process with hydrophilic polymers and surfactants. For a preliminary study of various additives, the equilibrium solubility of *trans*-resveratrol in aqueous solutions containing 1% additive were determined at 37 °C and are presented in Figure 1. The solubility of *trans*-resveratrol was 55.3 μg/mL in water, 52.3 μg/mL in pH 1.2 buffer solution, 53.7 μg/mL in pH 4.0 buffer solution, and 51.1 μg/mL in pH 6.8 buffer solution. These results indicate that the solubility of *trans*-resveratrol is very poor in aqueous solution and is similar to previously reported data [31]. Among the hydrophilic polymers, *trans*-resveratrol showed the highest solubility in HPMC (6 cp). Interestingly, the solubility of *trans*-resveratrol increased with increasing viscosity of HPMC, while *trans*-resveratrol solubility showed the opposite trend in PVP. Surfactants dramatically increased *trans*-resveratrol solubility via micelle formation. The most effective surfactant tested was poloxamer 407, followed by TPGS, tween, and gelucire 44/14. The solubility of *trans*-resveratrol with poloxamer 407 was approximately 20-fold the solubility of *trans*-resveratrol alone. Based on the solubility tests, HPMC (6 cp) was selected as the polymer for nanoparticle construction and poloxamer 407, TPGS, and gelucire 44/14 were evaluated as surfactants for preparation of *trans*-resveratrol-loaded composite nanoparticles using an SAS process.

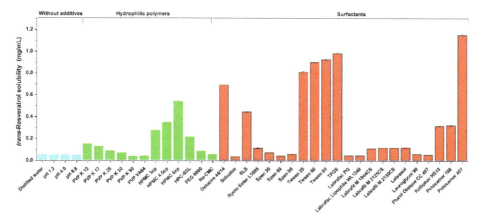

Figure 1. Solubility of *trans*-resveratrol in aqueous solutions containing 1% of various additives at 37 °C. Note: PVP = polyvinylpyrrolidone; HPMC = hydroxylpropylmethyl cellulose; SLS = sodium lauryl sulfate; CMC = sodium carboxymethylcellulose; HPC-SSL = hydroxylpropyl cellulose; TPGS = D-α-Tocopherol polyethylene glycol 1000 succinate; PEG = polyethylene glycol.

3.1. Preparation and Characterization of Trans-Resveratrol Composite Nanoparticles

In this study, we prepared composite nanoparticles containing 1:4 and 1:5 ratios of *trans*-resveratrol/HPMC using a SAS process and evaluated the molecular dispersion of *trans*-resveratrol within composite nanoparticles based on previously reported formulations [32]. In addition, the effects of surfactants (poloxamer 407, TPGS, and gelucire 44/14) on the physicochemical properties and in vivo performance of *trans*-resveratrol-loaded composite nanoparticles was evaluated at a 1:4:1 ratio of *trans*-resveratrol/HPMC/surfactants. In our previous work, nanoparticle agglomeration was observed in composite nanoparticles with high levels of low melting point surfactants (poloxamer 407, TPGS, and gelucire 44/14) prepared using an SAS process [33–35]. As shown in Figure 2 and Table 1, micronized *trans*-resveratrol morphology includes needle-shaped particles with mean particle sizes of 2.6 μm, while *trans*-resveratrol-HPMC nanoparticles are spherical particles with sizes of 180–190 nm and specific surface areas of 60–64 m^2/g. No significant differences were observed between *trans*-resveratrol-HPMC nanoparticles composed of 1:4 and 1:5 ratios of *trans*-resveratrol/HPMC. However, surfactant addition to nanoparticles increased mean particle size and reduced specific surface areas. In particular, *trans*-resveratrol/HPMC/TPGS nanoparticles with mean particle sizes of 293.4 nm exhibited fusion and aggregations of nanoparticles with specific surface areas of 36.4 m^2/g due to the low melting temperature of TPGS (37 °C). Nevertheless, mean particle sizes of all the composite nanoparticles produced were less than 300 nm. *Trans*-resveratrol was successfully incorporated into composite nanoparticles and the encapsulation efficiency exceeded 97% for all formulations, indicating that *trans*-resveratrol was not degraded during the SAS process (Table 1). In addition, SAS process yields were above 80% for all formulations.

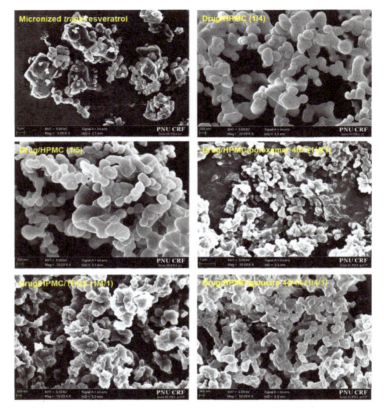

Figure 2. SEM images of *trans*-resveratrol composite nanoparticles.

Table 1. Composition, encapsulation efficiency, mean particle size, and specific surface area of *trans*-resveratrol composite nanoparticles.

Formulation	Encapsulation Efficiency (%)	Mean Particle Size (nm)	Specific Surface Area (m^2/g)
Micronized *trans*-resveratrol	-	2631.4 ± 203.1	2.98 ± 0.3
Drug/HPMC/poloxamer 407	98.9 ± 0.6	258.5 ± 19.5	40.2 ± 1.6
Drug/HPMC/TPGS	99.1 ± 0.8	293.4 ± 16.9	36.4 ± 1.1
Drug/HPMC/gelucire 44/14	97.2 ± 1.2	291.7 ± 15.2	33.1 ± 1.3
Drug/HPMC (1:5)	98.2 ± 0.9	188.2 ± 6.9	60.4 ± 2.3
Drug/HPMC (1:4)	99.9 ± 0.3	181.5 ± 8.5	63.2 ± 1.9

Data are expressed as the mean ± standard deviation (n = 4).

The crystallinity and dispersion of *trans*-resveratrol in composite nanoparticles was analyzed using modulated DSC and PXRD. In DSC thermograms (Figure 3A), the melting temperature and fusion enthalpy of raw *trans*-resveratrol are 268.96 °C and 270.96 J/g, respectively, in good agreement with previously reported data [36]. The complete disappearance of a single, sharp melting endotherm for *trans*-resveratrol was observed for composite nanoparticles, indicating that *trans*-resveratrol exists in an amorphous or molecularly dispersed state within composite nanoparticles. Furthermore, we also analyzed the glass transition temperature (T_g) using modulated DSC measurements. If small molecule drugs act as a plasticizer, the T_g value of the polymer should decrease with increasing drug content in polymer–drug blends. As shown in reversing heat flow versus temperature thermograms (Figure 3B), the T_g value of the composite nanoparticles decreased with increasing *trans*-resveratrol level, and all composite nanoparticles exhibited one T_g value. From these results, *trans*-resveratrol appears dispersed at the molecular level within the composite nanoparticles. In addition, samples of micronized *trans*-resveratrol and composite nanoparticles were characterized using PXRD to assess the preparation of amorphous composites of *trans*-resveratrol (Figure 3C). Micronized *trans*-resveratrol exhibited characteristic peaks at 2θ, similar to previously reported results [37] However, characteristic peak patterns for *trans*-resveratrol were not observed for all composite nanoparticle preparations, indicating that *trans*-resveratrol is molecularly dispersed within composite nanoparticles.

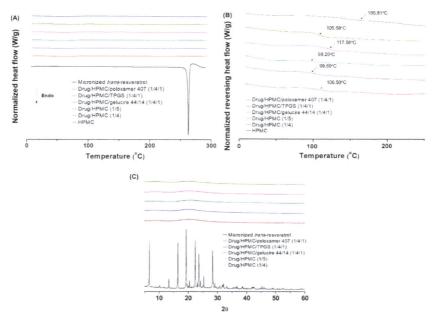

Figure 3. Differential scanning calorimetry thermograms of (**A**) heat flow versus temperature and (**B**) reversing heat flow versus temperature, and powder X-ray diffraction patterns (**C**) of *trans*-resveratrol composite nanoparticles.

The kinetic solubility of *trans*-resveratrol composite nanoparticles was determined in distilled water at 37 °C. As shown in Figure 4, the degree of solubility of *trans*-resveratrol in composite nanoparticles was dramatically increased compared to that of micronized *trans*-resveratrol. In particular, the solubility of *trans*-resveratrol/HPMC/poloxamer 407 (1:4:1) nanoparticles at 24 h was significantly higher (~7.2x) than that of micronized *trans*-resveratrol. The solubility values of *trans*-resveratrol-loaded composite nanoparticles at 24 h as ranked by the SNK test were as follows: drug/HPMC/poloxamer 407 (1:4:1) > drug/HPMC/TPGS (1:4:1) > drug/HPMC/gelucire 44/14 (1:4:1) > drug/HPMC (1:5) = drug/HPMC (1:4) > micronized *trans*-resveratrol. The maximum solubility of *trans*-resveratrol from composite nanoparticles was rapidly reached and maintained for at least 24 h through high inhibition of *trans*-resveratrol crystallization by HPMC [32,38].

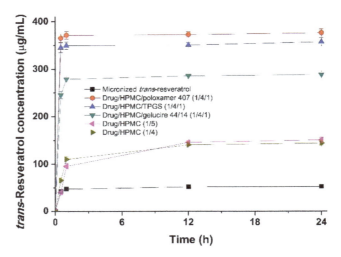

Figure 4. Kinetic solubility profiles of *trans*-resveratrol composite nanoparticles in distilled water at 37 °C.

3.2. Use of Trans-Resveratrol Composite Nanoparticles for Oral Delivery

To compare *trans*-resveratrol flux of different composite nanoparticles prepared by the SAS process and establish correlations between in vitro flux data and in vivo pharmacokinetic data of *trans*-resveratrol, in vitro flux measurements and in vivo pharmacokinetic experiments of *trans*-resveratrol in SD rats were carried out using the oral delivery of *trans*-resveratrol composite nanoparticles. As shown in Figure 5A, the concentration of *trans*-resveratrol from composite nanoparticles in the donor cell increased with increasing solubility of composite nanoparticles. Higher concentrations of *trans*-resveratrol in the receiver cell were observed by increasing the driving force of *trans*-resveratrol through the membrane. In addition, the concentration of *trans*-resveratrol over time in the donor cell indicated maximum dissolution of *trans*-resveratrol from the composite nanoparticles within 2 min, indicating fast dissolution from multiple HPMC/surfactant combinations. Flux (*J*) was obtained from the slope of the concentration of *trans*-resveratrol vs. time profile in the range of 30 to 240 min in Figure 5B, and is presented in Table 2. The system reached steady state within 30 min after dissolution–permeation measurements. As shown in Table 2, *trans*-resveratrol/HPMC/poloxamer 407 (1:4:1) nanoparticles exhibited the highest flux of 0.792 μg/min/cm^2, which was 3.0-fold higher than the flux of micronized *trans*-resveratrol. Flux ranks based on the SNK test for *trans*-resveratrol-loaded composite nanoparticles were as follows: drug/HPMC/poloxamer 407 (1:4:1) > drug/HPMC/TPGS (1:4:1) > drug/HPMC/gelucire 44/14 (1:4:1) > drug/HPMC (1:5) = drug/HPMC (1:4) > micronized *trans*-resveratrol. The trend of increased flux of composite nanoparticles is similar to the trend for kinetic solubility of *trans*-resveratrol. In addition, the very fast dissolution of *trans*-resveratrol from HPMC/surfactant nanoparticles was observed in the donor cell.

The increased solubility and flux of *trans*-resveratrol by composite nanoparticles increases the oral bioavailability of *trans*-resveratrol [39,40]. As shown in Figure 6, HPMC/surfactant composite nanoparticles have rapid absorption rates and significantly higher exposure 4 h after oral administration compared to micronized *trans*-resveratrol. C_{max} ranks based on the SNK test for *trans*-resveratrol-loaded composite nanoparticles were: drug/HPMC/poloxamer 407 (1:4:1) = drug/HPMC/TPGS (1:4:1) > drug/HPMC/gelucire 44/14 (1:4:1) > drug/HPMC (1:5) = drug/HPMC (1:4) > micronized *trans*-resveratrol. The $AUC_{0-12\,h}$ ranked by the SNK test for *trans*-resveratrol-loaded composite nanoparticles were as follows: drug/HPMC/poloxamer 407 (1:4:1) > drug/HPMC/TPGS (1:4:1) > drug/HPMC/gelucire 44/14 (1:4:1) > drug/HPMC (1:5) = drug/HPMC (1:4) > micronized *trans*-resveratrol. Composition ranks for flux data, thus, agreed with ranks for $AUC_{0-12\,h}$. Greater increases in C_{max} and $AUC_{0\rightarrow12\,h}$

were observed for composite nanoparticles compared to micronized *trans*-resveratrol. In particular, C_{max} and $AUC_{0\to12\,h}$ of *trans*-resveratrol/HPMC/poloxamer 407 (1:4:1) nanoparticles were 9.7-fold and 3.0-fold higher, respectively, than those of micronized *trans*-resveratrol, which may be due to the rapid metabolism of *trans*-resveratrol [41].

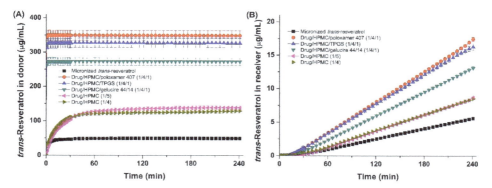

Figure 5. In vitro dissolution (**A**) and permeation profiles (**B**) of *trans*-resveratrol composite nanoparticles. Data are expressed as the mean ± standard deviation (*n* = 4).

Table 2. In vitro flux data and in vivo pharmacokinetic data for *trans*-resveratrol composite nanoparticles.

Formulation	In Vitro Flux (µg/cm²/min)	In Vivo Pharmacokinetic Data		
		$AUC_{0\to12\,h}$ (ng·h/mL)	C_{max} (ng/mL)	T_{max} (h)
Micronized *trans*-resveratrol	0.266 ± 0.006	163.4 ± 33.1	43.4 ± 6.3	1.0 ± 0.5
Drug/HPMC/poloxamer 407	0.792 ± 0.013 [a,b,c,d]	493.6 ± 45.4 [a,b,c,d]	420.3 ± 63.4 [a,b,c]	0.5 ± 0.1
Drug/HPMC/TPGS	0.761 ± 0.016 [a,b,c]	434.3 ± 30.1 [a,b,c]	368.7 ± 72.1 [a,b]	0.5 ± 0.2
Drug/HPMC/gelucire 44/14	0.618 ± 0.012 [a,b]	365.4 ± 45.0 [a,b]	268.7 ± 78.9 [a]	0.7 ± 0.4
Drug/HPMC (1:5)	0.419 ± 0.012 [a]	256.6 ± 53.8 [a]	133.5 ± 36.7	0.9 ± 0.3
Drug/HPMC (1:4)	0.409 ± 0.004 [a]	245.9 ± 76.9 [a]	132.2 ± 27.3	0.7 ± 0.2

Note: [a] $p < 0.05$ vs. micronized trans-resveratrol; [b] $p < 0.05$ vs. drug/HPMC (1:5); [c] $p < 0.05$ vs. drug/HPMC/Gelucire 44/14 (1:4:1); [d] $p < 0.05$ vs. drug/HPMC/TPGS (1:4:1). Data are expressed as the mean ± standard deviation (*n* = 4 or 6). $AUC_{0\to12\,h}$, the area under the plasma concentration versus time curve; $Cmax$, the maximum plasma concentration of trans-resveratrol; T_{max}, the time required to reach C_{max}.

To further investigate correlations between in vitro flux data and in vivo pharmacokinetic data for *trans*-resveratrol, linear regression analysis was used. A good correlation was observed between relative in vitro flux, relative in vivo C_{max}, and in vivo $AUC_{0\to12\,h}$ of composite nanoparticles and micronized *trans*-resveratrol ($R^2 > 0.989$). In fact, in vitro flux data reasonably represent in vivo pharmacokinetic data of *trans*-resveratrol, as previously reported for other poorly water-soluble compounds [33–35]. *Trans*-resveratrol/HPMC/poloxamer 407 (1:4:1) nanoparticles show 3.0-fold higher flux enhancement and 9.7-fold higher C_{max} compared to micronized *trans*-resveratrol. However, the 3.0-fold in vivo increase in $AUC_{0\to12\,h}$ for *trans*-resveratrol/HPMC/poloxamer 407 (1:4:1) nanoparticles relative to micronized *trans*-resveratrol more closely follows the in vitro flux enhancement. Similar to a previously reported method [28], we performed regression analysis between the total amount of *trans*-resveratrol absorbed at 240 min in flux measurements and in in vivo $AUC_{0-12\,h}$. At 240 min, the amounts of permeated *trans*-resveratrol in the receiver cell were 89.6 ± 2.0 µg for micronized *trans*-resveratrol, 136.4 ± 2.2 µg for drug/HPMC (1:4), 139.1 ± 2.9 µg for drug/HPMC (1:5), 209.7 ± 4.2 µg for drug/HPMC/gelucire 44/14 (1:4:1), 260.9 ± 8.2 µg for drug/HPMC/TPGS (1:4:1), and 279.5 ± 5.3 µg for drug/HPMC/poloxamer 407 (1:4:1), with the same ranks observed for in vivo $AUC_{0\to12\,h}$. As shown in Figure 7C, there is good positive linear correlation between the total amount of *trans*-resveratrol absorbed at 240 min in flux measurements and in vivo $AUC_{0\rightarrow 12\,h}$ data obtained from plasma concentration–time profiles in rats ($R^2 > 0.990$). In fact, in vitro flux data can predict the in vivo pharmacokinetic data for *trans*-resveratrol. In addition, experimental conditions for flux measurements can be modified to establish proportional linear relationships between in vitro and in vivo data.

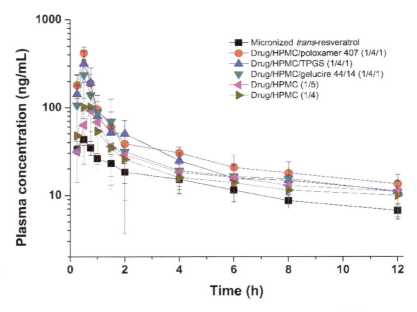

Figure 6. Plasma concentration versus time profiles of *trans*-resveratrol after oral administration of composite nanoparticles to Sprague–Dawley (SD) rats. Data are expressed as the mean ± standard deviation ($n = 6$).

Generally, the oral absorption of poorly water-soluble compounds can be accounted for by including the solubilization and dissolution processes as rate-limiting steps [42]. In this study, we demonstrated that a highly supersaturated solution generated using HPMC/surfactant nanoparticles is able to diffuse *trans*-resveratrol through membranes and enhance the flux of *trans*-resveratrol to receiver cells in in vitro flux measurement studies. Consequently, the enhanced flux of *trans*-resveratrol induces higher driving forces in the gastrointestinal epithelial membrane, resulting in enhanced oral delivery of *trans*-resveratrol in in vivo pharmacokinetic studies of rats [43,44]. Taken together, our results suggest that *trans*-resveratrol-loaded composite nanoparticles prepared using an SAS process are useful for orally delivering *trans*-resveratrol in a manner that allows fast absorption in the initial phase, resulting in higher overall exposure.

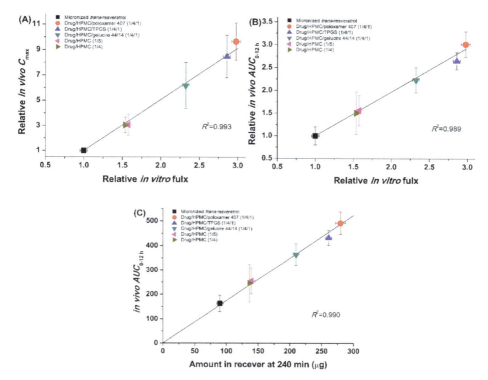

Figure 7. Correlations between in vitro flux data and in vivo pharmacokinetic data of *trans*-resveratrol: (**A**) in vitro flux vs. in vivo C_{max} of composite nanoparticles relative to micronized *trans*-resveratrol; (**B**) in vitro flux vs. in vivo $AUC_{0-12\,h}$ of composite nanoparticles relative to micronized *trans*-resveratrol; (**C**) total absorbed *trans*-resveratrol at 240 min in flux measurements vs. in vivo $AUC_{0\to 12\,h}$.

3.3. Utilization of Trans-Resveratrol Composite Nanoparticles for Skin Delivery

To investigate the application of *trans*-resveratrol-loaded composite nanoparticles for skin delivery, we performed ex vivo permeation studies using skin from rats. The cumulative amount of permeated *trans*-resveratrol (Q) per unit area of skin ($\mu g/cm^2$) was determined using HPLC analysis. To calculate flux, defined as the rate of diffusion of a substance through a permeable membrane, Q versus time profiles were generated (Figure 8). The steady state flux (J_{ss}) of *trans*-resveratrol was calculated based on the slope of the linear portion of the Q versus time profiles. The permeation of *trans*-resveratrol-loaded composite nanoparticles was higher compared to permeation of micronized *trans*-resveratrol (Figure 8). In particular, the steady state flux (J_{ss}) of *trans*-resveratrol/HPMC/poloxamer 407 (1:4:1) nanoparticles was significantly higher (~14.9-fold) than that of micronized *trans*-resveratrol. The steady state flux (J_{ss}) ranks for *trans*-resveratrol-loaded composite nanoparticles based on the SNK test are: drug/HPMC/poloxamer 407 (1:4:1) > drug/HPMC/TPGS (1:4:1) > drug/HPMC/gelucire 44/14 (1:4:1) > drug/HPMC (1:5) = drug/HPMC (1:4) > micronized *trans*-resveratrol. Skin penetration of *trans*-resveratrol has been confirmed in multiple studies [7,10,11]. The enhancement of solubility and dissolution rates of *trans*-resveratrol by composite nanoparticles produced using the SAS process has been shown to enhance *trans*-resveratrol penetration. Interestingly, the steady state flux (J_{ss}) of *trans*-resveratrol from composite nanoparticles containing poloxamer 407 and TPGS was much higher than that of composite nanoparticles containing gelucire 44/14, with approximately 2.2- and 1.9-fold increases, respectively, compared to micronized *trans*-resveratrol. This increase in skin permeation is likely a result of the penetration-enhancing properties of surfactants, such as poloxamer 407 and TPGS [45,46]. Surfactants can potentially solubilize stratum corneum lipids, and thus enhance penetration [47,48]. Surfactants have well-known effects on permeability characteristics of several biological membranes, including membranes in the skin, and thus can enhance the penetration of skin by other compounds present in formulations [49]. In addition, the enhanced permeation may be due to the permeation of solute-containing nanoparticles through shunt routes, such as hair follicles [50]. However, further mechanistic studies are needed of *trans*-resveratrol-loaded composite nanoparticles.

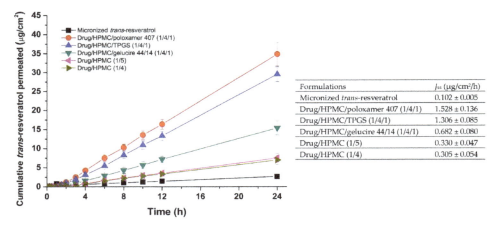

Figure 8. Cumulative ex vivo skin permeation profiles and flux (J_{ss}) data for *trans*-resveratrol composite nanoparticles. Data are presented as means ± standard deviation ($n = 6$).

4. Conclusions

In our study, we designed HPMC/surfactants nanoparticles using the supercritical antisolvent (SAS) process in order to increase solubility and dissolution properties of *trans*-resveratrol for effective oral and skin delivery. Spherical composite nanoparticles with a mean size smaller than 300 nm were successfully produced using an SAS process. Among the formulations tested, *trans*-resveratrol/HPMC/poloxamer 407 (1:4:1) nanoparticles show the highest flux of 0.792 μg/min/cm^{-2}, exhibit rapid absorption, and have significantly higher exposure 4 h after oral administration than micronized *trans*-resveratrol. Good correlations between in vitro flux and in vivo pharmacokinetic data were also observed. The increased solubility and flux of *trans*-resveratrol generated using HPMC/surfactant nanoparticles increased driving forces on the gastrointestinal epithelial membrane and rat skin, resulting in enhanced oral and skin delivery of *trans*-resveratrol. HPMC/surfactant nanoparticles produced by the SAS process are, therefore, promising formulations enhancing *trans*-resveratrol absorption via oral administration for use in healthcare products, as well for skin application of cosmetic products.

Author Contributions: Conceptualization, E.-S.H., J.-S.K., S.-J.H., and M.-S.K.; formal analysis, E.-S.H., W.-Y.S., S.-K.L., J.-S.J., J.-S.K., I.-h.B., D.H.C., H.P., and M.-S.K.; funding acquisition, M.-S.K.; investigation, E.-S.H., W.-Y.S., S.-K.L., and M.-S.K.; methodology, E.-S.H., W.-Y.S., S.-K.L., J.-S.J., J.-S.K., I.-h.B., D.H.C., H.P., and M.-S.K.; software, D.H.C., S.-J.H., and M.-S.K.; supervision, M.-S.K.; validation, E.-S.H.; writing—original draft, E.-S.H. and M.-S.K.; writing—review and editing, J.-S.K., I.-h.B., D.H.C., H.P., S.-J.H., and M.-S.K.

Funding: This research was supported by Basic Science Research Program through the National Research Foundation of Korea (NRF), funded by the Ministry of Science, ICT, and Future Planning (NRF- 2017R1C1B1006483).

Conflicts of Interest: The authors declare no conflict of interest.

References

1. Amri, A.; Chaumeil, J.C.; Sfar, S.; Charrueau, C. Administration of resveratrol: What formulation solutions to bioavailability limitations? *J. Control. Release* **2012**, *158*, 182–193. [CrossRef] [PubMed]
2. Zhao, Y.; Shi, M.; Ye, J.-H.; Zheng, X.-Q.; Lu, J.-L.; Liang, Y.-R. Photo-induced chemical reaction of trans-resveratrol. *Food Chem.* **2015**, *171*, 137–143. [CrossRef] [PubMed]
3. Ramírez-Garza, S.L.; Laveriano-Santos, E.P.; Marhuenda-Muñoz, M.; Storniolo, C.E.; Tresserra-Rimbau, A.; Vallverdú-Queralt, A.; Lamuela-Raventós, R.M. Health Effects of Resveratrol: Results from Human Intervention Trials. *Nutrients* **2018**, *10*, 1892. [CrossRef] [PubMed]
4. Na, J.-I.; Shin, J.-W.; Choi, H.-R.; Kwon, S.-H.; Park, K.-C. Resveratrol as a Multifunctional Topical Hypopigmenting Agent. *Int. J. Mol. Sci.* **2019**, *20*, 956. [CrossRef]
5. Fonseca, J.; Moradi, F.; Valente, A.J.F.; Stuart, J.A. Oxygen and Glucose Levels in Cell Culture Media Determine Resveratrol's Effects on Growth, Hydrogen Peroxide Production, and Mitochondrial Dynamics. *Antioxidants* **2018**, *7*, 157. [CrossRef]
6. Chimento, A.; De Amicis, F.; Sirianni, R.; Sinicropi, M.S.; Puoci, F.; Casaburi, I.; Saturnino, C.; Pezzi, V. Progress to Improve Oral Bioavailability and Beneficial Effects of Resveratrol. *Int. J. Mol. Sci.* **2019**, *20*, 1381. [CrossRef]
7. Marier, J.F.; Vachon, P.; Gritsas, A.; Zhang, J.; Moreau, J.P.; Ducharme, M.P. Metabolism and disposition of resveratrol in rats: Extent of absorption, glucuronidation, and enterohepatic recirculation evidenced by a linked-rat model. *J. Pharmacol. Exp. Ther.* **2002**, *302*, 369–373. [CrossRef]
8. Berman, A.Y.; Motechin, R.A.; Wiesenfeld, M.Y.; Holz, M.K. The therapeutic potential of resveratrol: A review of clinical trials. *Npj Precis. Oncol.* **2017**, *1*, 35. [CrossRef]
9. Walle, T. Bioavailability of resveratrol. *Ann. N. Y. Acad. Sci.* **2011**, *1215*, 9–15. [CrossRef]
10. Mamadou, G.; Charrueau, C.; Dairou, J.; Nzouzi, N.L.; Eto, B.; Ponchel, G. Increased intestinal permeation and modulation of presystemic metabolism of resveratrol formulated into self-emulsifying drug delivery systems. *Int. J. Pharm.* **2017**, *521*, 150–155. [CrossRef]
11. Intagliata, S.; Modica, M.N.; Santagati, L.M.; Montenegro, L. Strategies to Improve Resveratrol Systemic and Topical Bioavailability: An Update. *Antioxidants* **2019**, *8*, 244. [CrossRef] [PubMed]
12. Hajjar, B.; Zier, K.-I.; Khalid, N.; Azarmi, S.; Löbenberg, R. Evaluation of a microemulsion-based gel formulation for topical drug delivery of diclofenac sodium. *J. Pharm. Investig.* **2018**, *48*, 351–362. [CrossRef]
13. Jadhav, P.; Bothiraja, C.; Pawar, A. Resveratrol-piperine loaded mixed micelles: Formulation, characterization, bioavailability, safety and in vitro anticancer activity. *RSC Adv.* **2016**, *6*, 112795–112805. [CrossRef]
14. Zu, Y.; Zhang, Y.; Wang, W.; Zhao, X.; Han, X.; Wang, K.; Ge, Y. Preparation and in vitro/in vivo evaluation of resveratrol-loaded carboxymethyl chitosan nanoparticles. *Drug Deliv.* **2016**, *23*, 981–991. [CrossRef]
15. Dhakar, N.K.; Matencio, A.; Caldera, F.; Argenziano, M.; Cavalli, R.; Dianzani, C.; Zanetti, M.; López-Nicolás, J.M.; Trotta, F. Comparative Evaluation of Solubility, Cytotoxicity and Photostability Studies of Resveratrol and Oxyresveratrol Loaded Nanosponges. *Pharmaceutics* **2019**, *11*, 545. [CrossRef]
16. Gartziandia, O.; Lasa, A.; Pedraz, J.L.; Miranda, J.; Portillo, M.P.; Igartua, M.; Hernández, R.M. Preparation and Characterization of Resveratrol Loaded Pectin/Alginate Blend Gastro-Resistant Microparticles. *Molecules* **2018**, *23*, 1886. [CrossRef]
17. Shimojo, A.A.M.; Fernandes, A.R.V.; Ferreira, N.R.E.; Sanchez-Lopez, E.; Santana, M.H.A.; Souto, E.B. Evaluation of the Influence of Process Parameters on the Properties of Resveratrol-Loaded NLC Using 2^2 Full Factorial Design. *Antioxidants* **2019**, *8*, 272. [CrossRef]
18. Aguiar, G.P.S.; Arcari, B.D.; Chaves, L.M.P.C.; Magro, C.D.; Boschetto, D.L.; Piato, A.L.; Lanza, M.; Oliveira, J.V. Micronization of trans-resveratrol by supercritical fluid: Dissolution, solubility and in vitro antioxidant activity. *Ind. Crops Prod.* **2018**, *112*, 1–5. [CrossRef]
19. Calvo-Castro, L.A.; Schiborr, C.; David, F.; Ehrt, H.; Voggel, J.; Sus, N.; Behnam, D.; Bosy-Westphal, A.; Frank, J. The oral bioavailability of trans-resveratrol from a grapevine-shoot extract in healthy humans is significantly increased by micellar solubilization. *Mol. Nutr. Food Res.* **2018**, *62*, 1701057. [CrossRef]
20. Spogli, R.; Bastianini, M.; Ragonese, F.; Iannitti, R.G.; Monarca, L.; Bastioli, F.; Nakashidze, I.; Brecchia, G.; Menchetti, L.; Codini, M.; et al. Solid dispersion of resveratrol supported on magnesium dihydroxide (resv@mdh) microparticles improves oral bioavailability. *Nutrients* **2018**, *10*, 1925. [CrossRef]

21. He, H.; Zhang, Q.; Wang, J.-R.; Mei, X. Structure, physicochemical properties and pharmacokinetics of resveratrol and piperine cocrystals. *CrystEngComm* **2017**, *19*, 6154–6163. [CrossRef]
22. Singh, D.; Bedi, N.; Tiwary, A.K. Enhancing solubility of poorly aqueous soluble drugs: Critical appraisal of techniques. *J. Pharm. Investig.* **2018**, *48*, 509–526. [CrossRef]
23. Ma, X.; Williams, R.O., III. Polymeric nanomedicines for poorly soluble drugs in oral delivery systems: An update. *J. Pharm. Investig.* **2018**, *48*, 61–75.
24. Abuzar, S.M.; Hyun, S.-M.; Kim, J.-H.; Park, H.J.; Kim, M.-S.; Park, J.-S.; Hwang, S.-J. Enhancing the solubility and bioavailability of poorly water-soluble drugs using supercritical antisolvent (SAS) process. *Int. J. Pharm.* **2018**, *538*, 1–13. [CrossRef] [PubMed]
25. Hu, K.; McClements, D.J. Fabrication of biopolymer nanoparticles by antisolvent precipitation and electrostatic deposition: Zein-alginate core/shell nanoparticles. *Food Hydrocoll.* **2015**, *44*, 101–108. [CrossRef]
26. Ha, E.-S.; Kim, J.-S.; Lee, S.-K.; Sim, W.-Y.; Jeong, J.-S.; Kim, M.S. Solubility and modeling of telmisartan in binary solvent mixtures of dichloromethane and (methanol, ethanol, *n*-propanol, or *n*-butanol) and its application to the preparation of nanoparticles using the supercritical antisolvent technique. *J. Mol. Liq.* **2019**, *295*, 111710. [CrossRef]
27. Ha, E.-S.; Kim, J.-S.; Lee, S.-K.; Sim, W.-Y.; Jeong, J.-S.; Kim, M.S. Equilibrium solubility and solute-solvent interactions of carvedilol (Form I) in twelve mono solvents and its application for supercritical antisolvent precipitation. *J. Mol. Liq.* **2019**, *294*, 111622. [CrossRef]
28. Tsinman, K.; Tsinman, O.; Lingamaneni, R.; Zhu, S.; Riebesehl, B.; Grandeury, A.; Juhnke, M.; Van Eerdenbrugh, B. Ranking Itraconazole formulations based on the flux through artificial lipophilic membrane. *Pharm. Res.* **2018**, *35*, 161. [CrossRef]
29. Das, S.; Ng, K.Y. Quantification of trans-resveratrol in rat plasma by a simple and sensitive high performance liquid chromatography method and its application in pre-clinical study. *J. Liq. Chromatogr. Relat. Technol.* **2011**, *34*, 1399–1414. [CrossRef]
30. Alonso, C.; Martı, M.; Barba, C.; Carrer, V.; Rubio, L.; Coderch, L. Skin permeation and antioxidant efficacy of topically applied resveratrol. *Arch. Dermatol. Res.* **2017**, *309*, 423–431. [CrossRef]
31. Zupančič, Š.; Lavrič, Z.; Kristl, J. Stability and solubility of trans-resveratrol are strongly influenced by pH and temperature. *Euro. J. Pharm. Biopharm.* **2015**, *93*, 196–204. [CrossRef] [PubMed]
32. Wegiel, L.A.; Mosquera-Giraldo, L.I.; Mauer, L.J.; Edgar, K.J.; Taylor, L.S. Phase Behavior of Resveratrol Solid Dispersions Upon Addition to Aqueous media. *Pharm. Res.* **2015**, *32*, 3324–3337. [CrossRef] [PubMed]
33. Ha, E.-S.; Choo, G.-H.; Baek, I.-H.; Kim, M.-S. Formulation, Characterization, and in Vivo Evaluation of Celecoxib-PVP Solid Dispersion Nanoparticles Using Supercritical Antisolvent Process. *Molecules* **2014**, *19*, 20325–20339. [CrossRef] [PubMed]
34. Kim, M.-S.; Ha, E.-S.; Kim, J.-S.; Baek, I.; Yoo, J.-W.; Jung, Y.; Moon, H.R. Development of megestrol acetate solid dispersion nanoparticles for enhanced oral delivery by using a supercritical antisolvent process. *Drug Des. Dev. Ther.* **2015**, *9*, 4269–4277. [CrossRef]
35. Ha, E.S.; Choo, G.H.; Baek, I.H.; Kim, J.S.; Cho, W.; Jung, Y.S.; Jin, S.E.; Hwang, S.J.; Kim, M.S. Dissolution and bioavailability of lercanidipine-hydroxypropylmethyl cellulose nanoparticles with surfactant. *Int. J. Biol. Macromol.* **2015**, *72*, 188–222. [CrossRef]
36. Ansari, K.A.; Vavia, P.R.; Trotta, F.; Cavalli, R. Cyclodextrin-Based Nanosponges for Delivery of Resveratrol: In Vitro Characterisation, Stability, Cytotoxicity and Permeation Study. *AAPS PharmSciTech* **2011**, *12*, 279–286. [CrossRef]
37. Aguiar, G.P.S.; Boschetto, D.L.; Chaves, L.M.P.C.; Arcari, B.D.; Piato, A.L.; Oliveira, J.V.; Lanza, M. Trans-resveratrol micronization by SEDS technique. *Ind. Crops Prod.* **2016**, *89*, 350–355. [CrossRef]
38. Liu, X.; Feng, X.; Williams, R.O., III; Zhang, F. Characterization of amorphous solid dispersions. *J. Pharm. Investig.* **2018**, *48*, 19–41. [CrossRef]
39. Chang, C.W.; Wong, C.Y.; Wu, Y.T.; Hsu, M.C. Development of a solid dispersion system for improving the oral bioavailability of resveratrol in rats. *Eur. J. Drug Metab. Pharmacokinet.* **2017**, *42*, 239–249. [CrossRef]
40. Singh, S.K.; Makadia, V.; Sharma, S.; Rashid, M.; Shahi, S.; Mishra, P.R.; Wahajuddin, M.; Gayen, J.R. Preparation and in-vitro/in-vivo characterization of trans-resveratrol nanocrystals for oral administration. *Drug Deliv. Transl. Res.* **2017**, *7*, 395–407. [CrossRef]

41. Maier-Salamon, A.; Hagenauer, B.; Wirth, M.; Gabor, F.; Szekeres, T.; Jager, W. Increased transport of resveratrol across monolayers of the human intestinal Caco-2 cells is mediated by inhibition and saturation of metabolites. *Pharm. Res.* **2006**, *23*, 2107–2115. [CrossRef] [PubMed]
42. Ahsan, M.N.; Verma, P.R.P. Enhancement of in vitro dissolution and pharmacodynamic potential of olanzapine using solid SNEDDS. *J. Pharm. Investig.* **2018**, *48*, 269–278. [CrossRef]
43. Goo, B.; Sim, W.-Y.; Ha, E.-S.; Kim, M.-S.; Cho, C.-W.; Hwang, S.-J. Preparation of Spray-dried Emulsion of Sirolimus for Enhanced Oral Bioavailability. *Bull. Korean Chem. Soc.* **2018**, *39*, 1215–1218. [CrossRef]
44. Baek, I.H.; Kim, M.S. Improved supersaturation and oral absorption of dutasteride by amorphous solid dispersions. *Chem. Pharm. Bull.* **2012**, *60*, 1468–1473. [CrossRef] [PubMed]
45. Yang, C.; Wu, T.; Qi, Y.; Zhang, Z. Recent advances in the application of vitamin E TPGS for drug delivery. *Theranostics* **2018**, *8*, 464–485. [CrossRef]
46. Herpin, M.J.; Smyth, H.D.C. Super-heated aqueous particle engineering (SHAPE): A novel method for the micronization of poorly water soluble drugs. *J. Pharm. Investig.* **2018**, *48*, 135–142. [CrossRef]
47. Som, I.; Bhatia, K.; Yasir, M. Status of surfactants as penetration enhancers in transdermal drug delivery. *J. Pharm. Bioallied Sci.* **2012**, *4*, 2–9.
48. Rastogi, V.; Yadav, P.; Verma, N.; Verma, A. Preparation and characterization of transdermal mediated microemulsion delivery of T4 bacteriophages against *E.coli* bacteria: A novel anti-microbial approach. *J. Pharm. Investig.* **2018**, *48*, 394–407. [CrossRef]
49. Kim, J.-S.; Kim, M.-S.; Baek, I.-H. Enhanced Bioavailability of Tadalafil after Intranasal Administration in Beagle Dogs. *Pharmaceutics* **2018**, *10*, 187. [CrossRef]
50. Teichmann, A.; Otberg, N.; Jacobi, U.; Sterry, W.; Lademann, J. Follicular penetration: Development of a method to block the follicles selectively against the penetration of topically applied substances. *Skin Pharmacol. Physiol.* **2006**, *19*, 216–223. [CrossRef]

© 2019 by the authors. Licensee MDPI, Basel, Switzerland. This article is an open access article distributed under the terms and conditions of the Creative Commons Attribution (CC BY) license (http://creativecommons.org/licenses/by/4.0/).

Article

Gelatin-Based Hydrogels for the Controlled Release of 5,6-Dihydroxyindole-2-Carboxylic Acid, a Melanin-Related Metabolite with Potent Antioxidant Activity

Maria Laura Alfieri [1], Giovanni Pilotta [2], Lucia Panzella [1], Laura Cipolla [2,*] and Alessandra Napolitano [1,*]

1. Department of Chemical Sciences, University of Naples Federico II, I-80126 Naples, Italy; marialaura.alfieri@unina.it (M.L.A.); panzella@unina.it (L.P.)
2. Department of Biotechnology and Biosciences, University of Milano-Bicocca, I-20126 Milan, Italy; g.pilotta@campus.unimib.it
* Correspondence: laura.cipolla@unimib.it (L.C.); alesnapo@unina.it (A.N.)

Received: 27 February 2020; Accepted: 16 March 2020; Published: 18 March 2020

Abstract: The ability of gelatin-based hydrogels of incorporating and releasing under controlled conditions 5,6-dihydroxyindole-2-carboxylic acid (DHICA), a melanin-related metabolite endowed with marked antioxidant properties was investigated. The methyl ester of DHICA, MeDHICA, was also tested in view of its higher stability, and different solubility profile. Three types of gelatin-based hydrogels were prepared: pristine porcine skin type A gelatin (HGel-A), a pristine gelatin cross-linked by amide coupling of lysines and glutamic/aspartic acids (HGel-B), and a gelatin/chitosan blend (HGel-C). HGel-B and HGel-C differed in the swelling behavior, showed satisfactorily high mechanical strength at physiological temperatures and well-defined morphology. The extent of incorporation into all the gelatins tested using a 10% *w/w* indole to gelatin ratio was very satisfactory ranging from 60 to 90% for either indoles. The kinetics of indole release under conditions of physiological relevance was evaluated up to 72 h. The highest values were obtained with HGel-B and HGel-C for MeDHICA (90% after 6 h), and an appreciable release was observed for DHICA reaching 30% and 40% at 6 h for HGel-B and HGel-C, respectively. At 72 h, DHICA and MeDHICA were released at around 30% from HGel-A at pH 7.4, with an increase up to 40% at pH 5.5 in the case of DHICA. DHICA incorporated into HGel-B proved fairly stable over 6 h whereas the free compound at the same concentration was almost completely oxidized. The antioxidant power of the indole loaded gelatins was monitored by chemical assays and proved unaltered even after prolonged storage in air, suggesting that the materials could be prepared in advance with respect to their use without alteration of their efficacy.

Keywords: 5,6-dihydroxyindole-2-carboxylic acid; gelatin; cross-linked hydrogel; antioxidant activity; controlled release

1. Introduction

5,6-Dihydroxyindole-2-carboxylic acid (DHICA) is a key intermediate in the tyrosinase catalyzed oxidation of tyrosine leading to eumelanins, the main epidermal human pigments responsible for skin photoprotection. The levels of DHICA are dictated by the activity of the enzyme dopachrome tautomerase or tyrosinase related protein-2, that favors the non-decarboxylative rearrangement pathway of dopachrome, determining a significant incorporation of this indole in natural eumelanins and its presence at nM levels in blood and body fluids in the form of two methylated metabolites, 6-hydroxy-5-methoxyindole-2-carboxylic acid and 5-hydroxy-6-methoxyindole-2-carboxylic acid [1]. In addition to DHICA, other melanin precursors like tyrosine and 3,4-dihydroxyphenylalanine (DOPA) have been shown to exert important role acting as inducers and regulators of the melanogenic apparatus and of MSH receptors [2,3].

In the last decade, several evidences have accumulated indicating that DHICA may exert an antioxidant and protective function *per se* unrelated to pigment synthesis. Early studies showed that DHICA inhibits lipid peroxidation in vitro [4]. Subsequent works indicated that DHICA is oxidized by nitric oxide and efficiently inhibits H_2O_2-Fe(II)/EDTA (Fenton)-induced oxidation processes [5,6]. Moreover, DHICA exhibits excellent triplet quenching properties [7]. DHICA has also an intense absorption maximum at 313 nm, in the erythemigenic UVB region, and exhibits efficient excited state relaxation mechanisms of potential relevance to UV dissipation [8].

The antioxidant profile characterized in different in vitro assays suggested that it may act as a diffusible protective mediator under oxidative stress conditions [9]. In addition, studies on primary cultures of human keratinocytes disclosed its remarkable protective and differentiating effects [10]. At micromolar concentrations, DHICA induced: (a) time- and dose-dependent reduction of cell proliferation without concomitant toxicity; (b) enhanced expression of early and late differentiation markers; (c) increased activities and expression of antioxidant enzymes; and (d) decreased cell damage and apoptosis following UVA exposure. Similarly to DHICA, other structurally related compounds were reported to ensure effective protection of cutaneous homeostasis from hostile environmental factors [11].

All together these results suggest a high potential of DHICA for applicative purposes including treatment of disorders and pathological conditions affecting skin and mucous membranes. Severe limitations in this perspective stem from the ease of this compound to undergo oxidation with subsequent loss of its properties. In addition, proper formulation allowing for vehiculation through the skin and a controlled release would greatly add to the beneficial effects prolonging the action and taking the bioavailable concentrations relatively low.

In recent years, several natural compounds have been tested for the topical treatment of skin disorders by use of a variety of transcutaneous delivery systems including lipophilic nanoparticles like liposomes [12], solid lipid nanoparticles [13,14], nanostructured lipid carriers, monoolein aqueous dispersions [15,16], ethosomes [17,18], and lecithin organogels [19,20]. Speed up of wound healing process by a nanohydrogel embedding an antioxidant compound like baicalin has been described that exhibited optimal performance for a complete skin restoration and inhibition of specific inflammatory markers [21].

A variety of hydrophilic delivery systems have also been explored such as gelatin, the product of collagen hydrolysis, as it offers several advantages including the historical safe use in a wide range of medical applications, low costs, inherent electrostatic binding properties, and proteolytic degradability. In addition, gelatin versatility allows the design of different carrier systems, spanning from micro or nanoparticles, to fibers and hydrogels. Hydrogel based scaffolds are largely applied in the field of tissue engineering since their mechanical features can be tuned to the tissue being repaired and because they offer 3D networks able to support cell growth, differentiation, and migration.

Several reports have described the ability of gelatin hydrogels to adsorb bioactive molecules and/or drugs into the polymer network, thus allowing their controlled release, e.g., for pain treatment and wound healing and tissue regeneration applications [22,23].

In order to design gelatin-based systems, to proper tuning the mechanical properties, swelling behavior, thermal properties, and other physiochemical properties [24,25], gelatin is usually cross-linked by chemical, enzymatic or physical methods and many different approaches have been proposed in the last years. As for the chemical cross-linking, one may rely directly on the chemistry of amino acid side chains by coupling reactions (i.e., amide bond formation) or cross-linking by homo- or heterofunctional cross-linkers (i.e., glutaraldehyde [26], genipin [27], triazolinedione [28]) or convert them into suitable functionalities in order to exploit different chemistries (i.e., photopolymerizable acrylates [29,30] and thiol-ene chemistry [31]).

In addition, gelatin-based hydrogels may be further improved by the use of blends in combination with natural or synthetic polymers such as polyglutammic acid [32,33], chitosan [34], or polyvinyl alcohol [35].

Within this framework in the present work, we have investigated the ability of gelatin-based hydrogels of incorporating and releasing under controlled conditions DHICA. The methyl ester of DHICA, MeDHICA, was also tested in view of its higher stability and different solubility profile [36]. Three different type of gelatin-based hydrogels were prepared as drug release systems: (a) pristine porcine skin type A gelatin (HGel-A); (b) pristine gelatin cross-linked by amide coupling of lysines and glutamic/aspartic acids (HGel-B), and (c) a gelatin/chitosan blend (HGel-C).

These gelatin-based hydrogels could have different potential applications e.g., for topical uses or as scaffolds for cellular growth. In all cases, a satisfactory loading and a smooth release at physiological pH of DHICA and its methyl ester were observed while chemical assays confirmed the antioxidant power of the indole loaded gelatin hydrogels.

2. Materials and Methods

Gelatin type A from porcine skin (gel strength ~300 g bloom), 4-(4,6-dimethoxy-1,3,5-triazin-2-yl)-4-methylmorpholinium-chloride (DMTMM), chitosan, glacial acetic acid, 2,2-diphenyl-1-picrilhydrazyl (DPPH), ferric chloride (III) hexahydrate, and 2,4,6-tris(2-pyridyl)-s-triazine (TPTZ) were purchased from Sigma Aldrich (Milano, Italy) and used without any further purification. Phosphate buffer saline (PBS) 10× was purchased from VWR. DHICA and MeDHICA were prepared according to a procedure previously developed [36,37].

The UV–Vis spectra were recorded on a Jasco V-730 Spectrophotometer (Lecco, Italy).

HPLC analyses were performed on an Agilent 1100 binary pump instrument (Agilent Technologies, Milan, Italy) equipped with a SPD-10AV VP UV–visible detector using an octadecylsilane-coated column, 250 mm × 4.6 mm, 5 μm particle size (Phenomenex Sphereclone ODS, Bologna, Italy) at 0.7 mL/min. Detection wavelength was set at 300 nm. Eluant system: 1% formic acid:acetonitrile, 85:15 v/v.

2.1. Hydrogels Preparation

2.1.1. Pristine Gelatin Hydrogel (HGel-A)

A total of 0.2 g of gelatin was dissolved and mixed in 2 mL (10% w/v) of PBS (pH = 7.4) at 37 °C. After 5 min, DHICA or MeDHICA pre-dissolved in the minimal amount of DMSO were added to the gelatin solution up to 1, 5, 10% w/w with respect to gelatin in the case of DHICA, or 10% w/w in the case of MeDHICA, and continuously stirred until the solution appeared homogeneous. The hydrogels were set for gelation for 12 h at 4 °C and then washed with 5 mL PBS (pH = 7.4) for 30 min to remove not incorporated indole compounds.

2.1.2. Cross-Linked Pristine Gelatin Hydrogel (HGel-B)

For preparation of HGel-B (10% w/v), gelatin (1 g) was initially dispersed in 10 mL of PBS, pH = 7.4, at 45 °C with continuous stirring till complete dissolution (1.5 h). DMTMM was then added (44 mg, 0.16 mmol dissolved in 100 μL of PBS, roughly 20% mmol of gelatin carboxyl groups), and the solution was kept under stirring at 45 °C for 30/45 s. Finally, the solution was poured in a 24 multiwell plate (1 mL per well), plugged, and rested till gelation (5–10 min); the gels were then kept at 37 °C for 2 h and finally freeze-dried.

2.1.3. Gelatin-Chitosan Blend (HGel-C)

For preparation of HGel-C (gelatin/chitosan 8:1 w/w), 0.8 g of gelatin was initially dispersed in 8 mL (10% w/v gelatin solution) of PBS, pH = 7.4, and maintained at 45 °C under stirring till complete dissolution (1.5 h). Chitosan (0.1 g) was dissolved in 0.833 mL of 0.5 M acetic acid till complete dissolution (3 h). Gelatin solution was poured into the chitosan solution and kept under stirring at 45 °C. After 24 h, DMTMM (18 mg, 0.064 mmol dissolved in 70 µL of PBS, roughly 10% mmol based on the gelatin carboxyl groups) was added. The solution was poured in cylindric molds (17 mm in diameter with total volume of 2.2 mL, 1.7 mL/mold), plugged, and rested till gelation (5–10 min); the gels were then kept at 37 °C for 2 h. Finally, the gels were dried at 4 °C.

2.2. Determination of Swelling Degree of HGel-B and HGel-C

Washed hydrogels were swollen in distilled water up to 5 h and the swollen weight was recorded at 10 min intervals, after dabbing the hydrogels with a filter paper before weighing. Totally, three replicas were run. The degree of swelling (SD_i) was calculated as the following:

$$SD_i = [(Mw_i - Md)/Md] \times 100\% \tag{1}$$

where Mw_i is the swollen weight and Md is the dry weight [38].

2.3. Fourier Transform-Infrared Spectroscopy (FT-IR)

FT-IR analyses of loaded and unloaded gelatins HGel-B and HGel-C were done in the Attenuated Total Reflectance (ATR) mode using a Thermo Fisher Nicolet 5700 spectrophotometer equipped with a Smart Performer accessory (Rodano, Italy) mounting a ZnSe crystal for the analysis of solid samples.

2.4. Scanning Electron Microscopy (SEM)

The surface morphology of gelatins HGel-B and HGel-C were examined using a SEM Tabletop Microscope-1000 and a Field Emission SEM (FEI corporate, Hillsboro, OR, USA). The solid samples were mounted on a stub using double-sided adhesive tape before being coated with gold.

2.5. Loading of DHICA/MeDHICA to HGel-B and HGel-C Gelatins

HGel-B and HGel-C were swelled in distilled water for 1 h or 4 h, respectively. After that DHICA or MeDHICA, pre-dissolved in the minimal amount of DMSO, were added to the gelatin solution in PBS 1× (pH = 7.4) up to 10% w/w with respect to gelatin. In the case of HGel-B, DHICA was also used at 5% w/w with respect to gelatin. The optimum loading time was determined by UV–Vis monitoring of the remaining indole in the solution.

2.6. Kinetics of DHICA/MeDHICA Release

2.6.1. HGel-A

The kinetics of release of the indoles in PBS 1× at pH 7.4 or 5.5 was evaluated at room temperature by UV–Vis analysis over 72 h by refreshing the medium every 1 h in the first 6 h and then every 24 h.

2.6.2. HGel-B and HGel-C

The kinetics of release of indole componds (DHICA 5, 10% or MeDHICA 10% from HGel-B and DHICA/MeDHICA 10% from HGel-C) were determined in PBS 1× (pH = 7.4) at 37 °C by UV–Vis analysis over time. The medium was repeatedly refreshed every hour over 6 h.

2.7. Stability of DHICA/MeDHICA

Then, 10% HGel-B incorporating DHICA was immersed in PBS 1× at 37 °C and the release of the indole was monitored over 6 h without medium refreshing. The decay of free DHICA in the PBS solution at 37 °C at the same concentration (based on the estimated incorporation of DHICA in the gelatin as described above) was monitored by HPLC analysis.

2.8. Antioxidant Properties of the DHICA/MeDHICA Loaded Gelatins

2.8.1. DPPH Assay

The assay was performed as previously described [39]. Briefly, HGel-B and HGel-C, loaded with DHICA or MeDHICA at 10% w/w, were immersed at 37 °C (10 mg/mL) in PBS medium. Then, 60 µL aliquots were periodically withdrawn over 6 h and added to 200 µM DPPH solution in methanol (2 mL) with rapid mixing. The reaction was followed by spectrophotometric analysis measuring the absorbance at 515 nm after 10 min. Values are expressed as DPPH decay over time. The experiments were run in triplicates. In other experiments DHICA or MeDHICA loaded gelatins that had been taken in air over one week were immersed in 200 µM DPPH solution (using a 0.04 w/v ratio) and the antioxidant power was evaluated by UV–Vis recording spectra over time up to 7 days. In control experiments the DPPH assay was run on the materials soon after loading of the indoles.

2.8.2. Ferric Reducing/Antioxidant Power (FRAP) Assay

The assay was performed as previously described [40]. Briefly, the ferric reducing/antioxidant power (FRAP) reagent was prepared by sequentially mixing 0.3 M acetate buffer (pH = 3.6), 10 mM TPTZ in 40 mM HCl, and 20 mM ferric chloride in water, at a 10:1:1 $v/v/v$ ratio. To a solution of the FRAP reagent, 12 µL aliquots of the PBS medium, in which 10% HGel-B and HGel-C had been immersed at 37 °C, were added and rapidly mixed. After 10 min, the absorbance at 593 nm was measured. The assay was repeated on the samples over 6 h.

2.9. Statistical Analysis

In all the experiments, each sample was tested in three independent analyses, each carried out in triplicate. The results are presented as the mean ± SD values obtained.

3. Results and Discussion

3.1. Loading of Indole Compounds in Gelatin and Release Kinetics

In the initial experiments, porcine skin gelatin type A was dissolved at 10% w/v concentration in PBS at pH 7.4 at 37 °C (HGel-A) and DHICA or MeDHICA previously dissolved in the minimal amount of DMSO, were added under stirring to a 10% w/w concentration with respect to gelatin (10% w/w DHICA or MeDHICA/HGel-A). The solutions were set for gelation for 12 h at 4 °C and then washed with PBS to remove not incorporated indole compounds (HGel-A, Figure 1) [32].

Figure 1. Preparation of gelatin and loading of indole compounds.

UV–Vis spectrophotometric analysis of the indoles (λ_{max} 320 nm) in the washings allowed to estimate an extent of incorporation into the gelatins of 62 ± 1.7% in the case of DHICA and even higher, up to 80 ± 1.3%, for MeDHICA. Using DHICA at 1 or 5% w/w concentration with respect to gelatin, the extent of incorporation proved to be 48 and 59%, respectively.

The kinetics of release of the indoles at physiological pH and 25 °C was then evaluated over 72 h, by refreshing the medium every hour in the first 6 h and then every 24 h. For either indoles, the release was smooth over the observation period reaching values around 30% of the incorporated indole for the 10% *w/w* DHICA or MeDHICa/HGel-A (Figure 2, panel A). In the case of DHICA, the release was comparable for the 5% *w/w* DHICA/HGel-A and up to 60% after 72 h for the 1% *w/w* DHICA/HGel-A (Figure 2, panel B).

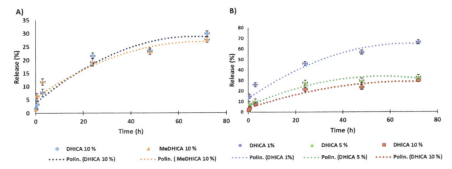

Figure 2. Release kinetics of 5,6-dihydroxyindole-2-carboxylic acid (DHICA) or the methyl ester of DHICA (MeDHICA) at 10% *w/w* (panel **A**), and DHICA 1, 5, 10% *w/w* (panel **B**) from pristine porcine skin type A gelatin (HGel-A) in PBS at pH 7.4 with refreshing of the medium over 72 h. Reported are the mean ± SD values of three experiments and the polynomial regression that fits the data.

The release of the indoles at 10% *w/w* in HGel-A was also monitored at pH 5.5, a pH value of relevance for topical delivery. A sustained release was observed for DHICA reaching 40% of the incorporated indole at 72 h, whereas this was much lower for MeDHICA (Figure 3).

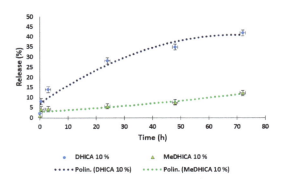

Figure 3. Kinetics of release of DHICA or MeDHICA at 10% *w/w* from HGel-A in PBS at pH 5.5 with refreshing of the medium over 72 h. Reported are the mean ± SD values of three experiments and the polynomial regression that fits the data.

The satisfactorily high kinetics of release for DHICA also at lower pH highlights the potential use of DHICA loaded gelatin hydrogels for epidermal drug delivery in topical uses, e.g., in wound healing to ameliorate the associated inflammatory state. It is worth noting that low pHs are also found in tumor environments. Given the observed gelatin releasing ability, the proposed system may have potential as pH-dependent targeted drug delivery system in anti-cancer therapy. On the contrary, MeDHICA release is negatively affected by low pH. A possible explanation can be found in the higher solubility of the compound at pH 7.4, due to the partially ionized phenol groups. In the case of DHICA this would in part be counterbalanced by the hydrophilic character of the carboxyl group that at pH 5.5 is also ionized to a significant extent based on the pKa value of 4.25 reported for DHICA carboxyl group [41].

However, the relatively low melting temperature of gelatin hydrogels would pose severe limitations to other possible applications suggested by the biological activity of DHICA, but implying exposure to physiological temperatures. Based on this consideration, in further experiments the possibility to get a tougher material that could remain unaltered even after prolonged exposure to physiological temperature or higher was explored by two different strategies: (i) chemical cross-linking; (ii) chitosan-gelatin blends.

3.2. Preparation of Cross-linked Gelatins and Gelatin-Chitosan Blends

DMTMM was used as the coupling agent for gelatin cross-linking [42]. DMTMM is a zero-length coupling agent promoting the activation of carboxyl groups for subsequent amide or ester formation. Like the N-(3-dimethylaminopropyl)-N'-ethylcarbodiimide hydrochloride and N-hydroxysuccinimide (EDC/NHS) system for amide formation representing the standard method for zero-length cross-linking between amino and carboxyl group ligation, DMTMM is water soluble and active towards the desired reaction in a water environment. Recently, it was demonstrated that DMTMM provides better yields than the EDC/NHS system, even in the absence of pH control, that is otherwise fundamental for EDC/NHS conjugation. [42] Thus, we envisaged DMTMM as a suitable coupling agent both for pristine gelatin and for the preparation of the hydrogel composed of gelatin and chitosan, since chitosan requires acidic pH to be solubilized.

For preparation of HGel-B, the coupling agent/gelatin ratio had firstly to be optimised, starting from a 10% w/v gelatin solution in PBS. DMTMM was used in a 5, 10, 20% molar ratio with respect to total gelatin free carboxyl groups (78–80 mmol of free carboxyl groups/g of protein). The 20% DMTMM proved the only condition affording thermally stable hydrogels at 37 °C.

In particular, HGel-B with a 1:10 or 1:5 cross-linker agent to gelatin to molar ratios were prepared and tested for DHICA/MeDHICA incorporation and release.

HGel-C was obtained by reaction of gelatin with chitosan ($\geq$ 75% deacetylation) in the presence of DMTMM coupling agent. 5:1 and 8:1 gelatin/chitosan w/w ratio were tested, using 10, 20, or 30% DMTMM molar ratio based on gelatin free carboxyl groups. Best conditions affording a stable hydrogel was the 8:1 gelatin/chitosan ratio with 10% of DMTMM.

Both HGel-B and HGel-C proved stable at 37 °C, but also at higher temperatures up to 100 °C.

3.3. Characterization of the HGel-B and HGel-C Gelatins

For both materials the swelling profile was determined in water at 25 °C. The swelling behavior expressed as swelling degree shown in Figure 4 appears rather different for the two gelatins. In the case of HGel-B the uptake of water is very rapid with respect to what observed for HGel-C. For this latter the swelling degree was found to be 800% after 6 h, whereas that of HGel-B reached 400% at 1 h.

The surface morphology of HGel-B and HGel-C was investigated using scanning electron microscopy (SEM, Figure 5). The two hydrogels showed quite different features. In particular, HGel-B had a rough wrinkled surface with some holes (Figure 5A), while the HGel-C exhibited a smooth membranous phase consisting of dome shaped orifices, microfibrils, and crystallite.

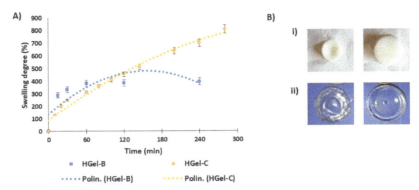

Figure 4. (**A**) Swelling degree over the time and (**B**) images (before and after swelling in water) of HGel-B (i) and HGel-C (ii). Reported are the mean ± SD values of three experiments and the polynomial regression that fits the data.

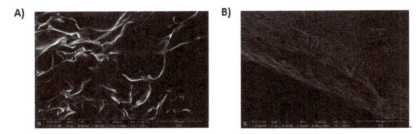

Figure 5. SEM images of HGel-B (**A**, scale bar 500 μM) and HGel-C (**B**, scale bar 400 μM).

FT-IR spectra taken in the ATR mode for HGel-B and HGel-C are shown in Figure 6. In either cases the characteristic bands of gelatin that is the amide I and amide II stretching at 1637 and 1540 cm^{-1}, respectively, and in the case of HGel-B the NH stretching band (band A) is well apparent [43]. The spectrum of HGel-C is dominated by the intense OH stretching band at around 3300 cm^{-1} in accord with the higher content of water of this sample while the well defined band at 1087 cm^{-1} (bridge C–O–C stretch) would be evidence for the presence of chitosan [44].

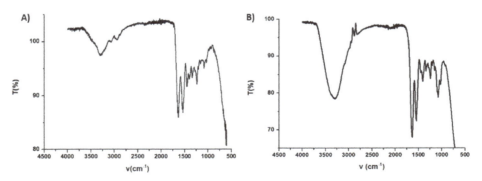

Figure 6. Attenuated Total Reflectance (ATR)-IR spectra of HGel-B (**A**) and HGel-C (**B**).

3.4. Loading and Release of DHICA and MeDHICA from HGel-B and HGel-C

The uptake of DHICA and MeDHICA for HGel-B is shown in Figure 7 panel A. For the HGel-B loaded at 10% *w/w* DHICA the uptake into the hydrogel scaffold was around 40% in the first 30 min with an increase up to 90% incorporation at equilibrium (4 h). The loading of MeDHICA 10% *w/w* was faster reaching 90% values in 2 h. On the other hand, the HGel-B in the presence of 5% *w/w* DHICA reaches 50% incorporation as the maximum value, while no appreciable loading could be detected using lower DHICA or MeDHICA to gelatin ratios (data not shown).

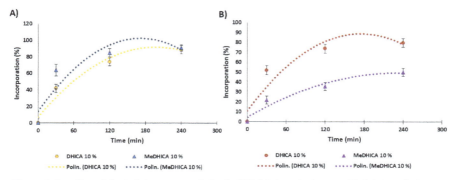

Figure 7. Loading of the indole compounds in the HGel-B (panel **A**) or in HGel-C (panel **B**) over time. Reported are the mean ± SD values of three experiments and the polynomial regression that fits the data.

DHICA loading into HGel-C starting from a 10% *w/w* DHICA gelatin ratio appears very fast reaching a 50% incorporation in the first 30 min and up to 80% at equilibrium (4 h); on the contrary, MeDHICA incorporation was less effective, featuring only a 50% loading after 4 h (Figure 7, panel B). Again, these different behaviors may be ascribed to the free carboxy group of DHICA, that may give acid-base interactions with chitosan rich in free amino groups. These favorable interactions are not allowed with the esterified carboxy group in MeDHICA.

The release kinetics of indole derivatives was determined in PBS at 37 °C (Figure 8, panel A). In the case of 10% *w/w* HGel-B, the observed release of DHICA was very smooth with 15% values at 1 h up to 30% at 6 h with medium refreshing every 1 h, and no significant changes for longer times up to 24 h. The release of DHICA is even lower (23% at 6 h) with 5% *w/w* HGel-B loaded with the indole, (data not shown). MeDHICA in 10% *w/w* HGel-B is released rapidly (57% after 1 h), up to 90% after 4 h with repeated medium refreshing. Much slower release kinetics was observed for DHICA in the case of HGel-C with an initial value of 8% after 1 h which increases up to 54% after 6 h with medium refreshing. On the other hand, MeDHICA is released to 35% after 1 h with a steep increase up to 90% in the first 4 h after repeated medium refreshing (Figure 8, panel B).

In summary (Table 1), loading and release values seem to be dictated mainly by the interactions between the gel and the indoles carboxy groups, and to a lesser extent by the acidic phenolic functionalities. The lower incorporation of DHICA in HGel-A is likely due to the unfavorable interactions of the carboxylate group of DHICA with the carboxylate groups of the acidic aminoacids of gelatin. Such an effect would be decreased with HGel-B and HGel-C, where gelatin carboxy groups are capped by the cross-linking reaction and by the increase of basic amino groups from chitosan, respectively. On the contrary, MeDHICA, missing the ionizable and strongly hydrophilic carboxy group does not experience unfavorable interactions with pristine gelatin (HGel-A) and HGel-B as indicated by the high loading extent, but has a lower affinity for the hydrophilic HGel-C, as indicated also by the higher extent of release with respect to DHICA.

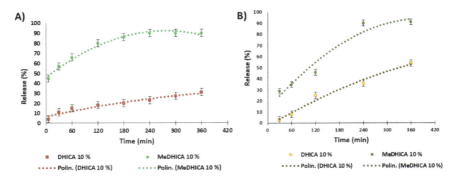

Figure 8. Release kinetics of the incorporated indoles from HGel-B (**A**) and HGel-C (**B**) in PBS with refreshing of the medium at 37 °C over 6 h. Reported are the mean ± SD values of three experiments and the polynomial regression that fits the data.

Table 1. Overview of the percentage values of loading and release of the indole compounds from the gelatin hydrogels investigated.

	Loading (%, at Equilibrium)			Release (%)		
	HGel-A	HGel-B	HGel-C	[a] HGel-A	[b] HGel-B	[b] HGel-C
DHICA (10% w/w)	62 ± 1.7	90 ± 2.2	80 ± 2.2	30 ± 0.57	30 ± 1.6	54 ± 1.6
MeDHICA (10% w/w)	80 ± 1.3	88 ± 1.4	50 ± 2.1	28 ± 0.8	90 ± 1.4	90 ± 1.2

[a] Release evaluated in PBS 1× pH 7.4 at room temperature; [b] Release evaluated in PBS 1× pH 7.4 at 37 °C.

3.5. Assessment of the Stability of DHICA in the HGel-B and -C Gelatins

One of the advantages that should be offered by incorporation of melanin related indoles into a biopolymer like gelatin is the increase of the stability to aerial oxidation in aqueous neutral media of physiological relevance. This issue is more critical in the case of DHICA with respect to MeDHICA given its higher proness to oxidation as shown in our previous studies [36].

Therefore, in subsequent experiments the kinetics of decay of free DHICA in PBS at 37 °C was evaluated by HPLC analysis over the time period used for monitoring the release from cross-linked gelatins. HGel-B loaded with DHICA at 10% was immersed in PBS at 37 °C and the release was again monitored over 6 h without medium refreshing leading to a release of 10% of the incorporated compound. The decay of free DHICA in the PBS solution at the same concentration estimated based on the incorporation shown in Figure 7, panel A, was monitored by HPLC analysis. Figure 9 (panel A) shows that under these latter conditions DHICA was consumed to 70% over 6 h being oxidized to dark melanin, whereas the indole entrapped into the gelatin and hence released slowly into solution was preserved from oxidation to a remarkable extent (Figure 9, panel B).

3.6. Evaluation of the Antioxidant Properties

Considering the remarkable antioxidant activity of the compounds under investigation, the antioxidant properties of the DHICA/MeDHICA released in PBS at 37 °C from the 10% HGel-B and HGel-C were also evaluated by the DPPH and FRAP assays. Briefly, aliquots of the medium were withdrawn over time and the DPPH decay after 10 min was evaluated spectrophotometrically (Figure 10, panels A and B). A similar procedure was followed to evaluate the ferric reducing antioxidant power of the medium containing the 10% HGel-B and HGel-C (Figure 10, panels C and D). As expected, the reducing potency increased over time as a result of the progressive release of the indoles from the hydrogels further confirming the observed stability of the indole compounds incorporated into the hydrogels.

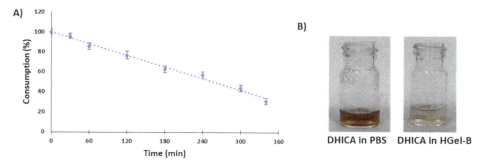

Figure 9. Panel **A**: Decay of free DHICA in the PBS solution monitored by HPLC analysis. Panel **B**: Appearance of the free DHICA solution in PBS (left) vs. the 10% *w/w* DHICA/HGel-B containing medium (right) over 6 h. Reported are the mean ± SD values of three experiments.

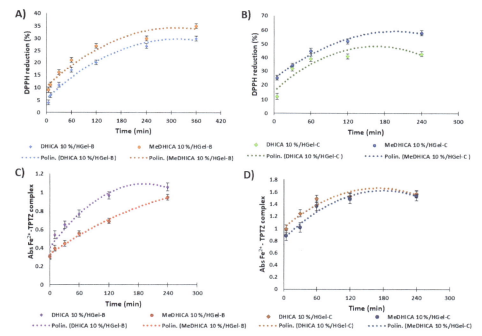

Figure 10. 2,2-diphenyl-1-picrilhydrazyl (DPPH) reduction properties of DHICA or MeDHICA 10% HGel-B (panel **A**) and 10% HGel-C (panel **B**) over time; increase of the absorbance at 593 nm due to the Fe^{2+} TPTZ complex induced by DHICA or MeDHICA 10% HGel-B (panel **C**) and 10% HGel-C (panel **D**). Reported are the mean ± SD values of three experiments and the polynomial regression that fits the data.

The antioxidant potency of HGel-C appeared higher than that of HGel-B in either assays. This result can be interpreted considering the higher release of the indoles in terms of rate and extent from HGel-C with respect to HGel-B. Previous studies on the antioxidant effect of the two indoles have shown an around 2.5-fold higher activity for DHICA with respect to MeDHICA in the DPPH assay [36]. This result allows to rationalize the effect on DPPH consumption observed for DHICA and MeDHICA incorporated into HGel-B (28 and 32%, respectively), in the light of the concentration of the indoles actually released in incubation medium after 6 h, that is 9 µM and 19 µM for DHICA and MeDHICA, respectively. This holds also for HGel-C with a 40 and 55% DPPH consumption for DHICA and MeDHICA, respectively, considering that DHICA and MeDHICA are present in solution at concentrations of 13 μM and 31 μM, respectively. Given the temperature conditions of these experiments HGel-A could not be tested.

The antioxidant potency of DHICA or MeDHICA incorporated into the gelatin hydrogels persisted even after prolonged storage of the material. This issue was demonstrated by DPPH assay applied to indole loaded HGel-B and HGel-C that have been exposed to air over one week. The reducing ability of the materials immersed in the 200 μM DPPH solution at a 0.04 w/v ratio at 25 °C increased over time reaching an almost complete consumption of the reagent after 7 days (Figure 11). The extent of incorporation affected the antioxidant effects. In agreement with the higher incorporation of DHICA in HGel-C the antioxidant activity of the loaded gelatin appeared slightly higher than that of MeDHICA/HGel-C, even if the higher solubility of MeDHICA in the assay medium (methanol) would have favored it compared to DHICA. Control experiments proved the stability of the DPPH reagent over the period of the assay. Moreover, the DPPH consumption profile of the loaded gelatins that had not been subjected to storage proved closely similar to that observed for the gelatins exposed to air over one week. For instance, in the case of HGel-B loaded with DHICA, a DPPH consumption of 86% is observed after 7 days. A DPPH consumption of 90% after seven days was obtained also for HGel-A that could be tested under these experimental conditions.

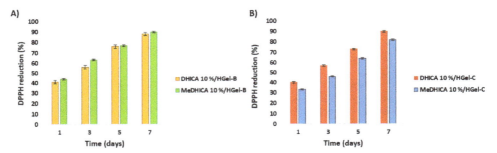

Figure 11. Reduction of DPPH by HGel-B (**A**) or HGel-C (**B**) loaded with DHICA or MeDHICA at 10%. Reported are the mean ± SD values of three experiments.

4. Conclusions

Gelatin based hydrogels are currently the focus of much interest in different fields of biomedical relevance. Even the raw material from different sources including fish skin has been described to have a potential for cell therapies for cutaneous wound, [24] whereas crosslinked gelatin hydrogels or gelatin blends hydrogel have been developed for three-dimensional (3D) cell culture platform [45]. The possibility to use gelatin based hydrogels as carrier for the controlled release of biologically active unstable compounds further adds to the opportunities offered by this system. In this frame the present work explored the potential of gelatin from porcine skin and cross-linked gelatin (HGel-B) or chitosan/gelatin blends (HGel-C) to incorporate DHICA, a melanin-related metabolite endowed with marked antioxidant and cell differentiation stimulation activity and its methyl ester, a more stable compound with a remarkable antioxidant profile. Both indole compounds were effectively incorporated into the pristine gelatin hydrogels and were smoothly released in PBS at pH 7.4, whereas only DHICA was released to a significant extent at pH 5.5, an observation of relevance for the implementation of devices for topical applications. Different incorporation and release profiles where exhibited by the two indoles for HGel-B and HGel-C. The higher affinity of DHICA for HGel-C likely reflects the higher hydrophilic character of the indole and of the cross-linked gelatins itself containing substantial amounts of chitosan as evidenced by FT-IR analysis. More interestingly, gelatin-based hydrogels offer protection to DHICA against air oxidation. Indeed, the cross-linked gelatins incorporating the indoles exhibited a marked antioxidant activity that persisted overtime and was not lost even after prolonged storage of the material. All together these results highlight gelatins as easily accessible biocompatible materials that could warrant a sustained release of labile bioactives under physiologically relevant conditions. The variability of the loading extent and kinetics of release of the three gelatin types investigated would suggest that proper modification of the raw material can allow tuning of the time scale of delivery for different therapeutical applications.

Author Contributions: Conceptualization, A.N., L.C., L.P.; investigation, M.L.A., G.P.; writing—original draft, A.N.; writing—review editing, L.P., L.C.; supervision, A.N., L.C. All authors have read and agreed to the published version of the manuscript.

Funding: This research received no external funding.

Acknowledgments: SEM and ATR/FT-IR analyses were carried out by M.L.A. at the Institut de Science et d'Ingénierie Supramoléculaires (University of Strasbourg, France) in the laboratories of the research group lead by Luisa De Cola. We acknowledge the access to these facilities and Mari Carmen Ortega for analysis of data and helpful discussion.

Conflicts of Interest: The authors declare no conflict of interest.

References

1. Ito, S.; Wakamatsu, K.; d'Ischia, M.; Napolitano, A.; Pezzella, A. Structure of melanins. In *Melanins and Melanosomes*; Borovansky, J., Riley, P.A., Eds.; Wiley-VCH Verlag GmbH: Weinheim, Germany, 2011; pp. 167–185.
2. Slominski, A.; Zmijewski, M.A.; Pawelek, J. L-tyrosine and L-dihydroxyphenylalanine as hormone-like regulators of melanocyte functions. *Pigment. Cell Melanoma Res.* **2012**, *25*, 14–27. [CrossRef]
3. Slominski, A.; Paus, R. Are L-tyrosine and L-dopa hormone-like bioregulators? *J. Theor. Biol.* **1990**, *143*, 123–138. [CrossRef]
4. Memoli, S.; Napolitano, A.; d'Ischia, M.; Misuraca, G.; Palumbo, A.; Prota, G. Diffusible melanin-related metabolites are potent inhibitors of lipid peroxidation. *Biochim. Biophys. Acta* **1997**, *1346*, 61–68. [CrossRef]
5. Novellino, L.; d'Ischia, M.; Prota, G. Nitric oxide-induced oxidation of 5,6-dihydroxyindole and 5,6-dihydroxyindole-2-carboxylic acid under aerobic conditions: Non-enzymatic route to melanin pigments of potential relevance to skin (photo)protection. *Biochim. Biophys. Acta* **1998**, *1425*, 27–35. [CrossRef]
6. Novellino, L.; Napolitano, A.; Prota, G. 5,6-Dihydroxyindoles in the Fenton reaction: A model study of the role of melanin precursors in oxidative stress and hyperpigmentary processes. *Chem. Res. Toxicol.* **1999**, *12*, 985–992. [CrossRef]
7. Zhang, X.; Erb, C.; Flammer, J.; Nau, W.M. Absolute rate constants for the quenching of reactive excited states by melanin and related 5,6-dihydroxyindole metabolites: Implications for their antioxidant activity. *Photochem. Photobiol.* **2000**, *71*, 524–533. [CrossRef]
8. Gauden, M.; Pezzella, A.; Panzella, L.; Neves-Petersen, M.T.; Skovsen, E.; Petersen, S.B.; Mullen, K.M.; Napolitano, A.; d'Ischia, M.; Sundström, V. Role of solvent, pH, and molecular size in excited-state deactivation of key eumelanin building blocks: Implications for melanin pigment photostability. *J. Am. Chem. Soc.* **2008**, *130*, 17038–17043. [CrossRef]
9. Panzella, L.; Napolitano, A.; d'Ischia, M. Is DHICA the key to dopachrome tautomerase and melanocyte functions? *Pigment. Cell Melanoma Res.* **2011**, *24*, 248–249. [CrossRef]
10. Kovacs, D.; Flori, E.; Maresca, V.; Ottaviani, M.; Aspite, N.; Dell'Anna, M.L.; Panzella, L.; Napolitano, A.; Picardo, M.; d'Ischia, M. The eumelanin intermediate 5,6-dihydroxyindole-2-carboxylic acid is a messenger in the cross-talk among epidermal cells. *J. Invest. Derm.* **2012**, *132*, 1196–1205. [CrossRef]
11. Slominski, A.T.; Zmijewski, M.A.; Semak, I.; Kim, T.K.; Janjetovic, Z.; Slominski, R.M.; Zmijewski, J.W. Melatonin, mitochondria, and the skin. *Cell Mol. Life Sci.* **2017**, *74*, 3913–3925. [CrossRef]
12. Ha, E.-S.; Sim, W.Y.; Lee, S.-K.; Jeong, J.-S.; Kim, J.-S.; Baek, I.; Choi, D.H.; Park, H.; Hwang, S.-J.; Kim, M.-S. Preparation and evaluation of resveratrol-loaded composite nanoparticles using a supercritical fluid technology for enhanced oral and skin delivery. *Antioxidants* **2019**, *8*, 554. [CrossRef] [PubMed]

13. Permana, A.D.; Tekko, I.A.; McCrudden, M.T.C.; Anjani, Q.K.; Ramadon, D.; McCarthy, H.O.; Donnelly, R.F. Solid lipid nanoparticle-based dissolving microneedles: A promising intradermal lymph targeting drug delivery system with potential for enhanced treatment of lymphatic filariasis. *J. Control. Release* **2019**, *316*, 34–52. [CrossRef] [PubMed]

14. Shimojo, A.A.M.; Fernandes, A.R.V.; Ferreira, N.R.E.; Sanchez-Lopez, E.; Santana, M.H.A.; Souto, E.B. Evaluation of the influence of process parameters on the properties of resveratrol-loaded NLC using 2^2 full factorial design. *Antioxidants* **2019**, *8*, 272. [CrossRef] [PubMed]

15. Fang, C.L.; Al-Suwayeh, S.A.; Fang, J.Y. Nanostructured lipid carriers (NLCs) for drug delivery and targeting. *Recent Pat. Nanotechnol.* **2013**, *1*, 41–55. [CrossRef]

16. Cortesi, R.; Cappellozza, E.; Drechsler, M.; Contado, C.; Baldisserotto, A.; Mariani, P.; Carducci, F.; Pecorelli, A.; Esposito, E.; Valacchi, G. Monoolein aqueous dispersions as a delivery system for quercetin. *Biomed. Microdevices* **2017**, *19*, 1–11. [CrossRef]

17. Paliwal, S.; Tilak, A.; Sharma, J.; Dave, V.; Sharma, S.; Yadav, R.; Patel, S.; Verma, K.; Tak, K. Flurbiprofen loaded ethosomes—transdermal delivery of anti-inflammatory effect in rat model. *Lipids Health Dis.* **2019**, *18*, 133. [CrossRef]

18. Zhang, Z.; Wo, Y.; Zhang, Y.; Wang, D.; He, R.; Chen, H.; Cui, D. *In vitro* study of ethosome penetration in human skin and hypertrophic scar tissue. *Nanomedicine* **2012**, *8*, 1026–1033. [CrossRef]

19. Raut, S.; Bhadoriya, S.S.; Uplanchiwar, V.; Mishra, V.; Gahane, A.; Jain, S.K. Lecithin organogel: A unique micellar system for the delivery of bioactive agents in the treatment of skin aging. *Acta Pharm. Sin. B* **2012**, *2*, 8–15. [CrossRef]

20. Alsaab, H.; Bonam, S.P.; Bahl, D.; Chowdhury, P.; Alexander, K.; Boddu, S.H. Organogels in drug delivery: A special emphasis on pluronic lecithin organogels. *J. Pharm Pharm Sci.* **2016**, *19*, 252–273. [CrossRef]

21. Manconi, M.; Manca, M.L.; Caddeo, C.; Cencetti, C.; Meo, C.D.; Zoratto, N.; Nacher, A.; Fadda, A.M.; Matricardi, P. Preparation of gellan-cholesterol nanohydrogels embedding baicalin and evaluation of their wound healing activity. *Eur. J. Pharm. Biopharm.* **2018**, *127*, 244–249. [CrossRef]

22. Foox, M.; Zilberman, M. Drug delivery from gelatin-based systems. *Expert Opin. Drug Deliv.* **2015**, *12*, 1547–1563. [CrossRef] [PubMed]

23. Kempen, D.H.; Lu, L.; Hefferan, T.E.; Creemers, L.B.; Maran, A.; Classic, K.L.; Dhert, W.J.; Yaszemski, M.J. Retention of *in vitro* and *in vivo* BMP-2 bioactivities in sustained delivery vehicles for bone tissue engineering. *Biomaterials* **2008**, *29*, 3245–3252. [CrossRef] [PubMed]

24. Huang, C.Y.; Wu, T.C.; Hong, Y.H.; Hsieh, S.L.; Guo, H.R.; Huang, R.H. Enhancement of cell adhesion, cell growth, wound healing, and oxidative protection by gelatins extracted from extrusion-pretreated Tilapia (Oreochromis sp.) fish scale. *Molecules* **2018**, *23*, 2406. [CrossRef] [PubMed]

25. Satapathy, M.K.; Nyambat, B.; Chiang, C.W.; Chen, C.H.; Wong, P.C.; Ho, P.H.; Jheng, P.R.; Burnouf, T.; Tseng, C.L.; Chuang, E.Y. A gelatin hydrogel-containing nano-organic PEI-Ppy with a photothermal responsive effect for tissue engineering applications. *Molecules* **2018**, *23*, 1256. [CrossRef] [PubMed]

26. Fan, H.Y.; Duquette, D.; Dumont, M.J.; Simpson, B.K. Salmon skin gelatin-corn zein composite films produced via crosslinking with glutaraldehyde: Optimization using response surface methodology and characterization. *Int. J. Biol. Macromol.* **2018**, *120*, 263–273. [CrossRef] [PubMed]

27. Manickam, B.; Sreedharan, R.; Elumalai, M. 'Genipin'—the natural water soluble cross-linking agent and its importance in the modified drug delivery systems: An overview. *Curr Drug Deliv.* **2014**, *11*, 139–145. [CrossRef] [PubMed]

28. Guizzardi, R.; Vaghi, L.; Marelli, M.; Natalello, A.; Andreosso, I.; Papagni, A.; Cipolla, L. Gelatin-based hydrogels through homobifunctional triazolinediones targeting, tyrosine residues. *Molecules* **2019**, *24*, 589. [CrossRef]

29. Nguyen, A.H.; McKinney, J.; Miller, T.; Bongiorno, T.; McDevitt, T.C. Gelatin methacrylate microspheres for controlled growth factor release. *Acta Biomater.* **2014**, *13*, 101–110. [CrossRef]

30. Occhetta, P.; Visone, R.; Russo, L.; Cipolla, L.; Moretti, M.; Rasponi, M. VA-086 methacrylate gelatine photopolymerizable hydrogels: A parametric study for highly biocompatible 3D Cell embedding. *J. Biomed. Mater. Res. A* **2015**, *103*, 2109–2117. [CrossRef]

31. Russo, L.; Sgambato, A.; Visone, R.; Occhetta, P.; Moretti, M.; Rasponi, M.; Nicotra, F.; Cipolla, L. Gelatin hydrogels via thiol-ene chemistry. *Mon. Fur. Chem.* **2015**, *147*, 587–592. [CrossRef]

32. Garcia, J.P.D.; Hsieh, M.-F.; Doma, B.T.; Peruelo, D.C.; Chen, I.-H.; Lee, H.-M. Synthesis of Gelatin-γ-Polyglutamic Acid-Based Hydrogel for the *in vitro* controlled release of epigallocatechin gallate (EGCG) from Camellia sinensis. *Polymers* **2014**, *6*, 39–58. [CrossRef]
33. Connell, L.S.; Gabrielli, L.; Mahony, O.; Russo, L.; Cipolla, L.; Jones, J.R. Functionalizing natural polymers with alkoxysilane coupling agents: Reacting 3-glycidoxypropyl trimethoxysilane with poly(γ-glutamic acid) and gelatin. *Polym. Chem.* **2017**, *8*, 1095–1103. [CrossRef]
34. Liu, F.; Antoniou, J.; Li, Y.; Yi, J.; Yokoyama, W.; Ma, J.; Zhong, F. Preparation of gelatin films incorporated with tea polyphenol nanoparticles for enhancing controlled-release antioxidant properties. *J. Agric. Food Chem.* **2015**, *63*, 3987–3995. [CrossRef] [PubMed]
35. Rodríguez-Rodríguez, R.; Espinosa-Andrews, H.; Velasquillo-Martínez, C.; García-Carvajal, Z.Y. Composite hydrogels based on gelatin, chitosan and polyvinyl alcohol to biomedical applications: A review. *Int. J. Polym. Mater. Po.* **2020**, *69*, 1–20. [CrossRef]
36. Micillo, R.; Iacomino, M.; Perfetti, M.; Panzella, L.; Koike, K.; D'Errico, G.; d'Ischia, M.; Napolitano, A. Unexpected impact of esterification on the antioxidant activity and (photo)stability of a eumelanin from 5,6-dihydroxyindole-2-carboxylic acid. *Pigm. Cell Melanoma Res.* **2018**, *31*, 475–483. [CrossRef] [PubMed]
37. D'Ischia, M.; Wakamatsu, K.; Napolitano, A.; Briganti, S.; Garcia-Borron, J.-C.; Kovacs, D.; Meredith, P.; Pezzella, A.; Picardo, M.; Sarna, T.; et al. Melanins and melanogenesis: Methods, standards, protocols. *Pigm. Cell Melanoma Res.* **2013**, *26*, 616–633. [CrossRef]
38. Zhao, Q.S.; Ji, Q.X.; Xinh, K.; Li, X.Y.; Liu, C.S.; Chen, X.G. Preparation and characteristics of novel porous hydrogel films based on chitosan and glycerophosphate. *Carbohydr. Polym.* **2009**, *76*, 410–416. [CrossRef]
39. Goupy, P.; Dufour, C.; Loonis, M.; Dangles, O. Quantitative kinetic analysis of hydrogen transfer reactions from dietary polyphenols to the DPPH radical. *J. Agric. Food Chem.* **2003**, *51*, 615–622. [CrossRef]
40. Benzie, I.F.F.; Strain, J.J. The ferric reducing ability of plasma (FRAP) as a measure of "antioxidant power": The FRAP assay. *Anal. Biochem.* **1996**, *239*, 70–76. [CrossRef]
41. Charkoudian, L.K.; Franz, K.J. Fe(III)-coordination properties of neuromelanin components: 5,6-dihydroxyindole and 5,6-dihydroxyindole-2-carboxylic acid. *Inorg. Chem.* **2006**, *4*, 3657–3664. [CrossRef]
42. D'Este, M.; Eglin, D.; Alini, M. A systematic analysis of DMTMM vs EDC/NHS for ligation of amines to hyaluronan in water. *Carbohydr. Polym.* **2014**, *108*, 239–246. [CrossRef] [PubMed]
43. Sandra, H.; La Ode, S.; Widya, F. Differentiation of bovine and porcine gelatin based on spectroscopic and electrophoretic analysis. *J. Food Pharm.Sci.* **2013**, *1*, 68–73.
44. Kumar, S.; Joonseok, K. Physiochemical, optical and biological activity of chitosan-chromone derivative for biomedical applications. *Int. J. Mol. Sci.* **2012**, *13*, 6102–6116. [CrossRef] [PubMed]
45. Pepelanova, I.; Kruppa, K.; Scheper, T.; Lavrentieva, A. Gelatin-methacryloyl (GelMA) hydrogels with defined degree of functionalization as a versatile toolkit for 3D cell culture and extrusion bioprinting. *Bioengineering* **2018**, *5*, 55. [CrossRef]

© 2020 by the authors. Licensee MDPI, Basel, Switzerland. This article is an open access article distributed under the terms and conditions of the Creative Commons Attribution (CC BY) license (http://creativecommons.org/licenses/by/4.0/).

Article

A Melanin-Related Phenolic Polymer with Potent Photoprotective and Antioxidant Activities for Dermo-Cosmetic Applications

Davide Liberti, Maria Laura Alfieri, Daria Maria Monti *, Lucia Panzella * and Alessandra Napolitano

Department of Chemical Sciences, University of Naples "Federico II", Via Cintia 4, I-80126 Naples, Italy; davide.liberti@unina.it (D.L.); marialaura.alfieri@unina.it (M.L.A.); alesnapo@unina.it (A.N.)
* Correspondence: mdmonti@unina.it (D.M.M.); panzella@unina.it (L.P.);
  Tel.: +39-081-679150 (D.M.M.); +39-081-674131 (L.P.)

Received: 26 February 2020; Accepted: 23 March 2020; Published: 25 March 2020

Abstract: Eumelanins, the dark variant of skin pigments, are endowed with a remarkable antioxidant activity and well-recognized photoprotective properties that have been ascribed to pigment components derived from the biosynthetic precursor 5,6-dihydroxyindole-2-carboxylic acid (DHICA). Herein, we report the protective effect of a polymer obtained starting from the methyl ester of DHICA (MeDHICA-melanin) against Ultraviolet A (UVA)-induced oxidative stress in immortalized human keratinocytes (HaCaT). MeDHICA-melanin was prepared by aerial oxidation of MeDHICA. At concentrations as low as 10 µg/mL, MeDHICA-melanin prevented reactive oxygen species accumulation and partially reduced glutathione oxidation in UVA-irradiated keratinocytes. Western blot experiments revealed that the polymer is able to induce the translocation of nuclear factor erythroid 2–related factor 2 (Nrf-2) to the nucleus with the activation of the transcription of antioxidant enzymes, such as heme-oxygenase 1. Spectrophotometric and HPLC analysis of cell lysate allowed to conclude that a significant fraction (ca. 7%), consisting mainly of the 4,4′-dimer of MeDHICA (ca. 2 µM), was internalized in the cells. Overall these data point to the potential use of MeDHICA-melanin as an antioxidant for the treatment of skin damage, photoaging and skin cancers.

Keywords: melanins; 5,6-dihydroxyindole-2-carboxylic acid; antioxidant; photoprotection; UVA; HaCaT cells; reactive oxygen species; glutathione; Nrf-2

1. Introduction

Melanins are the primary determinants of skin, hair and exoskeletal pigmentation in mammals, birds and insects [1–6], and their importance has been growing over the past few years not only from a biological point of view, related to their role in the human body, but also for the exploitation of their unique properties in biomedicine or in the cosmetic and health sectors [7–16]. These pigments are biosynthesized in melanocytes, starting with the oxidation of tyrosine to dopaquinone catalyzed by the enzyme tyrosinase (Figure 1) [1,17,18]. Dopaquinone may then undergo cyclization leading, after a further oxidation step, to 5,6-dihydroxyindole (DHI) or 5,6-dihydroxyindole-2-carboxylic acid (DHICA) whose oxidative polymerization ultimately leads to melanin pigments brown or dark in color, known as eumelanins [1,9,18]. On the other hand, entrapment of dopaquinone by cysteine, a process which is under genetic control, gives rise to isomeric cysteinyldopas whose polymerization is responsible for the biosynthesis of the reddish-brown pigments known as pheomelanins, typical of the red hair phenotype [1,19,20] (Figure 1).

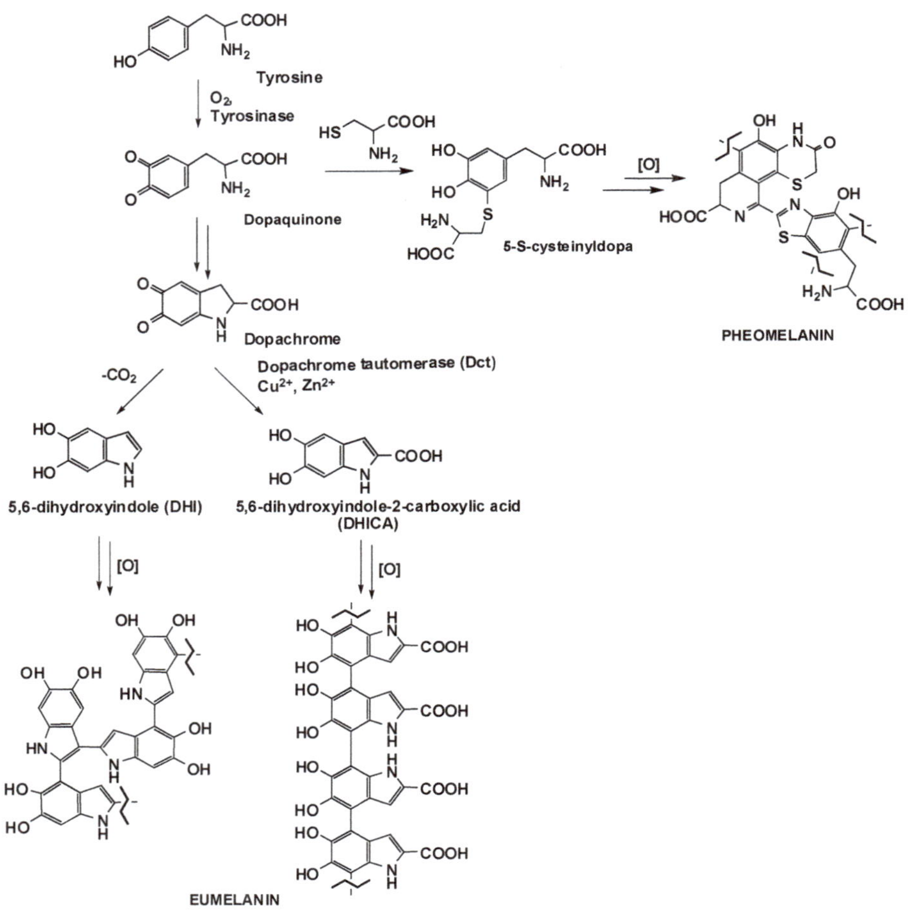

Figure 1. Biosynthetic pathways leading to eumelanins and pheomelanins.

Traditionally, eumelanins have been attributed a role as antioxidant and photoprotective agents in dark-skinned phenotypes, whereas pheomelanins have been implicated in the enhanced susceptibility to skin cancer of individuals belonging to the red-hair phenotype due to their photosensitizing and pro-oxidant properties [1,9,21–26]. Among eumelanins, bionspired synthetic pigments obtained by oxidative polymerization of DHICA have shown remarkable antioxidant properties, and have been proposed as a plausible explanation for the high content of DHICA-related units in natural eumelanins [27,28]. Indeed, DHICA-melanin is able to act as a potent hydroxyl radical scavenger in the Fenton reaction [29] and has been found to act as an efficient antioxidant and radical scavenger also in the 2,2-diphenyl-1-picrylhydrazyl (DPPH), 2,2'-azinobis(3-ethylbenzothiazoline-6-sulfonic acid) (ABTS) and nitric oxide scavenging assays [30]. DHICA-melanin also exhibited inhibition properties against in vitro lipid peroxidation [31], whereas silica/DHICA-melanin hybrid nanoparticles exerted protecting effects against hydrogen peroxide-induced cytotoxicity [32]. More recently, the higher antioxidant activity of DHICA-melanin compared to DHI-melanin has also been confirmed by the Folin–Ciocalteu assay [33].

The antioxidant properties of DHICA-melanin seem to also play a role in the maintenance of immune hyporesponsiveness to melanosomal proteins of relevance for the onset of autoimmune vitiligo [34]. Recently, a natural pigment isolated from marine *Aspergillus nidulans* and tentatively identified as a DHICA-melanin exhibited protective effects against Ultraviolet B (UVB)-induced oxidative stress in cellular and mice models [35,36]. UV radiations are known to be very harmful for the human skin, as they induce reactive oxygen species (ROS) production. Both Ultraviolet A (UVA) and UVB are able to induce DNA damages [37,38]. UVB radiations induce DNA dimerization reactions between adjacent pyrimidine bases, whereas UVA radiations, weakly absorbed by DNA, can excite endogenous chromophores, leading to mispairing of DNA bases with consequent translation of mutated proteins [39].

In this context, the remarkable antioxidant properties of DHICA-melanin and its chromophoric characteristics, allowing significant absorption in the UVA region [1,9,30], would suggest its use in dermo-cosmetic formulations with photoprotective action.

Yet, full exploitation of DHICA-melanin has so far been hampered by the low solubility in lipophilic or hydroalcoholic solvents usually employed in cosmetics, and the relatively high susceptibility to (photo)degradation [40,41]. On these bases, we recently developed a variant of DHICA-melanin which was obtained by oxidative polymerization of the methyl ester of DHICA (MeDHICA-melanin) and shown to consist of a collection of intact oligomers from the dimer up to the heptamer by MALDI-MS analysis (Figure 2) [42]. The material was characterized by an intense and broad absorption band centered at 330 nm and proved to be soluble in water miscible organic solvents. Moreover, MeDHICA-melanin retained the antioxidant properties of DHICA-melanin, proving indeed even more active. It was also stable to prolonged oxidation or exposure to a solar simulator [42].

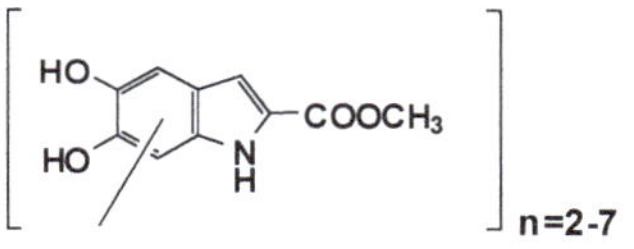

Figure 2. Structure proposed for the methyl ester of 5,6-dihydroxyindole-2-carboxylic acid (MeDHICA)-melanin based on MALDI-MS analysis [42].

Here, we report the protective effect of MeDHICA-melanin, prepared by aerial oxidation of MeDHICA, on oxidative photodamage of immortalized human keratinocytes (HaCaT) induced by UVA-exposure. Keratinocytes represent the most exposed cellular layer in the epidermis, functioning as a protective barrier from environmental stimuli, pathogens and radiation, and it is now generally recognized that molecules endowed with antioxidant activity, especially polyphenols, can strengthen the barrier function of keratinocytes from photoaging [43,44].

2. Materials and Methods

2.1. Reagents

MeDHICA was prepared as described in [42]. MeDHICA-melanin was prepared by aerial oxidation of MeDHICA in phosphate buffer at pH 8.5, as previously reported [42]. Phosphate buffer saline (PBS), Dulbecco's modified Eagle's medium (DMEM), fetal bovine serum (HyClone), 3-(4,5-dimethylthiazol-2-yl)-2,5-diphenyltetrazolium bromide (MTT), 2′,7′-dichlorodihydrofluorescein diacetate (H_2DCFDA), L-glutamine, trypsin-EDTA, Triton, 5,5′-dithiobis-2-nitrobenzoic acid (DTNB), thiobarbituric acid (TBA), and Bradford reagent were from Sigma-Aldrich (St. Louis, MI, USA). Bicinchoninic acid (BCA) protein assay kit was from Thermo Scientific (Waltham, MA, USA). Antibodies against nuclear factor erythroid 2–related factor 2 (Nrf-2) and heme oxygenase 1 (HO-1) were from Cell Signal Technology (Danvers, MA, USA). Antibodies against B-23 and β-actin and the chemiluminescence detection system (SuperSignal® West Pico) were from Thermo Fisher Scientific (Waltham, MA, USA).

2.2. Cell Culture

Human immortalized keratinocytes (HaCaT) were from Innoprot (Derio, Spain). Cells were cultured in DMEM, supplemented with 10% fetal bovine serum, 2 mM L-glutamine and antibiotics in a 5% CO_2 humidified atmosphere at 37 °C. Every 48 h, cells were refreshed in a ratio 1:5. The culture medium was removed and cells were rinsed with PBS and then detached with trypsin-EDTA. After centrifugation (5 min at 1000 rpm), cells were diluted in fresh medium.

2.3. Analysis of Cell Viability

Cells were seeded in 96-well plates (100 µL/well) at a density of 2.5×10^3 cells/cm^2. 24 h after seeding, cells were incubated with increasing concentrations (0.1, 1, 5 and 10 µg/mL) of MeDHICA-melanin. Mother solutions of MeDHICA-melanin were prepared in DMSO at a concentration of 0.2 mg/mL and proper aliquots were added to the incubation medium to get the desired concentration. After 24 h and 48 h incubation, cell viability was assessed by the MTT assay. The MTT reagent, dissolved in DMEM without phenol red, was added to the cells (0.5 mg/mL). After 4 h at 37 °C, the culture medium containing MTT was removed and the resulting formazan salt was dissolved in 2-propanol containing 0.01 M HCl (100 µL/well). Absorbance values of blue formazan were determined at 570 nm using an automatic plate reader (Microbeta Wallac 1420, Perkin Elmer, Milano, Italy). Cell survival was expressed as the percentage of viable cells in the presence of MeDHICA-melanin compared to the controls, represented by untreated cells and cells supplemented with identical volumes of DMSO, in order to exclude a possible effect of DMSO on cell viability.

2.4. UVA irradiation and H_2DCFDA Assay

To evaluate the protective effect of MeDHICA-melanin against oxidative stress, cells were plated at a density of 3.5×10^4 cells/cm^2 (in 60 mm eukaryotic cell plates), pre-incubated in the presence of increasing concentration (0.1–10 µg/mL) of MeDHICA-melanin for different lengths of time (from 5 to 120 min) and stressed by UVA light for 10 min (100 J/cm^2) [45]. Then, cells were incubated with the cell permeable, redox-sensitive fluorophore H_2DCFDA at a concentration of 20 µM for 30 min at 37 °C. Cells were then washed with cold PBS 2 times, detached by trypsin, centrifuged at 1000 rpm for 10 min and resuspended in PBS containing 30 mM glucose, 1 mM $CaCl_2$, and 0.5 mM $MgCl_2$ (PBS plus) at a cell density of 1×10^5 cells/mL. H_2DCFDA is nonfluorescent until it is hydrolyzed by intracellular esterases, and in the presence of ROS it is readily oxidized to the highly fluorescent 2′,7′-dichlorofluorescein (DCF). DCF fluorescence intensity was measured at an emission wavelength of 525 nm with excitation wavelength set at 488 nm using a Perkin-Elmer LS50 spectrofluorometer (Perkin Elmer, Milano, Italy). Emission spectra were acquired at a scanning speed of 300 nm/min, with 5 slit widths for excitation and emission. ROS production was expressed as percentage of DCF fluorescence intensity of the sample under test, with respect to the untreated sample.

2.5. Western Blot Analysis

HaCaT cells were plated at a density of 3.5×10^4 cells/cm^2 (100 mm eukaryotic cell plates) in complete medium for 24 h and then treated with 10 µg/mL of MeDHICA-melanin for 5, 15, 30 and 60 min. To extract nuclear proteins, cells were first incubated with PBS buffer containing 0.1% Triton and protease inhibitors to extract cytosolic proteins. After centrifugation at 1200 rpm for 10 min, nuclear pellet was obtained and proteins were extracted by resuspending the pellet in radioimmunoprecipitation assay buffer (RIPA) buffer (150 mM NaCl, 1% NP-40, 0.1% SDS, proteases inhibitors in 50 mM Tris-HCl pH 8.0). Proteins were quantified by BCA protein assay kit. Western blotting was used to analyze 100 µg of proteins, as reported [46]. Nuclear factor erythroid 2–related factor 2 (Nrf-2) and heme oxygenase 1 (HO-1) levels were detected by using specific antibodies. To normalize protein intensity levels, specific antibodies against B-23 and β-actin were used for nuclear and cytosolic extracts, respectively. Signals were detected by using the chemiluminescence detection system.

2.6. Catalase Assay

HaCaT cells were plated at a density of 3.5×10^4 cells/cm^2 (in a 100 mm eukaryotic cell plate) in complete medium for 24 h and then treated with 10 µg/mL of MeDHICA-melanin for 60 min. At the end of the experiment, total cell lysate was obtained by resuspending each cell pellet in 50 µL of lysis buffer (100 mM Tris-HCl, 300 mM NaCl and 0.5% NP-40 at pH 7.4 with addition of inhibitors of proteases and phosphatases). Proteins were quantified by BCA protein assay kit. To measure catalase activity, a procedure reported in the existing literature was followed [47]. Briefly, cell lysates (50 µg of proteins) were incubated for 30 min at room temperature in 1 mL of hydrogen peroxide solution (50 mM potassium phosphate buffer, pH 7.0, 0.036% w/w H_2O_2). Then, the hydrogen peroxide concentration in solution was determined by measuring the absorbance at 240 nm. The percentage of peroxide removed was calculated as following:

$$\% \ H_2O_2 \ \text{reduced} = 1 - OD_{240nm} \ \text{sample}/OD_{240nm} \ \text{standard}$$

Standard is referred to the hydrogen peroxide solution in the absence of lysate and measured at 240 nm after 30 min of incubation.

2.7. Determination of Intracellular Glutathione (GSH) Levels

HaCaT cells were plated at a density of 3.5×10^4 cells/cm^2 (60 mm eukaryotic cell plates) in complete medium for 24 h and then treated with 10 µg/mL of MeDHICA-melanin for 60 min. At the end of the photoirradiation experiment, cells were lysed, and protein concentration was determined by the Bradford colorimetric assay. Proteins (50 µg) were incubated in the presence of 3 mM EDTA, 144 µM DTNB in 30 mM Tris-HCl at pH 8.2, and centrifuged at 13,000 rpm for 5 min at 4 °C. Supernatants were collected, and the absorbance was measured at 412 nm by using a multiplate reader (Bio-Rad, Hercules, CA, USA). GSH levels were expressed as % of the sample under test with respect to the untreated sample.

2.8. Analysis of Lipid Peroxidation Levels

HaCaT cells were seeded at a density of 3.5×10^4 cells/cm^2 (100 mm eukaryotic cell plates) in complete medium for 24 h and then treated with 10 µg/mL of MeDHICA-melanin for 120 min. After UVA irradiation, cells were kept at 37 °C for 90 min, before performing the thiobarbituric acid reactive substances (TBARS) assay as described [48]. Briefly, cells were detached and 5×10^4 cells suspended in 0.67% TBA containing 20% trichloroacetic acid (TCA) (1:1 v/v). After heating for 30 min at 100 °C, samples were centrifuged at 2500 rpm for 5 min at 4 °C, and supernatants spectrophotometrically analyzed at 532 nm.

2.9. Quantification of Internalized Melanin

HaCaT cells were plated at a density of 3.5×10^4 cells/cm^2 (100 mm eukaryotic cell plates) in complete medium for 24 h and then treated with 10 µg/mL of MeDHICA-melanin for 60 min. After treatment, total cell lysate was obtained, 50 µg of proteins (Bradford assay) were diluted in 1 mL of potassium phosphate buffer (50 mM, pH 7.0) and UV-vis spectra were recorded. The amount of MeDHICA-melanin internalized by the cells was determined by using a calibration curve obtained with pure MeDHICA-melanin. In particular, increasing concentrations (0.6–20 µg/mL) of MeDHICA-melanin, alone or in the presence of 50 µg of cell lysate, were used to record the UV-vis spectra. The calibration curve was built by plotting values of absorbance at 330 nm against MeDHICA-melanin concentration.

2.10. HPLC and LC-MS Analysis of Cell Lysate

HPLC analysis was performed on an instrument (Agilent 1100, Santa Clara, CA, USA) equipped with a binary pump and a SPD-10AV VP UV-vis detector set at 300 nm. The chromatographic separation was achieved on a Sphereclone octadecylsilane-coated column, 250 mm × 4.6 mm, 5 µm particle size

(Phenomenex, Torrance, CA, USA) at 0.7 mL/min using binary gradient elution conditions as follows: 0.1% formic acid (solvent A), acetonitrile (solvent B) from 35% to 70%, 0–45 min. LC-MS analyses were run on a LC-MS ESI-TOF 1260/6230DA Agilent instrument operating in positive ionization mode in the following conditions: Nebulizer pressure 35 psig; drying gas (nitrogen) 5 L/min, 325 °C; capillary voltage 3500 V; fragmentor voltage 175 V. An Eclipse Plus C18 column, 150 × 4.6 mm, 5 µm (Agilent), at a flow rate of 0.4 mL/min was used, using the same eluant as above. The cell lysate, obtained as described in Section 2.9, was lyophilized and subjected to acetylation treatment with acetic anhydride (500 µL) and pyridine (75 µL) overnight. After repeated washings with methanol to remove solvents, the residue was taken up in methanol and analyzed by HPLC and LC-MS. A control lysate sample obtained in the absence of MeDHICA-melanin was also analyzed.

2.11. Statistical Analysis

In all the experiments, each sample was tested in three independent analyses, each carried out in triplicate. The results are presented as mean of results obtained (mean ± SD) and compared by one-way ANOVA following Tukey's multiple comparison test using Graphpad Prism for Windows, version 6.01 (San Diego, CA, USA).

3. Results and Discussion

3.1. Biocompatibility of MeDHICA-Melanin on Keratinocytes

In order to assess the possible use of MeDHICA-melanin for cosmetic applications, its biocompatibility was tested on immortalized human keratinocytes (HaCaT), as these cells are normally present in the outermost layer of the skin. Increasing amounts of MeDHICA-melanin (from 0.1 to 10 µg/mL) were incubated with the cells for 24 and 48 h. At the end of each incubation, cell viability was assessed by the MTT assay. As shown in Figure 3, cell viability was not affected at any of the experimental conditions tested, neither after 24 h nor after 48 h incubation, thus suggesting that MeDHICA-melanin was fully biocompatible on HaCaT cells.

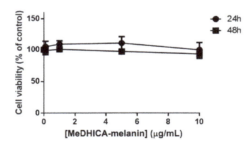

Figure 3. Effects of MeDHICA-melanin on HaCaT cells viability. Dose-response curves after 24 h (black circles) and 48 h (black squares) incubation of HaCaT cells with increasing concentration of MeDHICA-melanin (0.1–10 µg/mL). Cell viability was assessed by the MTT assay and cell survival expressed as percentage of viable cells in the presence of MeDHICA-melanin, with respect to control cells (i.e., cells grown in the absence of the melanin). The results shown are means ± SD of three independent experiments.

3.2. Inhibition of UVA-Induced Damage on HaCaT Cells by MeDHICA-Melanin

To assess the protective effect of MeDHICA-melanin against photoinduced oxidative stress, irradiation with UVA was chosen as a source of stress as this has been shown to induce many side effects on human skin [49]. A dose-response experiment was first performed to evaluate the optimal MeDHICA-melanin concentration to be used. HaCaT cells were incubated with increasing concentrations of MeDHICA-melanin (0.1–10 µg/mL) for 2 h prior to UVA irradiation treatment, and immediately after irradiation ROS production was evaluated by the H_2DCFDA assay.

As shown in Figure 4A, DCF fluorescence was significantly increased after UVA irradiation (2.3 fold increase, $p < 0.005$), whereas MeDHICA-melanin had no effect on ROS levels on non-irradiated cells. Interestingly, when cells were preincubated with MeDHICA-melanin prior to UVA exposure, ROS production was decreased in a dose dependent manner, and reached the levels observed in non-irradiated cells when the melanin was tested at 10 µg/mL ($p < 0.005$) (Figure 4A).

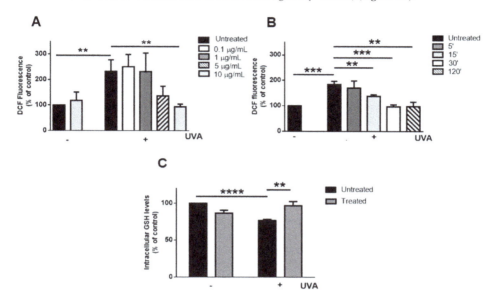

Figure 4. Antioxidant effects of MeDHICA-melanin on UVA-stressed HaCaT cells. (**A**) Dose-response analysis of intracellular ROS levels by 2′,7′-dichlorodihydrofluorescein diacetate (H_2DCFDA) assay. Cells were pre-incubated with increasing concentrations of MeDHICA-melanin for 2 h prior to UVA irradiation (100 J/cm^2) for 10 min. Cells were incubated with 0.1 µg/mL (white bars), 1 µg/mL (dark grey bars), 5 µg/mL (dashed bars) or 10 µg/mL (light grey bars) MeDHICA-melanin. Black bars refer to untreated cells. (**B**) Time-course analysis of intracellular ROS levels by H_2DCFDA assay. Cells were incubated for 5 min (dark grey bars), 15 min (light grey bars), 30 min (white bars) or 120 min (dashed bars) with MeDHICA-melanin before being irradiated by UVA. Black bars are referred to untreated cells. (**C**) Intracellular GSH levels evaluated by 5,5′-dithiobis-2-nitrobenzoic acid (DTNB) assay. Cells were pre-incubated with MeDHICA-melanin (10 µg/mL) for 1 h before UVA irradiation. Values are expressed as % with respect to control (i.e. untreated) cells. Data shown are means ± SD of three independent experiments. ** indicates $p < 0.005$, *** indicates $p < 0.0005$, **** indicates $p < 0.0001$.

The effect of the preincubation time with MeDHICA-melanin on ROS production was also evaluated (Figure 4B). A significant protection against UVA damage was observed already after 15 min of incubation with 10 µg/mL of MeDHICA-melanin ($p < 0.005$). These data confirmed the potent antioxidant activity of MeDHICA-melanin [42] also in a cellular model, further highlighting its potential as an active ingredient in cosmetic formulations when compared to other natural or synthetic materials, such as phenol-rich plant extracts or other melanin-related samples. As an example, 10-fold higher concentrations (100 µg/mL) and longer pre-incubation times (1 h) have been reported in the case of a water extract from red grapevine leaves containing high levels of polyphenols to observe an effect comparable to that of the present study on the decrease of ROS generation in HaCaT cells irradiated with lower doses of UVA (25 J/cm^2) [38]. Also, the activity of silymarin was much lower than that observed with MeDHICA-melanin: 30 min of pre-incubation with 250 μg/mL of the compound were able to reduce the ROS produced by irradiating HaCaT cells with 20 J/cm^2 UVA by only 30% [50].

Based on these promising results, subsequent experiments were carried using 10 μg/mL MeDHICA-melanin.

The intracellular levels of GSH were evaluated in view of the important role of this biomolecule in the cellular redox balance, and the decrease associated to oxidative stress [51]. Following UVA irradiation, a 25% decrease ($p < 0.0001$) of intracellular GSH levels was observed with respect to control cells, whereas GSH levels were unaltered in cells preincubated with MeDHICA-melanin (Figure 4C). Similar effects have been reported on UVA-irradiated HaCaT cells for phenol-rich extracts from *Eugenia uniflora* [52] and *Syzygium aqueum* [53] leaves, which, however, had to be tested at higher concentrations (50 μg/mL) and for a longer time (2 h) to show a protective effect.

The different behavior of all these samples compared to MeDHICA-melanin can be ascribed to differences in the chemical structures of the compounds tested, determining crucial variations not only in the intrinsic antioxidant activity, but also in the cell-permeation ability as well as in the UVA-interaction properties.

The protective effects of MeDHICA-melanin on HaCaT cells were further confirmed by analyzing the lipid peroxidation levels 90 min after irradiation (Figure S1). The results indicated that MeDHICA-melanin was able to keep lipid peroxidation unaltered. In fact, cells pretreated with MeDHICA-melanin and then exposed to UVA radiation showed a significantly lower intracellular level of lipid peroxidation when compared to untreated cells exposed to UVA (50% decrease, $p < 0.005$) (Figure S1). Notably, a lower protective effect against lipid peroxidation has been reported for the well-recognized antioxidant pterostilbene in 20 J/cm^2 UVA irradiated-HaCaT cells after 24 h pretreatment with 2.5 μg/mL of the compound [54]. However, a significant increase in lipid peroxidation levels was observed in cells incubated only with MeDHICA-melanin (2.4 fold increase), without photoirradiation, an intriguing observation that will be addressed in future studies. In any case, the overall results clearly indicate that MeDHICA-melanin is able to protect HaCaT cells from UVA-induced oxidative stress.

3.3. Induction of Nrf-2 Nuclear Translocation by MeDHICA-Melanin

The MeDHICA-melanin protective effect was analyzed at a molecular level by studying the involvement of Nrf-2. Under normal physiological conditions, the complex between Nrf-2 and Keap-1 keeps Nrf-2 in the cytosol and the protein is degraded through the proteasome. Oxidative stress, or small amounts of antioxidants, induce the dissociation between Keap-1 and Nrf-2, and the latter is translocated to the nucleus. Once in the nucleus, it binds to antioxidant responsive element (ARE) sequences and activates the transcription of several phase-II detoxifying enzymes, such as HO-1 and catalase [55]. Thus, cells were incubated with MeDHICA-melanin for 5, 15 and 30 min and then nuclear Nrf-2 levels were evaluated by Western blot analyses. As shown in Figure 5A, a significant increase of Nrf-2 nuclear levels was observed after 30 min of incubation (about 2 fold increase, $p < 0.005$). Nrf-2 activation was confirmed by measuring HO-1 levels (Figure 5B), which were found to significant increase after 60 min incubation of the cells with MeDHICA-melanin (about 2 fold increase, $p < 0.05$). Nrf-2 activation was also confirmed by measuring consumption of H_2O_2 added to the incubation medium, that could indirectly indicate the activity of catalase. As reported in Figure 5C, the levels of H_2O_2 detected in keratinocyte lysates were lower (33% decrease, $p < 0.05$) in the cells after incubation with MeDHICA-melanin with respect to the control sample.

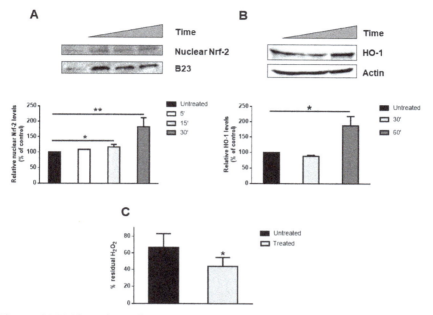

Figure 5. MeDHICA-melanin effects on Nrf-2 activation in HaCaT cells. Cells were incubated with MeDHICA-melanin (10 µg/mL) for different lengths of time, and (**A**) nuclear Nrf-2, or (**B**) cytosolic HO-1 proteins were analyzed by Western blotting. (**A**) HaCaT cells were incubated with MeDHICA-melanin for 5 min (white bars), 15 min (light grey bars) and 30 min (dark grey bars) and then nuclear proteins extracted to perform Western blot analysis of Nrf-2. Nrf-2 was quantified by densitometric analysis and normalized to B-23. (**B**) Western blot analysis for HO-1 performed on cytosolic proteins obtained from HaCaT cells after incubation with MeDHICA-melanin for 30 min (light grey bars) and 60 min (dark grey bars). HO-1 was quantified by densitometric analysis and normalized to β-Actin. (**C**) Cells were incubated with melanin (10 µg/mL) for 1 h and then 50 µg of cell lysate were incubated with 0.036% w/w H_2O_2. Hydrogen peroxide concentration in solution was determined by measuring the absorbance at 240 nm. Black bars are referred to control cells. Data shown are means ± SD of three independent experiments. * indicates $p < 0.05$ ** indicates $p < 0.005$.

3.4. Cellular Uptake of MeDHICA-Melanin

In order to verify whether MeDHICA-melanin was internalized in the cells, HaCaT cells were incubated with MeDHICA-melanin at 10 µg/mL for 60 min, after that the total cell lysate was obtained. UV–vis spectra of lysates from untreated and treated cells were recorded and the amount of melanin internalized by the cells was estimated to be about 7% using a measurement of the absorbance at 330 nm of treated cells lysate and a comparison with the calibration curve obtained by using pure MeDHICA-melanin (Figure 6). This result is in agreement with the by now well-established idea that melanin is internalized by cells to serve as a protective agent [56].

MeDHICA-melanin internalization was further corroborated by HPLC analysis of the cell lysate upon incubation with the pigment. To improve the chromatographic properties and the stability, the cell lysate obtained by HaCaT incubation with 10 µg/mL of MeDHICA-melanin was acetylated with acetic anhydride-pyridine overnight at room temperature and then analyzed by HPLC (Figure 7A).

Comparison of the elutographic properties with those of an authentic standard [42] allowed to identify the compound eluted at 19 min as the acetylated derivatives of the 4,4'-dimer of MeDHICA (ca. 2 µM). The identity of this product was further confirmed by LC-MS analysis ($[M+H]^+$ for acetylated dimer = 581 m/z) (Figure 7B,C).

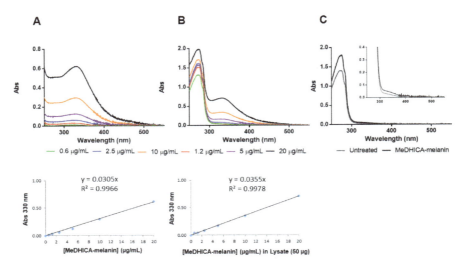

Figure 6. Quantification of internalized MeDHICA-melanin. Increasing concentrations (0.6–20 µg/mL) of MeDHICA-melanin, (**A**) alone, or (**B**) in the presence of 50 µg of cell lysate, were used to record the UV–vis spectra. Calibration curves built by plotting values of absorbance at 330 nm against MeDHICA-melanin concentration are also shown. (**C**) UV–vis spectra of untreated cells (grey line) and MeDHICA-melanin treated (black line) HaCaT cells.

4. Conclusions

UVA radiations are highly harmful as they can penetrate the skin, crossing the epidermis and reaching the dermis. They are responsible for a variety of physiopathological conditions, ranging from inflammation to premature skin aging and skin cancer development [45,57,58]. On the other hand, the antioxidant and photoprotective properties of eumelanins and model pigments from DHICA are well-established [29,30,41,42].

In this paper, MeDHICA-melanin, previously shown to possess marked in vitro antioxidant activity and favorable solubility properties for dermo-cosmetic applications [42], was demonstrated to exert protective effects on a cellular model of immortalized keratinocytes (HaCaT) exposed to UVA radiations. All the endpoint parameters of oxidative stress, i.e., ROS, lipid peroxidation, and intracellular oxidized glutathione levels, were significantly inhibited in cells incubated with MeDHICA-melanin at concentrations compatible with cell viability. Moreover, a comparison with related findings recently reported in the literature on natural phenols or phenol-rich extracts further highlighted the advantages of MeDHICA-melanin.

Similarly to other natural antioxidants [58], MeDHICA-melanin also proved to provide protection by activating the Nrf-2 pathway. Indeed, the nuclear translocation of Nrf-2 was effective, as demonstrated by the activation of downstream genes. To our knowledge, this is the first report on the correlation between the antioxidant activity of melanins and the Nrf-2 pathway. We also demonstrated that MeDHICA-melanin was able to enter keratinocytes, although, as expected, only

Low molecular weight components such as dimeric compounds were appreciably internalized after 1 h incubation. These data provide further confirmation to biological studies indicating that keratinocytes are able to internalize melanins as the result of a cross-talk with melanocytes, mediated by the protease-activated receptor 2 (PAR-2) and the Ras-related protein Rab11 [59,60]. Overall, our findings demonstrate that MeDHICA-melanin is able to enter the cells and activate the antioxidant system to protect skin cells from UVA-induced damage, encouraging its use as an effective component in dermo-cosmetic formulations for the treatment of skin damage, photoaging and skin cancers.

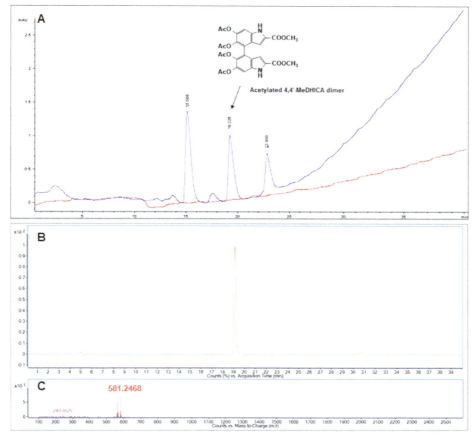

Figure 7. HPLC and LC-MS analysis of cell lysate. (**A**) HPLC profile of acetylated cell lysate (blue trace) and acetylated lysate from control cells (not incubated with MeDHICA-melanin) (red trace). (**B**) LC-MS extracted ion chromatogram (*m/z* 581). (**C**) MS spectrum of the compound eluted at 19 min.

Supplementary Materials: The following are available online at http://www.mdpi.com/2076-3921/9/4/270/s1, Figure S1: Lipid peroxidation levels evaluated by TBARS assay.

Author Contributions: Conceptualization, D.M.M., L.P. and A.N.; methodology, D.M.M., L.P. and A.N.; validation, D.M.M., L.P. and A.N.; investigation, D.L and M.L.A.; data curation, D.M.M., L.P. and A.N.; writing—original draft preparation, D.L., M.L.A, D.M.M and L.P.; writing—review and editing, D.M.M., L.P. and A.N. All authors have read and agreed to the published version of the manuscript.

Funding: This research was funded in part by Kao Corporation.

Conflicts of Interest: The authors declare no conflict of interest.

References

1. Micillo, R.; Panzella, L.; Koike, K.; Monfrecola, G.; Napolitano, A.; D'Ischia, M. "Fifty shades" of black and red or how carboxyl groups fine tune eumelanin and pheomelanin properties. *Int. J. Mol. Sci.* **2016**, *17*, 746. [CrossRef] [PubMed]
2. Galvan, I.; Solano, F. Bird integumentary melanins: Biosynthesis, forms, function and evolution. *Int. J. Mol. Sci.* **2016**, *17*, 520. [CrossRef] [PubMed]

3. Del Bino, S.; Duval, C.; Bernerd, F. Clinical and biological characterization of skin pigmentation diversity and its consequences on UV impact. *Int. J. Mol. Sci.* **2018**, *19*, 2668. [CrossRef] [PubMed]
4. Miyamura, Y.; Coelho, S.G.; Wolber, R.; Miller, S.A.; Wakamatsu, K.; Zmudzka, B.Z.; Ito, S.; Smuda, C.; Passeron, T.; Choi, W.; et al. Regulation of human skin pigmentation and responses to ultraviolet radiation. *Pigment Cell Res.* **2007**, *20*, 2–13. [CrossRef]
5. Ito, S.; Wakamatsu, K. Diversity of human hair pigmentation as studied by chemical analysis of eumelanin and pheomelanin. *J. Eur. Acad. Dermatol. Venereol.* **2011**, *25*, 1369–1380. [CrossRef]
6. Sugumaran, M.; Barek, H. Critical analysis of the melanogenic pathway in insects and higher animals. *Int. J. Mol. Sci.* **2016**, *17*, 1753. [CrossRef]
7. Meredith, P.; Sarna, T. The physical and chemical properties of eumelanin. *Pigment Cell Res.* **2006**, *19*, 572–594. [CrossRef]
8. Xie, W.; Pakdel, E.; Liang, Y.; Kim, Y.J.; Liu, D.; Sun, L.; Wang, X. Natural eumelanin and its derivatives as multifunctional materials for bioinspired applications: A review. *Biomacromolecules* **2019**, *20*, 4312–4331. [CrossRef]
9. Panzella, L.; Ebato, A.; Napolitano, A.; Koike, K. The late stages of melanogenesis: Exploring the chemical facets and the application opportunities. *Int. J. Mol. Sci.* **2018**, *19*, 1753. [CrossRef]
10. D'Ischia, M. Melanin-based functional materials. *Int. J. Mol. Sci.* **2018**, *19*, 228. [CrossRef]
11. Solano, F. Melanin and melanin-related polymers as materials with biomedical and biotechnological applications-cuttlefish ink and mussel foot proteins as inspired biomolecules. *Int. J. Mol. Sci.* **2017**, *18*, 1561. [CrossRef] [PubMed]
12. Huang, L.; Liu, M.; Huang, H.; Wen, Y.; Zhang, X.; Wei, Y. Recent advances and progress on melanin-like materials and their biomedical applications. *Biomacromolecules* **2018**, *19*, 1858–1868. [CrossRef] [PubMed]
13. Barra, M.; Bonadies, I.; Carfagna, C.; Cassinese, A.; Cimino, F.; Crescenzi, O.; Criscuolo, V.; D'Ischia, M.; Maglione, M.G.; Manini, P.; et al. Eumelanin-based organic bioelectronics: Myth or reality? *MRS Adv.* **2016**, *1*, 3801–3810. [CrossRef]
14. D'Ischia, M.; Wakamatsu, K.; Cicoira, F.; Di Mauro, E.; Garcia-Borron, J.C.; Commo, S.; Galvan, I.; Ghanem, G.; Kenzo, K.; Meredith, P.; et al. Melanins and melanogenesis: From pigment cells to human health and technological applications. *Pigment Cell Melanoma Res.* **2015**, *28*, 520–544. [CrossRef] [PubMed]
15. Longo, D.L.; Stefania, R.; Aime, S.; Oraevsky, A. Melanin-based contrast agents for biomedical optoacoustic imaging and theranostic applications. *Int. J. Mol. Sci.* **2017**, *18*, 1719. [CrossRef]
16. D'Ischia, M.; Napolitano, A.; Pezzella, A.; Meredith, P.; Buehler, M.J. Melanin biopolymers: Tailoring chemical complexity for materials design. *Angew. Chem. Int. Ed.* **2019**. [CrossRef]
17. Panzella, L.; Napolitano, A. Natural and bioinspired phenolic compounds as tyrosinase inhibitors for the treatment of skin hyperpigmentation: Recent advances. *Cosmetics* **2019**, *6*, 57. [CrossRef]
18. Ito, S.; Wakamatsu, K.; D'Ischia, M.; Napolitano, A.; Pezzella, A. Structure of Melanins. In *Melanins and Melanosomes*; Borovansky, J., Riley, P.A., Eds.; Wiley: Weinheim, Germany, 2011; pp. 167–185.
19. Ito, S.; Wakamatsu, K. Chemistry of mixed melanogenesis—Pivotal roles of dopaquinone. *Photochem. Photobiol.* **2008**, *84*, 582–592. [CrossRef]
20. Napolitano, A.; Panzella, L.; Leone, L.; D'Ischia, M. Red hair benzothiazines and benzothiazoles: Mutation-inspired chemistry in the quest for functionality. *Acc. Chem. Res.* **2013**, *46*, 519–528. [CrossRef]
21. Kang, M.; Kim, E.; Temocin, Z.; Li, J.; Dadachova, E.; Wang, Z.; Panzella, L.; Napolitano, A.; Bentley, W.E.; Payne, G. Reverse engineering to characterize redox properties: Revealing melanin's redox activity through mediated electrochemical probing. *Chem. Mater.* **2018**, *30*, 5814–5826. [CrossRef]
22. Kim, E.; Panzella, L.; Napolitano, A.; Payne, G.F. Redox activities of melanins investigated by electrochemical reverse engineering: Implications for their roles in oxidative stress. *J. Invest. Dermatol.* **2020**, *140*, 537–543. [CrossRef] [PubMed]
23. Napolitano, A.; Panzella, L.; Monfrecola, G.; D'Ischia, M. Pheomelanin-induced oxidative stress: Bright and dark chemistry bridging red hair phenotype and melanoma. *Pigment Cell Melanoma Res.* **2014**, *27*, 721–733. [CrossRef]
24. Morgan, A.M.; Lo, J.; Fisher, D.E. How does pheomelanin synthesis contribute to melanomagenesis? Two distinct mechanisms could explain the carcinogenicity of pheomelanin synthesis. *BioEssays* **2013**, *35*, 672–676. [CrossRef] [PubMed]

25. Abdel-Malek, Z.A.; Ito, S. Being in the red: A no-win situation with melanoma. *Pigment Cell Melanoma Res.* **2013**, *26*, 164–166. [CrossRef]

26. Ito, S.; Wakamatsu, K.; Sarna, T. Photodegradation of eumelanin and pheomelanin and its pathophysiological implications. *Photochem. Photobiol.* **2018**, *94*, 409–420. [CrossRef]

27. Panzella, L.; Napolitano, A.; D'Ischia, M. Is DHICA the key to dopachrome tautomerase and melanocyte functions? *Pigment Cell Melanoma Res.* **2011**, *24*, 248–249. [CrossRef]

28. Kovacs, D.; Flori, E.; Maresca, V.; Ottaviani, M.; Aspite, N.; Dell'Anna, M.L.; Panzella, L.; Napolitano, A.; Picardo, M.; D'Ischia, M. The eumelanin intermediate 5,6-dihydroxyindole-2-carboxylic acid is a messenger in the cross-talk among epidermal cells. *J. Invest. Dermatol.* **2012**, *132*, 1196–1205. [CrossRef]

29. Jiang, S.; Liu, X.M.; Dai, X.; Zhou, Q.; Lei, T.C.; Beermann, F.; Wakamatsu, K.; Xu, S.Z. Regulation of DHICA-mediated antioxidation by dopachrome tautomerase: Implication for skin photoprotection against UVA radiation. *Free Radic. Biol. Med.* **2010**, *48*, 1144–1151. [CrossRef]

30. Panzella, L.; Gentile, G.; D'Errico, G.; Della Vecchia, N.F.; Errico, M.E.; Napolitano, A.; Carfagna, C.; D'Ischia, M. Atypical structural and π-electron features of a melanin polymer that lead to superior free-radical-scavenging properties. *Angew. Chem. Int. Ed.* **2013**, *52*, 12684–12687. [CrossRef]

31. Blarzino, C.; Mosca, L.; Foppoli, C.; Coccia, R.; De Marco, C.; Rosei, M.A. Lipoxygenase/H_2O_2-catalyzed oxidation of dihydroxyindoles: Synthesis of melanin pigments and study of their antioxidant properties. *Free Radic. Biol. Med.* **1998**, *26*, 446–453. [CrossRef]

32. Silvestri, B.; Vitiello, G.; Luciani, G.; Calcagno, V.; Costantini, A.; Gallo, M.; Parisi, S.; Paladino, S.; Iacomino, M.; D'Errico, G.; et al. Probing the eumelanin-silica interface in chemically engineered bulk hybrid nanoparticles for targeted subcellular antioxidant protection. *ACS Appl. Mater. Interfaces* **2017**, *9*, 37615–37622. [CrossRef] [PubMed]

33. Cecchi, T.; Pezzella, A.; Di Mauro, E.; Cestola, S.; Ginsburg, D.; Luzi, M.; Rigucci, A.; Santato, C. On the antioxidant activity of eumelanin biopigments: A quantitative comparison between free radical scavenging and redox properties. *Nat. Prod. Res.* **2019**, 1–9. [CrossRef] [PubMed]

34. Liu, X.-M.; Zhou, Q.; Xu, S.-Z.; Wakamatsu, K.; Lei, T.-C. Maintenance of immune hyporesponsiveness to melanosomal proteins by DHICA-mediated antioxidation: Possible implications for autoimmune vitiligo. *Free Radic. Biol. Med.* **2011**, *50*, 1177–1185. [CrossRef] [PubMed]

35. Shanuja, S.K.; Iswarya, S.; Gnanamani, A. Marine fungal DHICA as a UVB protectant: Assessment under in vitro and in vivo conditions. *J. Photochem. Photobiol. B* **2018**, *179*, 139–148. [CrossRef]

36. Shanuja, S.K.; Iswarya, S.; Rajasekaran, S.; Dinesh, M.G.; Gnanamani, A. Pre-treatment of extracellular water soluble pigmented secondary metabolites of marine imperfect fungus protects HDF cells from UVB induced oxidative stress. *Photochem. Photobiol. Sci.* **2018**, *17*, 1229–1238. [CrossRef]

37. Baron, Y.; Corre, S.; Mouchet, N.; Vaulont, S.; Prince, S.; Galibert, M.D. USF-1 is critical for maintaining genome integrity in response to UV-induced DNA photolesions. *PLoS Genet.* **2012**, *8*, e1002470. [CrossRef]

38. Marabini, L.; Melzi, G.; Lolli, F.; Dell'Agli, M.; Piazza, S.; Sangiovanni, E.; Marinovich, M. Effects of *Vitis vinifera* L. leaves extract on UV radiation damage in human keratinocytes (HaCaT). *J. Photochem. Photobiol. B* **2020**, *204*, 111810. [CrossRef]

39. Aust, A.E.; Eveleigh, J.F. Mechanisms of DNA oxidation. *Proc. Soc. Exp. Biol. Med.* **1999**, *222*, 246–252. [CrossRef]

40. Szewczyk, G.; Zadlo, A.; Sarna, M.; Ito, S.; Wakamatsu, K.; Sarna, T. Aerobic photoreactivity of synthetic eumelanins and pheomelanins: Generation of singlet oxygen and superoxide anion. *Pigment Cell Melanoma Res.* **2016**, *29*, 669–678. [CrossRef]

41. Ito, S.; Kikuta, M.; Koike, S.; Szewczyk, G.; Sarna, M.; Zadlo, A.; Sarna, T.; Wakamatsu, K. Roles of reactive oxygen species in UVA-induced oxidation of 5,6-dihydroxyindole-2-carboxylic acid-melanin as studied by differential spectrophotometric method. *Pigment Cell Melanoma Res.* **2016**, *29*, 340–351. [CrossRef]

42. Micillo, R.; Iacomino, M.; Perfetti, M.; Panzella, L.; Koike, K.; D'Errico, G.; D'Ischia, M.; Napolitano, A. Unexpected impact of esterification on the antioxidant activity and (photo)stability of a eumelanin from 5,6-dihydroxyindole-2-carboxylic acid. *Pigment Cell Melanoma Res.* **2018**, *31*, 475–483. [CrossRef] [PubMed]

43. Zhao, P.; Alam, M.B.; Lee, S.H. Protection of UVB-induced photoaging by fuzhuan-brick tea aqueous extract via MAPKs/Nrf2-mediated down-regulation of MMP-1. *Nutrients* **2018**, *11*, 60. [CrossRef] [PubMed]

44. Kostyuk, V.; Potapovich, A.; Albuhaydar, A.R.; Mayer, W.; De Luca, C.; Korkina, L. Natural substances for prevention of skin photoaging: Screening systems in the development of sunscreen and rejuvenation cosmetics. *Rejuvenation Res.* **2018**, *2*, 91–101. [CrossRef] [PubMed]
45. Petruk, G.; Illiano, A.; Del Giudice, R.; Raiola, A.; Amoresano, A.; Rigano, M.M.; Piccoli, R.; Monti, D.M. Malvidin and cyanidin derivatives from açai fruit (*Euterpe oleracea* Mart.) counteract UV-A-induced oxidative stress in immortalized fibroblasts. *J. Photochem. Photobiol. B* **2017**, *172*, 42–51. [CrossRef]
46. Galano, E.; Arciello, A.; Piccoli, R.; Monti, D.M.; Amoresano, A. A proteomic approach to investigate the effects of cadmium and lead on human primary renal cells. *Metallomics* **2014**, *6*, 587–597. [CrossRef]
47. Beers, R.F.; Sizer, I.W. A spectrophotometric method for measuring the breakdown of hydrogen peroxide by catalase. *J. Biol. Chem.* **1952**, *195*, 133–140.
48. Del Giudice, R.; Petruk, G.; Raiola, A.; Barone, A.; Monti, D.M.; Rigano, M.M. Carotenoids in fresh and processed tomato (*Solanum lycopersicum*) fruits protect cells from oxidative stress injury. *J. Sci. Food Agric.* **2017**, *97*, 1616–1623. [CrossRef]
49. Haywood, R.; Rogge, F.; Lee, M. Protein, lipid, and DNA radicals to measure skin UVA damage and modulation by melanin. *Free Radic. Biol. Med.* **2008**, *44*, 990–1000. [CrossRef]
50. Fidrus, E.; Ujhelyi, Z.; Fehér, P.; Hegedűs, C.; Janka, E.A.; Paragh, G.; Vasas, G.; Bácskay, I.; Remenyik, E. Silymarin: Friend or foe of UV exposed keratinocytes? *Molecules* **2019**, *24*, 1652. [CrossRef]
51. Aquilano, K.; Baldelli, S.; Ciriolo, M.R. Glutathione: New roles in redox signaling for an old antioxidant. *Front. Pharmacol.* **2014**, *5*, 196. [CrossRef]
52. Sobeh, M.; El-Raey, M.; Rezq, S.; Abdelfattah, M.A.O.; Petruk, G.; Osman, S.; El-Shazly, A.M.; El-Beshbishy, H.A.; Mahmoud, M.F.; Wink, M. Chemical profiling of secondary metabolites of *Eugenia uniflora* and their antioxidant, anti-inflammatory, pain killing and anti-diabetic activities: A comprehensive approach. *J. Ethnopharmacol.* **2019**, *240*, 111939. [CrossRef] [PubMed]
53. Sobeh, M.; Mahmoud, M.F.; Petruk, G.; Rezq, S.; Ashour, M.L.; Youssef, F.S.; El-Shazly, A.M.; Monti, D.M.; Abdel-Naim, A.B.; Wink, M. *Syzygium aqueum*: A polyphenol- rich leaf extract exhibits antioxidant, hepatoprotective, pain-killing and anti-inflammatory activities in animal models. *Front. Pharmacol.* **2018**, *9*, 566. [CrossRef] [PubMed]
54. Deng, H.; Li, H.; Ho, Z.Y.; Dai, X.Y.; Chen, Q.; Li, R.; Liang, B.; Zhu, H. Pterostilbene's protective effects against photodamage caused by UVA/UVB irradiation. *Pharmazie* **2018**, *73*, 651–658. [PubMed]
55. Ma, Q. Role of Nrf2 in oxidative stress and toxicity. *Annu. Rev. Pharmacol. Toxicol.* **2013**, *53*, 401–426. [CrossRef]
56. Huang, Y.; Li, Y.; Hu, Z.; Yue, X.; Proetto, T.; Jones, Y.; Gianneschi, N.C. Mimicking melanosomes: Polydopamine nanoparticles as artificial microparasols. *ACS Cent. Sci.* **2017**, *3*, 564–569. [CrossRef]
57. Kulms, D.; Schwarz, T. Molecular mechanisms of UV-induced apoptosis. *Photodermatol. Photoimmunol. Photomed.* **2000**, *16*, 195–201. [CrossRef]
58. Hseu, Y.-C.; Chou, C.W.; Senthil Kumar, K.J.; Fu, K.T.; Wang, H.M.; Hsu, L.S.; Kuo, Y.H.; Wu, C.R.; Chen, S.C.; Yang, H.L. Ellagic acid protects human keratinocyte (HaCaT) cells against UV-A-induced oxidative stress and apoptosis through the upregulation of the HO-1 and Nrf-2 antioxidant genes. *Food Chem. Toxicol.* **2012**, *50*, 1245–1255. [CrossRef]
59. Moreiras, H.; Lopes-da-Silva, M.; Seabra, M.C.; Barral, D.C. Melanin processing by keratinocytes: A non-microbial type of host-pathogen interaction? *Traffic* **2019**, *20*, 301–304. [CrossRef]
60. Seiberg, M.; Paine, C.; Sharlow, E.; Andrade-Gordon, P.; Costanzo, M.; Eisinger, M.; Shapiro, S.S. The protease-activated receptor 2 regulates pigmentation via keratinocyte-melanocyte interactions. *Exp. Cell Res.* **2000**, *254*, 25–32. [CrossRef]

© 2020 by the authors. Licensee MDPI, Basel, Switzerland. This article is an open access article distributed under the terms and conditions of the Creative Commons Attribution (CC BY) license (http://creativecommons.org/licenses/by/4.0/).

Article

Spray Drying of Blueberry Juice-Maltodextrin Mixtures: Evaluation of Processing Conditions on Content of Resveratrol

César Leyva-Porras [1,2], María Zenaida Saavedra-Leos [3,*], Elsa Cervantes-González [3], Patricia Aguirre-Bañuelos [4], Macrina B. Silva-Cázarez [3] and Claudia Álvarez-Salas [1]

1. Facultad de Ingeniería, Universidad Autónoma de San Luis Potosí, Av. Dr. Manuel Nava 6, San Luis Potosí 78210, Mexico; cesar.leyva@cimav.edu.mx (C.L.-P.); claudia.salas@uaslp.mx (C.Á.-S.)
2. Centro de Investigación en Materiales Avanzados S.C. (CIMAV), Miguel de Cervantes # 120, Complejo Industrial Chihuahua, Chihuahua 31136, Mexico
3. Coordinación Académica Región Altiplano, Universidad Autónoma de San Luis Potosí, Carretera Cedral Km, 5+600 Ejido San José de las Trojes, Matehuala 78700, Mexico; elsa.cervantes@uaslp.mx (E.C.-G.); macrina.silva@uaslp.mx (M.B.S.-C.)
4. Facultad de Ciencias Químicas, Universidad Autónoma de San Luis Potosí, Av. Dr. Manuel Nava 6, San Luis Potosí 78210, Mexico; paguirreb@uaslp.mx
* Correspondence: zenaida.saavedra@uaslp.mx; Tel.: +52(488)125-0150

Received: 16 August 2019; Accepted: 24 September 2019; Published: 1 October 2019

Abstract: Resveratrol is an antioxidant abundant in red fruits, and one of the most powerful inhibiting reactive oxygen species (ROS) and oxidative stress (OS) produced by human metabolism. The effect of the spray drying processing conditions of blueberry juice (BJ) and maltodextrin (MX) mixtures was studied on content and retention of resveratrol. Quantitatively, analysis of variance (ANOVA) showed that concentration of MX was the main variable influencing content of resveratrol. Response surface plots (RSP) confirmed the application limits of maltodextrins based on their molecular weight, where low molecular weight MXs showed a better performance as carrying agents. After qualitatively comparing results for resveratrol against those reported for a larger antioxidant molecule (quercetin 3-D-galactoside), it was observed a higher influence of the number of active sites available for the chemical interactions, instead of stearic hindrance effects.

Keywords: resveratrol; analysis of variance (ANOVA); spray drying; blueberry juice-maltodextrins; conservation of antioxidants

1. Introduction

Human metabolism continuously produces reactive oxygen species (ROS) by normal respiration and cellular functions. These species may cause cellular damage by affecting the DNA, stimulating free radical chain reactions, and provoking more than one hundred diseases [1]. To counteract the reactive oxygen species (ROS), human body produces antioxidant compounds that may scavenge free radicals either by transferring free electrons or hydrogen atoms [2]. However, an over production of oxidative species and a weak immune system, may result in an imbalance in the body, developing oxidative stress (OS). This OS can generate changes in cell volume, and make other biomolecules (i.e., proteins, lipids and nucleic acids) to malfunction, leading to other major degenerative diseases such as cancer, diabetes, atherosclerosis, stroke, asthma, arthritis, dermatitis and aging, among others [3]. Thus, one approach to reduce OS is a regular consumption of antioxidants, which are found in fruits, vegetables, leafs and roots. There are various types of antioxidants in nature, among them are found flavonoids and stilbenoids structures such as quercetin 3-D-galactoside, resveratrol, myricetin and kaempferol [4,5]. Among these compounds, resveratrol (3,4′,5-trihydroxystilbene) is a polyphenolic compound that exists in the *cis*- and *trans*- isomeric forms, synthesized by plants as a phytoalexin in response to injury, fungal attack and exposure to UV light [6]. It is one of the most potent antioxidants against ROS and OS, and its intake is beneficial for human health, in the modulation of vascular cell function, suppression of platelet aggregation, reducing myocardial damage, inhibiting kinase activity, as anti-inflammatory and effective against the carcinogenesis [1,3]. Particularly, resveratrol is abundant in red wine, grape berry skins and seeds, berries, nuts and roots such as the Itadori plant (*Polygonum cuspidatum*) [7,8]. According to Tomé-Carneiro et al. the cardiovascular risk factor is reduced with a dose of 8 mg/day of resveratrol for one year [9]. In terms of red wine consumption, this dose would be equivalent to drinking 1–3 L of wine per day, depending on wine variety [10]. Additionally to a large volume ingestion of antioxidants containing products, utilization of resveratrol in the food industry is limited by factors such as low stability against oxidation, high photosensitivity, insolubility in water and short biological half-life (i.e., rapid metabolism and elimination) [11,12]. Thus, it is necessary to develop food products containing a high concentration of resveratrol, offering a reduced volume, while keeping the stability and bioavailability of antioxidants.

Both, flavor and high content of antioxidants in blueberry (*Vaccinium corymbosum*) juice (BJ) has triggered its consumption and popularity in regions such as Europe, Asia, North America and Latin America, but the short harvest season of the fruit and rapid perishability, limit its availability in the market [13]. In consequence, more than 50% of the total production must be processed in food products that may support a long shelf life such as juices, nectars, yogurts, marmalades, syrups and juice powders. Unfortunately, thermal degradation of antioxidants may occur during processing at temperatures higher than 60 °C. Spray drying of fruit juices is an alternative that offers a solution to this issue. In this sense, several studies have reported spray drying of different fruit juices with carrying agents. These agents are employed as aids in the drying process because of their relative high glass transition temperature (T_g). Fruit juices such as orange, pineapple, apple, watermelon, mango and blueberry, have been successfully dried with carrying agents such as inulin, maltodextrins, starch, Arabic gum and sodium alginate [14–20]. Maltodextrins are polysaccharides obtained from the acidic hydrolysis of starch, with a nutritional contribution of only 4 calories per gram. Besides, maltodextrins are commercially found in a wide range of molecular weight distributions (MWD), which lead to different thermal properties and potential applications [21]. Recently, Saavedra-Leos et al. used a set of four maltodextrins as carrying agents in spray drying of BJ [22]. Through analysis of variance (ANOVA) and response surface plots (RSP), they determined the effect of processing conditions, found an optimal set of experimental variables for drying the juice and conserving of quercetin 3-D-galactoside, and set the application limits of maltodextrins based on their MWD. Although there is a large number of works published on the subject of conservation of resveratrol, few have addressed the issue of optimization of processing conditions. For example, Lim, Ma and Dolan carried out a systematic experiment for testing the spray drying conditions of blueberry by-products. They found a yield of 94% when employing a ratio of blueberry solid to maltodextrin of 30:70, and an outlet temperature of 90 °C [19]. Jiménez-Aguilar et al. studied the spray drying conditions of blueberry and mesquite gum on color and degradation of anthocyanins, and found a direct relation between color and concentration of anthocyanins [23]. Tatar Turan et al. studied the effect of an ultrasonic nozzle on spray drying of blueberry juice with maltodextrin and Arabic gum as carrier agent and coating material, respectively [24]. They reported the optimal drying conditions as inlet temperature of 125 °C, ultrasonic power of 9 W, and feed pump rate of 8%. Correia et al. encapsulated the polyphenolics in wild blueberry with different protein-food ingredients, by two-way ANOVA and Tukey's test; they found that soy protein produced the highest total phenolic content [25]. Darniadi, Ho and Murray, also used ANOVA for comparing two methodologies in drying of blueberry juice: Foam-mat freeze-drying (FMFD) and spray drying. They concluded that the highest powder yield was achieved with FMFD methodology and a ratio of maltodextrin to whey protein of 1.5 [26]. Shimojo et al. employed a 2^2 full factorial experimental design for evaluating the process parameters on the properties of resveratrol loaded in nanostructured lipid carriers [27]. Additionally, other works have also studied the drying processing conditions of alternative drying methodologies [28–30].

Therefore, in the present work, ANOVA and RSP were employed for setting the optimal processing conditions of blueberry juice-maltodextrin (BJ-MX) mixtures. The effects of inlet temperature, concentration and type of maltodextrin, were studied on content, and retention of resveratrol, and the results were compared with those reported by Saavedra-Leos et al. (2019) for quercetin 3-D-galactoside. This work contributes to understanding the application limits of maltodextrins when used as carrying agents in the spray drying of sugar-rich systems, and for conservation of antioxidants.

2. Materials and Methods

2.1. Materials

Blueberry juice (BJ) was prepared using crushed fresh blueberry fruit (*Vaccinium corymbosum*), commercially available in a local market center (Costco Wholesale Corp., San Luis Potosí, Mexico). Prior to juice extraction, the fruits were stored in a refrigerator by 12 h, and crushed in a juice extractor Turmix E-17 (Guadalajara, Mexico). Juice and bagasse were stored in a glass container inside the fridge by 12 h, and after this period, were separated by vacuum filtration with paper filter Whatman No. 4, used in clarification of juices and wines. BJ was stored in darkness inside a refrigerator at 4 °C for avoiding degradation of antioxidants.

Four types of maltodextrins (MX) were employed as carrying agents, and identified according to their dextrose equivalent (DE) as DE of 7 (commercial grade maltodextrin, CM), DE of 10 (M10), DE of 20 (M20) and DE of 40 (M40). CM dry powder was purchased from INAMALT (Guadalajara, Mexico), while M10, M20 and M40 from INGREDION Mexico (Guadalajara, Mexico).

2.2. Experimental Design

Without needing many experimental runs, and keeping a confidence interval [31], D-optimal experimental design ensures an optimal selection of spray drying conditions for maximizing the content of antioxidants. Then, for this purpose, two independent continuous variables (inlet temperature (T) and maltodextrin concentration (C)), and one categorical independent variable (type of MX) were tested. The minimum and maximum levels of these variables are described in Table 1. Content of resveratrol was set as unique response variable. The D-optimal experimental design consisted of 25 experiments, necessary to achieve a quadratic model in the quantitative factors. From these 25 experiments, five runs were repeated, i.e., runs 8, 10, 11, 12 and 14 were repeated with runs 21, 22, 23, 24 and 25, respectively. All the experiments were executed randomly. Experimental design and RSP were carried out employing Matlab 2013a (Natick, MA, USA).

Table 1. Testing levels of variables in D-optimal experimental design.

Tested Variables	Testing Level	
	Minimum	Maximum
Inlet temperature (°C)	170	210
Maltodextrin concentration (wt%)	10	30
Dextrose equivalent	CM	M40

2.3. Spray Drying

Dehydration of blueberry juice was carried out in a Mini Spray Dryer B290 (Buchi, Switzerland), by feeding mixtures of blueberry juice and maltodextrin (BJ-MX) into the spray dryer at room temperature. Blends of BJ-MX were prepared by adding the necessary amount of maltodextrin in the juice, and by constant mechanical stirring. Hot air was employed as drying vehicle, at a volumetric flow rate of

$35\ m^3/h$, and constant pressure of 1.5 bar. The rest of processing conditions were varied according to the testing level of variables described above.

2.4. Content and Retention of Resveratrol

Quantification of antioxidant content in dry samples was carried out by dissolving 0.5 g of powder in 0.5 mL of phosphoric acid (10% *v/v* in water), and 3 mL of methanol used as an extracting solvent. In order to maximize the extraction of antioxidants, the solution was stirred for 5 min, and left resting 24 h in darkness. Solution was filtered in an Acrodisc filter (0.45 µm), and the filtered was diluted with 200 µL of methanol. A constant volume of 10 µL was injected in a high performance liquid chromatography (HPLC) instrument. Injections were carried out by triplicate. Content of resveratrol was quantified by HPLC with a Waters system (Waters Assoc. Milford, MA, USA), equipped with a binary pump, an auto-injector (model 717), and a dual wavelength absorbance detector (model 2487). The analyses were carried out at room temperature, and a pH of 3.0 in the solution. A constant flow rate of 1 mL/min solution of 50% acetonitrile-phosphoric acid was employed as the mobile phase. Detection was set at a wavelength of 306 nm. Chromatographic separation was done with an Agilent Zorbax C-18 column (75 mm × 4.6 mm DI 3.5 µm). All data were analyzed with the Empower Pro software Version 4.0 (Mildford, MA, USA).

Calibration curves were constructed employing a resveratrol HPLC grade standard (>99.0%, Sigma-Aldrich, Toluca, Mexico). A stock solution of 1000 µg/mL, and several aliquots (0.01, 1, 5, 10 and 20 µg/mL) were prepared as the calibration curve. Calibration curves were prepared the same day of injecting in the HPLC. Elution time of resveratrol was 2.5 min, while mobile phase eluted at 0.92 min. The intensity of resveratrol peak linearly increased with concentration. Content of resveratrol in the sample was determined from comparison with the calibration curve.

Resveratrol was evaluated in the original juice, and in dry powders. The content of antioxidants was expressed as micrograms of resveratrol per gram of blueberry juice powder (µg/g). Percent of retention (R) of resveratrol was determined according to equation (1):

$$R\ (\%) = \frac{Q_P \times 100}{Q_j}, \tag{1}$$

where Q_P is the content of resveratrol in dry powder (in ppm), and Q_J is the content of antioxidants in fresh juice (8.38 ppm).

2.5. Statistical Analysis

The effects of experimental variables (factors) and their interactions were evaluated with an analysis of variance (ANOVA). In ANOVA, the F-value indicates the effect of the independent variable on the response variable. Thus, if F is equal to one, the independent variable has no effect, while if $F > 1$ the independent variable has an effect and its effect is lager as F increases above 1. The p-value represents the probability of an F-value large enough for influencing the experiment; if this value is equal or lower than the significance level, then the assumption of the influence of the independent variables on the response variable is correct. The probability of rejecting the previous assumption even when it is true, is given by the significance level.

A quadratic model with second-order interactions and main effects were used to explain a relationship between the given continuous variables as indicated in Equation (2):

$$Z = \alpha_0 + \Sigma\alpha_i X_i + \Sigma\alpha_{ii}X_{ii}^2 + \Sigma\alpha_{ij}X_i X_j, \tag{2}$$

where, Z represents the response variable (content of resveratrol), X_i and X_j are the factors (temperature, and concentration of maltodextrin) and α_0, α_i, α_{ii} and α_{ij} are the linear regression coefficients of the model.

In the process of selecting a model, some parameters of the complete model were first adjusted with Equation (2). Based on a normality test of Anderson Darling for the response variable (R), a transformation of the response was made by the Box-Cox analysis when it was necessary to stabilize the variance. Then, for simplification, the model was hierarchically pruned, and used only with significant factors. Here we present the results obtained with the pruned model, and transformed into response variables.

3. Results and Discussion

3.1. Content and Retention of Resveratrol

From 25 experiments, 18 runs were successfully spray-dried, while in seven runs the microstructure collapsed. Since the results for resveratrol were discussed herein, results of physicochemical characterization and yield can be consulted in the work previously reported for quercetin 3-D-galactoside [22], since both works are parts of the same experiment. Table 2 shows the detailed experimental description of the 25 runs and results of content, and retention (R) of resveratrol at different processing conditions. Minimum and maximum values for content, and retention, of 18 spray-dried samples were 0.0–0.47 µg/g, and 0.0–10.24%, respectively. The corresponding average values for these two parameters were 0.28 µg/g, and 6.25%. The highest retention was obtained for run 10 with a value of 10.24%. Overall, average concentration of resveratrol in dried experiments diminished about 96%. This was calculated considering the initial concentration of resveratrol in fresh BJ, and the content of antioxidant determined in dried samples. On the other hand, content and retention values were relatively lower than those obtained for quercetin 3-D-galactoside [22]. Therefore, based on these observations, it is evident that resveratrol is less prone to interact chemically with maltodextrins, being more exposed to thermal degradation during the spray drying process. While some qualitative relations among the independent and categorical variables may be inferred from Table 2, the quantitative analysis of the effect of processing variables on content of resveratrol will be discussed in following section.

Table 2. Content and retention of resveratrol in spray drying of blueberry juice-maltodextrin (BJ-MX).

Run Identification	Factors			Resveratrol	
	T (°C)	C (wt%)	MD	Content (µg/g) [a]	Retention (wt%)
1	170	30	CM	0.23 ± 0.031	5.01
2	187	10	M10	0	0.00
3	170	30	M20	0.33 ± 0.108	7.19
4	210	10	M40	0	0.00
5	210	30	CM	0.21 ± 0.056	4.58
6	210	18	M10	0.39 ± 0.036	8.50
7	210	10	M20	0	0.00
8	170	10	M40	0	0.00
9	210	10	CM	0	0.00
10	170	22	M10	0.47 ± 0.282	10.24
11	170	10	M20	0	0.00
12	181	30	M40	0.29 ± 0.010	6.32
13	190	20	CM	0.29 ± 0.020	6.32
14	193	30	M10	0.22 ± 0.053	4.79
15	210	30	M20	0.27 ± 0.122	5.88
16	175	12.5	CM	0.28 ± 0.020	6.10
17	190	20	M10	0.28 ± 0.212	6.10
18	190	25	M20	0.31 ± 0.057	6.76
19	209	24	M40	0.26 ± 0.033	5.67
20	210	20	CM	0.30 ± 0.065	6.54

Antioxidants **2019**, *8*, 437

Table 2. *Cont.*

Run Identification	Factors			Resveratrol	
	T (°C)	C (wt%)	MD	Content (µg/g) [a]	Retention (wt%)
21	170	10	M40	0	0.00
22	193	30	M10	0.20 ± 0.019	4.36
23	170	10	M20	0	0.00
24	181	30	M40	0.25 ± 0.020	5.45
25	170	22	M10	0.3 ± 0.170	6.54

[a] Average and standard deviation values calculated from three repetitions.

3.2. ANOVA and Response Surface Plots Analysis (RSP)

Table 3 shows ANOVA results calculated for the content of resveratrol. ANOVA results showed that concentration (C) of MX was the variable with the most important effect, while the type of maltodextrin, and inlet temperature had a negligible effect. These observations were confirmed by the p-value at a significance level of 0.05. The interactions between the same variable (intra) showed that only the concentration (C^2) had an effect on content, but 2.4 times less than the single concentration (C). The rest of interactions between variables (inter and intra) such as T·C, T·MX, C·MX and T^2 showed a p-value higher than the significance level, thus their effects were negligible. Therefore, concentration was the independent variable with the larger effect on the content of resveratrol.

These observations indicated that concentration was the main processing variable affecting the content of resveratrol, while the type of maltodextrin and inlet temperature showed no effect on the response variables, suggesting that content of resveratrol was unresponsive to inlet temperatures exerted in the experimental design, and to differences in molecular weight distribution of MXs.

Table 3. ANOVA results determined for content of antioxidant in BJ-MX.

	Content				
Source	DF	SS [a]	MS [b]	F	p [*]
Model	14	1.069	0.076	15.978	0.0001
T	1	0.011	0.011	2.425	0.150
C	1	0.531	0.531	111.03	0.0001
MX	3	0.005	0.002	0.401	0.755
T·C	1	0.0005	0.0005	0.122	0.733
T·MX	3	0.021	0.007	1.516	0.269
C·MX	3	0.033	0.011	2.308	0.138
T^2	1	0.011	0.011	2.463	0.147
C^2	1	0.222	0.222	46.605	0.0001
Residual	10	0.047	0.004		
Total	24	1.117			

[a] Sum of squares; [b] mean squares; [*] calculated at a significance level of 0.05; T = inlet temperature, C = maltodextrin concentration, MX = type of maltodextrin.

Figure 1 shows RSP for the content of resveratrol as a function of the type of MX. The surfaces shape was similar regardless of the type of MX. In all cases, the highest content of resveratrol was observed at a concentration of MX of 23%. At higher or lower concentration values, content of resveratrol decreased rapidly. However, slight differences were observed with inlet temperature. CM showed a maximum content of resveratrol at 170 °C, and a decrement at 210 °C. M10 showed the opposite behavior, with a maximum value at 210 °C and a decrement at 170 °C. High molecular weight maltodextrins (M20 and M40) showed two maximum values at opposite temperatures. All of these observations allowed setting the optimal processing conditions for spray drying of BJ-MX. Highest values for content of resveratrol were obtained at a concentration of MX of 23%, and inlet temperature of 170 °C for CM or 210 °C for M10. In addition, RSP results confirmed the utilization limits of MXs, where low molecular weight MXs produced a higher content of resveratrol than high molecular weight MXs. These observations demonstrated that depending on the molecular weight distributions of MXs, these polysaccharides might be employed selectively as carrying agents in spray drying of diverse sugar-rich systems. Indeed, chemical interactions between maltodextrins and antioxidants (i.e., resveratrol) may be affected by other variables rather than by the degree of polymerization of maltodextrin. Other variables that may affect this interaction into a greater or lesser extent are the volumetric flow of solution injected into the dryer, wet bulb temperature (i.e., relative humidity of air), type of nozzle used (for example, regular versus ultrasonic) and adjuvant agents (i.e., soy protein and sodium alginate). However, in the present work, ANOVA was focused on two spray drying variables (concentration of maltodextrin and inlet temperature) and one categorical variable (type of maltodextrin). In this sense, Ameri and Maa, indicated that increasing the total content of solids in feed solution, increased the recovering of powders in spray drying [32]. Nadeem et al. concluded that yield was related to concentration of maltodextrin, rather than to drying temperature [33]. Caliskan and Gulsah found that increasing the concentration of maltodextrin resulted in an increment in the yield of dried powder [34]. Bhusari et al. attributed this behavior to an increment in the T_g of mixtures [35]. Peng et al. indicated that above 30% of the carrying agent was detrimental for product quality [36]. Saavedra-Leos et al. determined an inverse relation between T_g and DE of maltodextrins [21]. Through several works reported by the Saavedra-Leos group [14,22,37–39], we have observed that in some juices such as orange and blueberries, their physicochemical properties varied with the differences in the type and distribution of molecular weight of the carrier agent. In these studies, we have attributed this behavior mainly to: (i) Differences in the molecular weight distribution of the carrier agent, (ii) the arrangement of polymer chains (i.e., entangled or linear) and (iii) the type of molecule in these chains (for example, glucose for maltodextrin or fructose for inulin). Additionally, according to Darniadi, Ho and Murray, when mixing low molecular weight sugars and high molecular active compounds, the active compound tends to segregate in some extent to the surface of dried particles [26]. While this behavior prevents the particles sticking on the dryer walls, it also exposes the active ingredient to a faster degradation.

Table 4 shows the predictive equations for the content of resveratrol as a function of the type of MX. These equations were extrapolated from SRP and in consequence were only valid within the interval of conditions tested herein. Inlet temperature presented a negative effect and concentration a positive effect. From both processing variables, the extent of concentration was about 1.9 times larger than that of temperature. The values for interactions (T·C), and square of temperature (T^2) showed a relatively low positive value, but their numerical contributions were similar to that of concentration. Although the square of concentration (C^2) showed a negative effect, its value was still lower than that of single concentration, thus indicating a little contribution on the content of resveratrol. Interactions (T·C, T^2 and C^2) showed a constant value indicating that these interactions were insensitive with respect to the type of MX. Numerical calculations employing these equations supported the results found from RSP. CM showed the largest content value when using a temperature of 170 °C, while for M10 the higher content was obtained with a temperature of 210 °C.

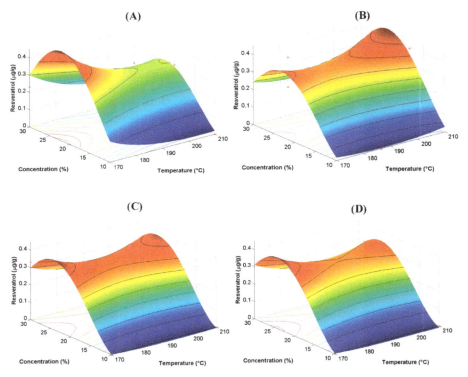

Figure 1. Response surface plots analysis (RSP) for the content of resveratrol as a function of the type of MX: (**A**) CM, (**B**) M10, (**C**) M20 and (**D**) M40. The symbols (*) in each RSP represent the center and start points for estimation of second order effects.

Table 4. Predicting equations extrapolated from SRP for the content of resveratrol in BJ-MX, as a function of type of MX.

Type Of MX	Content
CM	$\ln(Q + 0.01) = 5.8694 - 0.0639T + 0.1069C + 0.00003T \cdot C + 0.00015T^2 - 0.0025C^2$
M10	$\ln(Q + 0.01) = 4.6662 - 0.0583T + 0.1137C + 0.00003T \cdot C + 0.00015T^2 - 0.0025C^2$
M20	$\ln(Q + 0.01) = 4.8695 - 0.0602T + 0.1191C + 0.00003T \cdot C + 0.00015T^2 - 0.0025C^2$
M40	$\ln(Q + 0.01) = 5.0389 - 0.0611T + 0.1193C + 0.00003T \cdot C + 0.00015T^2 - 0.0025C^2$

Q = content of resveratrol (μg/g), T = inlet temperature (°C), C = concentration of maltodextrin (wt. %).

3.3. Effect of the Chemical Structure in the Content of Antioxidants

In this section we compared the results of the content of resveratrol reported herein, against those for quercetin 3-D-galacoside previously reported [22]. The effect of spray drying processing conditions showed similar behavior for both antioxidants, but the main difference relied on the values of the content of each antioxidant. In general, the content of quercetin 3-D-galacoside was 2–8 times higher than that of resveratrol. Conversely, according to Shrikanta, Kumar and Govindaswamy, the total content of polyphenols was 1–14 times higher than the total content of flavonoids in underutilized Indian fruits [8]. In this sense, Araujo-Díaz et al. reported a concentration of resveratrol 1.5 times higher than for 3-D-galactoside when employing maltodextrin in spray drying of blueberry juice [37]. Figure 2 shows a schematic representation of the chemical structure of both antioxidants, and a maltodextrin repetitive unit. Resveratrol is a stilbene with a C6-C2-C6 structure and three hydroxyls (OH^-). While quercetin is a flavonoid with a C6-C3-C6 structure, five hydroxyls, one alkoxy group (ether) and one carbonyl group (ketone); the galactoside refers to the galactose molecule containing four hydroxyls. On the other hand, maltodextrins are polysaccharide molecules consisting of glucose units linked by glycosidic α-(1-4) and α-(1-6) bonds, with a variable length of its polymeric chains expressed as DE [40]. The glucose molecule contains three hydroxyl groups, one alkoxy group (ether) and one hydroxyl group at each extreme of polymeric chains. All these functional groups are responsible for carrying out molecular chemical interactions such as hydrogen bonding, and Van der Waals interactions. In several works it has been reported that these chemical interactions (inter and intramolecular) are responsible of the adsorption of water on different carrying agents such as inulin and maltodextrins [21,38]. In these works, water adsorption was promoted with increasing the molecular weight of carrying agents. The set of maltodextrins employed as carrying agents in this work, was similar to that reported by Saavedra et al. and presented a degree of polymerization (DP) of: CM 2-12, M10 2-16, M20 2-21 and M40 2-30, units of glucose [21]. Although a high DP may indicate a larger number of hydroxyl groups exposed for chemical interactions, the polymeric chains in MXs may arrange in different configurations rather than linearly, forming entangled branches, thus reducing the availability of active sites. Additionally to the arrangement of polymeric chains, the stearic hindrance between the adsorbing molecules and MXs, is another aspect affecting the final content of antioxidants. Evidently, the size of quercetin 3-D-galactosie molecule is larger than that of resveratrol, suggesting that stearic hindrance was not the main factor influencing larger chemical interactions with MX, but the number of functional groups available such as hydroxyls, alkoxys and carbonyls. Based on these arguments, it is possible to infer that: (i) The availability of functional groups is the main cause of chemical interactions, since quercetin has more of these groups than resveratrol; hence its greater interaction with maltodextrin, and (ii) in maltodextrins, the availability of these functional groups is limited by branching and entangled of polymeric chains. For this reason, in both cases i.e., the content of resveratrol and quercetin, low molecular weight MXs presented higher antioxidant content than high molecular weight MXs, indicating that polymeric chains in these carrying agents are less branched and entangled, thus promoting more chemical interactions.

trans-resveratrol

Quercetin 3-D-galactoside

Maltodextrin repetitive unit $2 < n < 20$

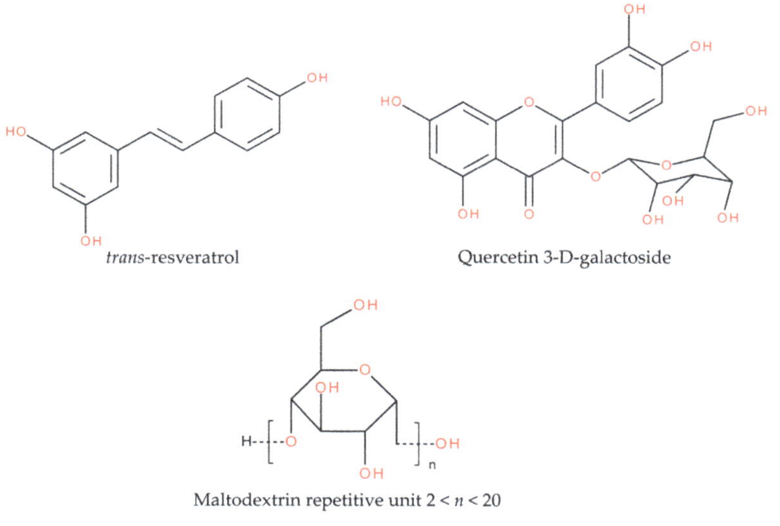

Figure 2. Schematic representation of chemical structures of resveratrol, quercetin 3-D-galactoside and the maltodextrin repetitive unit.

4. Conclusions

The effect of processing conditions in spray drying of blueberry juice and maltodextrin mixtures (BJ-MX) on content and retention of resveratrol was studied. Analysis of variance (ANOVA) showed that concentration (C) was the main variable influencing the content of resveratrol. Response surface plots (RSP) allowed observing that the content of resveratrol was also affected by the molecular weight distribution of MXs employed as carrying agents, where low molecular weight MX presented a higher content of resveratrol. With this study, it was possible to set the optimal processing conditions for spray drying of BJ-MX such as concentration of 23% of maltodextrin and temperature of 170 °C for CM, and 210 °C for M10. Additionally, the results reported herein were compared with those reported for content of quercetin 3-Q-galactoside, finding that quercetin 3-Q-galactoside was more readily to interact with MXs because of a higher availability of functional groups rather than by stearic hindrance effects.

Author Contributions: Conceptualization, C.L.-P. and M.Z.S.-L.; Data curation, E.C.-G.; Formal analysis, M.Z.S.-L. and P.A.-B.; Funding acquisition, M.Z.S.-L.; Investigation, C.L.-P., M.Z.S.-L. and P.A.-B.; Methodology, M.B.S.-C. and C.Á.-S.; Project administration, C.L.-P. and E.C.-G.; Software, E.C.-G.; Visualization, M.B.S.-C. and C.Á.-S.; Writing—original draft, C.L.-P. and M.Z.S.-L; Writing—review & editing, C.L.-P. and M.Z.S.-L.

Funding: This research received no external funding.

Acknowledgments: The postdoctoral support managed by the Academic core "Chemistry and food technology, CA-259", through the Program for Teacher Professional Development (PRODEP) is gratefully acknowledge.

Conflicts of Interest: The authors declare no conflict of interest.

References

1. Gülçin, İ. Antioxidant properties of resveratrol: A structure—Activity insight. *Innov. Food Sci. Emerg. Technol.* **2010**, *11*, 210–218. [CrossRef]
2. Naik, G.; Priyadarsini, K.; Satav, J.; Banavalikar, M.; Sohoni, D.; Biyani, M.; Mohan, H. Comparative antioxidant activity of individual herbal components used in Ayurvedic medicine. *Phytochemistry* **2003**, *63*, 97–104. [CrossRef]
3. Pandey, K.B.; Rizvi, S.I. Anti-oxidative action of resveratrol: Implications for human health. *Arab. J. Chem.* **2011**, *4*, 293–298. [CrossRef]
4. Faria, A.; Oliveira, J.; Neves, P.; Gameiro, P.; Santos-Buelga, C.; de Freitas, V.; Mateus, N. Antioxidant properties of prepared blueberry (*Vaccinium myrtillus*) extracts. *J. Agric. Food Chem.* **2005**, *53*, 6896–6902. [CrossRef] [PubMed]
5. Sinelli, N.; Spinardi, A.; Di Egidio, V.; Mignani, I.; Casiraghi, E. Evaluation of quality and nutraceutical content of blueberries (*Vaccinium corymbosum* L.) by near and mid-infrared spectroscopy. *Postharvest Biol. Technol.* **2008**, *50*, 31–36. [CrossRef]
6. Frémont, L. Biological effects of resveratrol. *Life Sci.* **2000**, *66*, 663–673. [CrossRef]
7. Burns, J.; Yokota, T.; Ashihara, H.; Lean, M.E.; Crozier, A. Plant foods and herbal sources of resveratrol. *J. Agric. Food Chem.* **2002**, *50*, 3337–3340. [CrossRef] [PubMed]
8. Shrikanta, A.; Kumar, A.; Govindaswamy, V. Resveratrol content and antioxidant properties of underutilized fruits. *J. Food Sci. Technol.* **2015**, *52*, 383–390. [CrossRef]
9. Tomé-Carneiro, J.; Gonzálvez, M.; Larrosa, M.; Yáñez-Gascón, M.J.; García-Almagro, F.J.; Ruiz-Ros, J.A.; García-Conesa, M.T.; Tomás-Barberán, F.A.; Espín, J.C. One-year consumption of a grape nutraceutical containing resveratrol improves the inflammatory and fibrinolytic status of patients in primary prevention of cardiovascular disease. *Am. J. Cardiol.* **2012**, *110*, 356–363. [CrossRef]
10. Kuršvietienė, L.; Stanevičienė, I.; Mongirdienė, A.; Bernatonienė, J. Multiplicity of effects and health benefits of resveratrol. *Medicina* **2016**, *52*, 148–155. [CrossRef]
11. Balanč, B.; Trifković, K.; Dordević, V.; Marković, S.; Pjanović, R.; Nedović, V.; Bugarski, B. Novel resveratrol delivery systems based on alginate-sucrose and alginate-chitosan microbeads containing liposomes. *Food Hydrocoll.* **2016**, *61*, 832–842. [CrossRef]
12. Intagliata, S.; Modica, M.N.; Santagati, L.M.; Montenegro, L. Strategies to Improve Resveratrol Systemic and Topical Bioavailability: An Update. *Antioxidants* **2019**, *8*, 244. [CrossRef] [PubMed]

13. Retamales, J.B.; Hancock, J.F. The Blueberry Industry. In *Blueberries*, 2nd ed.; Russell, R., Wilford, S., Eds.; Cabi: Boston, MA, USA, 2018; pp. 1–17.
14. Saavedra-Leos, M.Z.; Leyva-Porras, C.; Martínez-Guerra, E.; Pérez-García, S.A.; Aguilar-Martínez, J.A.; Álvarez-Salas, C. Physical properties of inulin and inulin-orange juice: Physical characterization and technological application. *Carbohydr. Polym.* **2014**, *105*, 10–19. [CrossRef] [PubMed]
15. Jittanit, W.; Niti-Att, S.; Techanuntachaikul, O. Study of spray drying of pineapple juice using maltodextrin as an adjunct. *Chiang Mai J. Sci.* **2010**, *37*, 498–506.
16. De Oliveira, M.A.; Maia, G.A.; De Figueiredo, R.W.; De Souza, A.C.R.; De Brito, E.S.; De Azeredo, H.M.C. Addition of cashew tree gum to maltodextrin-based carriers for spray drying of cashew apple juice. *Int. J. Food Sci. Tech.* **2009**, *44*, 641–645. [CrossRef]
17. Quek, S.Y.; Chok, N.K.; Swedlund, P. The physicochemical properties of spray-dried watermelon powders. *Chem. Eng. Process. Process. Intensif.* **2007**, *46*, 386–392. [CrossRef]
18. Cano-Chauca, M.; Stringheta, P.; Ramos, A.; Cal-Vidal, J. Effect of the carriers on the microstructure of mango powder obtained by spray drying and its functional characterization. *Innov. Food Sci. Emerg. Technol.* **2005**, *6*, 420–428. [CrossRef]
19. Lim, K.; Ma, M.; Dolan, K.D. Effects of spray drying on antioxidant capacity and anthocyanidin content of blueberry by-products. *J. Food Sci.* **2011**, *76*, H156–H164. [CrossRef]
20. Waterhouse, G.I.; Sun-Waterhouse, D.; Su, G.; Zhao, H.; Zhao, M. Spray-drying of antioxidant-rich blueberry waste extracts; interplay between waste pretreatments and spray-drying process. *Food Bioprocess Technol.* **2017**, *10*, 1074–1092. [CrossRef]
21. Saavedra-Leos, Z.; Leyva-Porras, C.; Araujo-Díaz, S.B.; Toxqui-Terán, A.; Borrás-Enríquez, A.J. Technological application of maltodextrins according to the degree of polymerization. *Molecules* **2015**, *20*, 21067–21081. [CrossRef]
22. Saavedra-Leos, M.Z.; Leyva-Porras, C.; López-Martínez, L.A.; González-García, R.; Martínez, J.O.; Compeán Martínez, I.; Toxqui-Terán, A. Evaluation of the Spray Drying Conditions of Blueberry Juice-Maltodextrin on the Yield, Content, and Retention of Quercetin 3-d-Galactoside. *Polymers* **2019**, *11*, 312. [CrossRef] [PubMed]
23. Jiménez-Aguilar, D.; Ortega-Regules, A.; Lozada-Ramírez, J.; Pérez-Pérez, M.; Vernon-Carter, E.; Welti-Chanes, J. Color and chemical stability of spray-dried blueberry extract using mesquite gum as wall material. *J. Food Compos. Anal.* **2011**, *24*, 889–894. [CrossRef]
24. Tatar Turan, F.; Cengiz, A.; Sandıkçı, D.; Dervisoglu, M.; Kahyaoglu, T. Influence of an ultrasonic nozzle in spray-drying and storage on the properties of blueberry powder and microcapsules. *J. Sci. Food Agric.* **2016**, *96*, 4062–4076. [CrossRef]
25. Correia, R.; Grace, M.H.; Esposito, D.; Lila, M.A. Wild blueberry polyphenol-protein food ingredients produced by three drying methods: Comparative physico-chemical properties, phytochemical content, and stability during storage. *Food Chem.* **2017**, *235*, 76–85. [CrossRef] [PubMed]
26. Darniadi, S.; Ho, P.; Murray, B.S. Comparison of blueberry powder produced via foam-mat freeze-drying versus spray-drying: Evaluation of foam and powder properties. *J. Sci. Food Agric.* **2018**, *98*, 2002–2010. [CrossRef]
27. Shimojo, A.A.; Fernandes, A.R.V.; Ferreira, N.R.; Sanchez-Lopez, E.; Santana, M.H.; Souto, E.B. Evaluation of the Influence of Process Parameters on the Properties of Resveratrol-Loaded NLC Using 22 Full Factorial Design. *Antioxidants* **2019**, *8*, 272. [CrossRef] [PubMed]
28. MacGregor, W. Effects of air velocity, air temperature, and berry diameter on wild blueberry drying. *Dry. Technol.* **2005**, *23*, 387–396. [CrossRef]
29. Pallas, L.A.; Pegg, R.B.; Shewfelt, R.L.; Kerr, W.L. The role of processing conditions on the color and antioxidant retention of jet tube fluidized bed–dried blueberries. *Dry. Technol.* **2012**, *30*, 1600–1609. [CrossRef]
30. Turan, F.T.; Cengiz, A.; Kahyaoglu, T. Evaluation of ultrasonic nozzle with spray-drying as a novel method for the microencapsulation of blueberry's bioactive compounds. *Innov. Food Sci. Emerg. Technol.* **2015**, *32*, 136–145. [CrossRef]
31. Munson-McGee, S.H. D-Optimal Experimental Designs for Uniaxial Expression. *J. Food Process Eng.* **2014**, *37*, 248–256. [CrossRef]
32. Ameri, M.; Maa, Y. Spray drying of biopharmaceuticals: Stability and process considerations. *Dry. Technol.* **2006**, *24*, 763–768. [CrossRef]

33. Nadeem, H.S.; Torun, M.; Özdemir, F. Spray drying of the mountain tea (*Sideritis stricta*) water extract by using different hydrocolloid carriers. *LWT Food Sci. Technol.* **2011**, *44*, 1626–1635. [CrossRef]
34. Caliskan, G.; Dirim, S.N. The effects of the different drying conditions and the amounts of maltodextrin addition during spray drying of sumac extract. *Food Bioprod. Process.* **2013**, *91*, 539–548. [CrossRef]
35. Bhusari, S.; Muzaffar, K.; Kumar, P. Effect of carrier agents on physical and microstructural properties of spray dried tamarind pulp powder. *Powder Technol.* **2014**, *266*, 354–364. [CrossRef]
36. Peng, Z.; Li, J.; Guan, Y.; Zhao, G. Effect of carriers on physicochemical properties, antioxidant activities and biological components of spray-dried purple sweet potato flours. *LWT Food Sci. Technol.* **2013**, *51*, 348–355. [CrossRef]
37. Araujo-Díaz, S.; Leyva-Porras, C.; Aguirre-Bañuelos, P.; Álvarez-Salas, C.; Saavedra-Leos, Z. Evaluation of the physical properties and conservation of the antioxidants content, employing inulin and maltodextrin in the spray drying of blueberry juice. *Carbohydr. Polym.* **2017**, *167*, 317–325. [CrossRef]
38. Leyva-Porras, C.; Saavedra-Leos, M.; López-Pablos, A.; Soto-Guerrero, J.; Toxqui-Terán, A.; Fozado-Quiroz, R. Chemical, thermal and physical characterization of inulin for its technological application based on the degree of polymerization. *J. Food Process Eng.* **2017**, *40*, e12333. [CrossRef]
39. Saavedra-Leos, M.Z.; Leyva-Porras, C.; Alvarez-Salas, C.; Longoria-Rodríguez, F.; López-Pablos, A.L.; González-García, R.; Pérez-Urizar, J.T. Obtaining orange juice—Maltodextrin powders without structure collapse based on the glass transition temperature and degree of polymerization. *CyTA J. Food* **2018**, *16*, 61–69. [CrossRef]
40. Valenzuela, C.; Aguilera, J.M. Effects of maltodextrin on hygroscopicity and crispness of apple leathers. *J. Food Eng.* **2015**, *144*, 1–9. [CrossRef]

© 2019 by the authors. Licensee MDPI, Basel, Switzerland. This article is an open access article distributed under the terms and conditions of the Creative Commons Attribution (CC BY) license (http://creativecommons.org/licenses/by/4.0/).

Article

Enhancing Thermal Stability and Bioaccesibility of Açaí Fruit Polyphenols through Electrohydrodynamic Encapsulation into Zein Electrosprayed Particles

Carol López de Dicastillo [1,2,*], Constanza Piña [1,2], Luan Garrido [1,2], Carla Arancibia [3] and María José Galotto [1,2]

[1] Food Packaging Laboratory (Laben-Chile), Department of Science and Food Technology, Faculty of Technology, University of Santiago de Chile, Obispo Umaña 050, 9170201 Santiago, Chile; constanza.pina@usach.cl (C.P.); luan.garrido@usach.cl (L.G.); maria.galotto@usach.cl (M.J.G.)
[2] Center for the Development of Nanoscience and Nanotechnology (CEDENNA), 9170124 Santiago, Chile
[3] Food Properties Research Group, Department of Science and Food Technology, Faculty of Technology, University of Santiago de Chile, Obispo Umaña 050, 9170201 Santiago, Chile; carla.arancibia@usach.cl
* Correspondence: analopez.dediscastillo@usach.cl

Received: 20 August 2019; Accepted: 1 October 2019; Published: 9 October 2019

Abstract: The açaí fruit *(Euterpe oleracea* Mart.) is well known for its high content of antioxidant compounds, especially anthocyanins, which provide beneficial health properties. The incorporation of this fruit is limited to food products whose processing does not involve the use of high temperatures due to the low thermal stability of these functional components. The objective of this work was the encapsulation of açaí fruit antioxidants into electrosprayed zein, a heat-resistant protein, to improve their bioavailability and thermal resistance. First, the hydroalcoholic açaí extract was selected due to its high polyphenolic content and antioxidant capacities, and, subsequently, it was successfully encapsulated in electrosprayed zein particles. Scanning electron microscopy studies revealed that the resulting particles presented cavities with an average size of 924 nm. Structural characterization by Fourier transform infrared spectroscopy revealed certain chemical interaction between the active compounds and zein. Encapsulation efficiency was approximately 70%. Results demonstrated the effectiveness of the encapsulated extract on protecting polyphenolic content after high-temperature treatments, such as sterilization (121 °C) and baking (180 °C). Bioaccesibility studies also indicated an increase of polyphenols presence after in vitro digestion stages of encapsulated açaí fruit extract in contrast with the unprotected extract.

Keywords: encapsulation; electrospinning; polyphenol; açaí (*Euterpe oleracea* Mart.); zein

---

1. Introduction

In the last decade, the market for nutraceuticals has grown enormously due to accentuated interest by consumers in their therapeutic effects for health disorders, including neurodegenerative and cardiovascular diseases [1,2]. A broad variety of active compounds from natural sources has been researched. The main active compounds from plants are polyphenols, secondary metabolites that make up a large family of substances, from simple molecules to complex structures [3]. Numerous studies have shown that certain fruit contains high levels of antioxidant active compounds. Specifically, one fruit with great antioxidant capacity is açaí (*Euterpe oleracea* Mart.), which has recently emerged as a promising source of energy, and nutritional and antimicrobial properties [4,5]. Preventing oxidative stress in human endothelial cells and the therapeutic effect on neurodegenerative diseases have emerged as bioactivities related to this fruit [6–9]. Açaí is a palm native to South America that grows mostly in the Amazon estuary, in the north of Brazil. Some studies have also revealed this fruit significantly reduces the risk of atherosclerosis through antioxidant and anti-inflammatory activities [10,11]. The principal flavonoid responsible for this anti-inflammatory activity is the flavone velutin [12]. Açaí fruit is also composed by high content of polyphenolic compounds, especially anthocyanins, as major components cyanidin-3-glucoside and cyanidin-3-rutinoside, and phenolic acids [13–15]. Açaí is commonly sold as dehydrated powder to be added to food, in dietary supplements or beverages. However, its current applications are limited to certain foods that do not include such high thermal processes as baking and cooking due to the low stability of the polyphenols, principally the anthocyanins. These molecules can be degraded at increased temperatures, leading to the loss of their functional properties [16]. Recently, a kinetic study of anthocyanin's açaí thermal degradation by Costa et col. (2018) revealed their degradations fitted a kinetic model of the first order [17]. On the other hand, it is also well known that for natural bioactive compounds to present real benefits, they must be available for absorption after the process of gastrointestinal digestion [18,19]. Thus, encapsulation of natural bioactive compounds is an interesting alternative for performing a double purpose of extending possibilities of incorporation into broader food matrices and enhancing bioavailability [20].

Although several procedures have been used to encapsulate active compounds, such as spray drying, lyophilization, and emulsification, these techniques present disadvantages, such as the complexity of the equipment, use of high temperatures, non-uniform conditions in the drying chamber and lack of particle size control [21]. In recent years, a technology that has received special attention is electrospinning and/or electrospraying [22]. Research studies have clearly shown electrospraying and electrospinning are techniques with functional advantages such as sustained release property, high encapsulation efficiency, and enhanced stability of encapsulated food bioactive compounds [22]. This technique consists of spinning polymeric solutions through high electric fields that exceed the forces of surface tension in the solution of charged polymers. At a certain voltage, fine jets of solution are expelled from the capillary to the collector plate. The solvent evaporates and the segments of fibers or particles are deposited randomly on a substrate. Depending on the specific conditions of polymer solution and the equipment, the process can result in a stretched jet or dispersion of droplets [23]. Several bioactive substances have been successfully encapsulated into electrosprayed particles by using a wide variety of natural polymers as encapsulating materials, depending on the compound to be encapsulated [24]. Among the edible materials, carbohydrates, lipids, and protein have gained the most interest. The latter have numerous advantages, such as increasing the bioavailability of the encapsulated compounds and high binding capacity with active compounds [25]. Corn zein protein has been shown to be a protein resistant to temperatures above 200 °C and has been used as an encapsulation material for some compounds, such as curcumin, improving stability against different values of pH and ultraviolet (UV) radiation [26,27]. Thus, this work presents the selection of a powerful açaí fruit extract, based on its highest phenolic content and antioxidant activities, to be further encapsulated into electrosprayed zein capsules. Although the limited use of açaí fruit in food formulations is evident due to its thermal instability, few works have developed alternatives to protect its active compounds. These encapsulated structures were morphological and structurally characterized and considered a suitable shell to impart thermal protection and enhance the bioavailability of phenolic compounds.

2. Materials and Methods

2.1. Test Materials and Reagents

Freeze-dried and milled organic açaí fruit was obtained from "Healthy Foods". Zein (Z 3625), 2,2-diphenyl-1-picrylhydrazyl (DPPH), 2,2'-azinobis(3-ethylbenzothiazoline-6-sulphonate) (ABTS), Folin–Ciocalteu phenol reagent, anhydrous sodium carbonate, gallic acid (GA), ferric 2,4,6-tripyridyl-s-triazine (TPTZ) and 6-hydroxy-2,5,7,8-tetramethylchroman-2-carboxylic acid (Trolox) were obtained from Sigma–Aldrich (Santiago, Chile). Lipase (L3126) and pancreatin (P1750) from porcine pancreas, pepsin (P6887) from porcine gastric mucosa, and porcine bile extract (B8631) were purchased from Sigma–Aldrich (Sigma–Aldrich S.A., USA). NaOH, HCl and different salts to prepare simulated

Antioxidants **2019**, *8*, 464

digestion fluids (KCl, KH_2PO_4, $NaHCO_3$, NaCl, $MgCl_2.(H_2O)_6$, $(NH_4)_2CO_3$, and $CaCl_2.(H_2O)_2$) were purchased from Merck (Merck KGaA S.A., Darmstadt, Germany).

2.2. Selection of Açaí Fruit Extract

2.2.1. Preparation of Açaí Fruit Extracts

Active compounds from açaí fruit were extracted using absolute ethanol, ethanol 50%, and distilled water in a 1:300 (solid (g):solvent (mL)) ratio to study the effect of the solvent polarity on the extraction capacity of the most relevant antioxidant compounds. The extractions were carried out at 40 °C for 3 h with an agitation of 150 rpm. Samples were centrifuged, filtered, and used for the antioxidant assays and the analysis of polyphenolic content (PC). Extracts obtained were named "Aç1, Aç2, Aç3" for extracts under ethanol, ethanol 50% and water, respectively.

2.2.2. Determination of Total Phenolic Content and Antioxidant Activity Studies

Total phenolic content (TPC) of the extracts was determined following the Folin–Ciocalteu method [28]. 100 µL of each extract was mixed with 3100 µL of distilled water and 200 µL of Folin–Ciocalteu reagent. The samples were taken to darkness for 5 min and 600 µL of anhydrous sodium carbonate at 20% (w/v) was added [29]. The samples were shaken and brought back into darkness for 2 h. The absorbance readings were performed at 765 nm. Results were expressed as mg of gallic acid equivalent (mg GAE) g^{-1} of dried açaí.

Antioxidant evaluation of extracts was carried out through three antioxidant assays: Trolox Equivalent Antioxidant Capacity (TEAC), 2,2-diphenyl-1-picrylhydrazil (DPPH), and Ferric Reducing Antioxidant Power (FRAP). All antioxidant results were expressed as mg Trolox g^{-1} dried açaí. Both TEAC and DPPH methods measure the antioxidant power of extracts by the percentage inhibition of $ABTS^{+\bullet}$ and $DPPH^{\bullet}$ radicals, respectively, via both single-electron transference (SET) and hydrogen-atom transference (HAT) mechanisms [30]. The cationic radical $ABTS^{+\bullet}$ was generated from an oxidation reaction of the ABTS reagent with potassium persulfate incubated in the dark at room temperature for 16 h. $ABTS^{+\bullet}$ working solution was obtained by dilution of the concentrated solution until an absorbance value of 1 at 734 nm. 3 mL of working $ABTS^{+\bullet}$ radical solution was mixed with 300 µL of each extract and three controls were prepared with the addition of 300 µL of water. DPPH radical-scavenging activity of açaí extracts was evaluated according to the method described by Okada and Okada with some modifications [31,32]. 5 mL of extracts were incubated with 0.5 mL of 6.4×10^{-4} DPPH solution for 30 min in the dark at room temperature, and absorbance was determined at 517 nm. FRAP assay measures the antioxidant activity through reduction of ferric 2,4,6-tripyridyl-s-triazine (TPTZ) to a colored product via SET mechanism. FRAP reagent was prepared by mixing 25 mL of 0.3 M acetate buffer (pH 3.6) with 2.5 mL of 10 mM TPTZ (2,4,6-tripyridyl-s-triazine) and 2.5 mL of 20 mM $FeCl_3$. 2850 µL of FRAP reagent was mixed with 150 µL of each extract and the absorbance was measured at 593 nm after 30 min of reaction at room temperature. The assays were performed in triplicate and results were expressed as mg Trolox/g fruit.

2.3. Encapsulation of Açaí Extract with Highest Phenolic Content

2.3.1. Determination of Zein–Açaí Extract Solution Properties

The açaí extract to be encapsulated (Aç2) was selected according to the highest concentration of active compounds by means of highest polyphenolic content and antioxidant capacities. Aç2 was obtained following the same procedure as described in Section 2.2.1 and subsequently, this extract was concentrated to a final concentration of 0.4 g dried açaí mL^{-1} using a rotary evaporator. This concentrated extract was named $AÇ_{CC}$. Electrospinning solutions were prepared with 2 mL of $AÇ_{CC}$, 8 mL of ethanol and zein was added at different concentrations (16, 18 and 20% $w\ v^{-1}$). The mixtures were gently stirred at room temperature for 1 h until homogeneous solutions were obtained

(ZN16-AÇ$_{CC}$, ZN18-AÇ$_{CC, and}$ ZN20-AÇ$_{CC}$, respectively). Additionally, three control solutions of zein using 80% ethanolic solution were prepared at the same concentrations without the extract to study the effect of the incorporation of açaí extract on the properties of the polymer solutions (ZN16, ZN18, and ZN20, respectively).

The zein–açaí extract and control zein solutions were characterized by determination of viscosity and conductivity. Viscosity was evaluated using the SC4-18 spindle at a deformation rate of 79.2 s^{-1}. In addition, the conductivity was measured using a conductivity meter from 0.01 to 1000 mS cm^{-1}. Both studies were performed in triplicate at room temperature.

2.3.2. Electrospinning Process of the Zein–Açaí Extract Solutions

The encapsulation was carried out using electrospinning equipment (Spraybase®power SupplyUnit, Maynooth, Ireland) with a vertical standard configuration equipped with a capillary connected to a high-voltage source. The technique was carried out at room temperature and 40% relative humidity. Initially, the purpose was to reveal the optimal concentration of zein to obtain electrosprayed capsules. In this process, the capillary was located 10 cm from the collector plate using a voltage of 13 kV. Each zein–açaí extract solution was introduced in a 5 mL syringe, which was expelled by the capillary during the process with an injection flow of 0.15 mL h^{-1}. The first samples were collected on a slide for easy observation by optical microscopy, and to obtain an initial and fast appreciation of the morphology of the electrospun structures. Once the zein concentration was fixed, the electrospinning parameters were studied to be able to fix the best conditions to obtain electrosprayed particles through a homogeneous and stable process. Flow rate and distance between capillary and collector were studied as follows: Samples S1, S2, and S3 with 10 cm distance and flow rates 0.3, 0.4, and 0.5 mL h^{-1}, respectively; and samples S4, S5, and S6 with 12 cm as distance and 0.3, 0.4, and 0.5 mL h^{-1}, respectively.

2.4. Characterization of Electrosprayed Açaí-Containing Capsules

2.4.1. Morphological Analysis

The morphologies and size distribution of electrosprayed zein ZN/AÇ$_{EXT}$ capsules resulting from processing samples S1–S6 were observed. Electrosprayed structures were previously coated with gold–palladium and analyzed by scanning electron microscopy (ZeissEVO MA10SEM, Oberkochen, Germany) at 20 kV. Average particle diameters were analyzed with Image analyzer software (Image J v 1.37) (Bethesda, MD, USA).

2.4.2. Structural Analysis

Functional chemical group analysis of the samples was performed through spectrometer equipment Bruker Alpha (Ettlingen, Karlsruhe, Germany) with transmission spectra accessory mode. Concentrated açaí extract was previously lyophilized for further analysis (AÇ$_{EXT}$). Electrosprayed zein particles without açaí (ZNe) were also analyzed to study possible chemical interactions between both components. Pellets with samples and potassium bromide were prepared by pressure and the spectra were obtained in a range from 4000 to 400 cm^{-1} with a resolution of 2 cm^{-1} and 64 scans.

2.4.3. Phenolic Loading Capacity (LC) and Encapsulation Efficiency (EE)

Loading capacity (%LC) was measured as the mass ratio between the total polyphenols content (PT) determined in the ZN/AÇ$_{EXT}$ capsules and the theoretical phenolic content incorporated during the preparation of the capsules. 0.015 g of ZN/AÇ$_{EXT}$ was dissolved with 1.2 mL of 80% ethanol for PT measurement. Samples were filtered through a 0.22 µm filter and the supernatant was analyzed following the Folin–Ciocalteu method described previously. It is worth mentioning that zein was also analyzed and did not present phenolic content interference. The encapsulation efficiency (%EE) was determined following the procedure of Idham, Muhamad & Sarmidi (2012) with some modifications [33].

This test consisted of the determination of total polyphenols (PT) and surface polyphenols (PS) of the encapsulated extract. 0.015 g of ZN/AÇ$_{EXT}$ was weighed and mixed with 1.2 mL of distilled water in an Eppendorf tube to measure PS. On the other hand, 0.015 g of ZN/AÇ$_{EXT}$ was mixed with 1.2 mL of 80% ethanol for PT measurement. In both cases, the samples were vortexed for 30 s. Finally, each sample was filtered through a 0.22 μm filter and the supernatants were analyzed following the Folin–Ciocalteu method described previously. %EE was calculated following Equation (1):

$$\% \, EE = (PT - PS) \cdot PT^{-1} \cdot 100 \tag{1}$$

2.5. Thermal Properties of Zein–Açaí Capsules

2.5.1. Thermal Stability Test

Thermogravimetric analyses (TGA) of açaí fruit (AÇ), lyophilized hydroalcoholic açaí extract (AÇ$_{EXT}$), and encapsulated (ZN/AÇ$_{EXT}$) were carried out using a Mettler Toledo Gas Controller GC20 Stare System (Schwerzenbach, Switzerland) TGA/DSC. Samples were heated from 30 to 600 °C at 10 °C/min under nitrogen atmosphere.

2.5.2. Stability of Total Phenolic Content of AÇ by Encapsulation

Phenolic content of AÇ, AÇ$_{EXT}$ and ZN/AÇ$_{EXT}$ samples were determined before (at room temperature) and subsequently exposed to high-temperature conditions. These parameters were selected in order to simulate common heat-treatment processes that these active capsules could suffer when being used as nutraceuticals incorporated into food: (A) sterilization process: autoclaving at 121 °C and 15 psi for 15 min; and (B) baked process: 180–185 °C for 25 min.

2.6. In Vitro Digestion

Dispersions of açaí (AÇ), lyophilized hydroalcoholic açaí extract (AÇ$_{EXT}$), and encapsulated (ZN/AÇ$_{EXT}$) were subjected to in vitro digestibility assays to evaluate their bioaccesibilities through the analysis of phenolic content after a gastric and intestinal phase. In vitro gastric and intestinal phases were prepared based on the consensus of the protocol for simulating static digestion method described by the COST Action InfoGest [34]. Simulated gastric fluid (SGF) consisted of a stock gastric solution (6.9 mM KCl, 0.9 mM KH$_2$PO$_4$, 25 nm NaHCO$_3$, 47.2 mM NaCl, 0.12 mM MgCl$_2$*6H$_2$O, 0.5 mM 2NH$_4$CO$_3$), water, 0.3 M CaCl$_2$, 1.0 M HCl, lecithin, and pepsin. At this stage, the different dispersions were mixture with SGF in a 1:1 proportion and incubated at 37 °C for 90 min with continuous agitation at 200 rpm, where pH of the mixture was monitored and controlled to a value of 2 through the addition of 0.5 M HCl, and by using an automatic titrator (902 Titrando, Metrohm, USA). Simulated intestinal fluid (SIF) consisted of a stock gastric solution (6.8 mM KCl, 0.8 mM KH$_2$PO$_4$, 85 nM NaHCO$_3$, 38.4 mM NaCl, 0.33 mM MgCl$_2$.6H$_2$O), water, 0.3 M CaCl$_2$, 1.0 M NaOH, bile, pancreatin, and lipase. At this stage, the mixtures obtained from the gastric phase were mixed with SIF in a 1:1 proportion and incubated at 37 °C for 120 min with continuous agitation at 200 rpm. During this second stage, pH was monitored and maintained at pH 7 by adding 0.5 M NaOH. The resulting aliquots after each stage of in vitro digestion were collected and frozen at −18 °C, and phenolic content were evaluated following the Folin–Ciocalteu method. All experiments were carried out in duplicate.

2.7. Statistical Analysis

One-way analyses of variance were carried out. The software SPSS version 11.5 (SPSS Inc., Chicago, IL, USA) was used. Differences in pairs of mean values were evaluated by the Tukey b-test at a confidence interval of 95%. Data were represented as the average ± standard deviation.

3. Results and Discussion

3.1. Evaluation of Polyphenolic Content and Antioxidant Capacity of Açaí Extracts

Total phenolic content (TPC) and antioxidant capacities of açaí fruit extracts measured through Folin–Ciocalteu, TEAC, DPPH, and FRAP methods are listed in Table 1. The extraction of natural active compounds was highly influenced by the solubility of these compounds in the extractive solvent. In this case, hydroalcoholic extraction achieved the greatest performance by extracting the highest amount of phenolic and antioxidant compounds with different polarities related to both ethanol and water [35]. A better solvation of the compounds, as a result of the hydrogen bond interactions between the polar sites of the antioxidant molecules and both solvents, in comparison to whether each solvent is used separately for extraction, has been demonstrated [36]

Table 1. Polyphenolic content and antioxidant capacities results of açaí fruit extracts.

Extract	TPC (mg GAE/g)	TEAC (mg Trolox/g)	DPPH (mg Trolox/g)	FRAP (mg Trolox/g)
Aç1	23.8 [a] ± 0.2	26.1 [a] ± 0.9	14.1 [a] ± 0.1	32.8 ± 0.7
Aç2	43.4 [c] ± 0.2	130.1 [b] ± 0.9	62.7 [c] ± 0.5	68.4 [c] ± 0.3
Aç3	35.4 [b] ± 0.4	122.4 [b] ± 2.0	51.2 [b] ± 1.2	57.2 [b] ± 0.6

Letters a–c indicate significant differences among the extracts of the same method.

The Folin–Ciocalteu method is broadly used to measure the content of total phenolic compounds in plant products. It is based on the fact that the phenolic compounds react with the Folin–Ciocalteu reagent at basic pH, giving rise to a blue coloration that can be easily spectrophotometrically determined [37]. The resulting TPC values indicated that the type of solvent directly affected the total number polyphenols extracted. Other works have shown lower TPC values for açaí fruit pulp extracts, e.g., 31.7 ± 0.6 and 26.7 ± 0.5 mg GAE g^{-1} açaí when extractions were carried out by using 1% acetic acid aqueous solution and a solvent mixture acetone/water 70:30, respectively [36,38]. Açaí fruit has been revealed to be the fruit with the highest total polyphenolic content, followed by murtilla, calafate, and maqui, whose TPC values were 34.9, 33.9, and 31.2 mg GAE g^{-1}, respectively, according to the antioxidant database directed by Speisky et al. [39]. Nevertheless, a reliable comparison of TPC values between studies is very difficult to achieve because the lack of standardization of this assay and the extraction conditions can imply several orders of significant difference in detected phenols [30].

The highest phenolic content and antioxidant capacity of açaí fruit pulp was obtained for the Aç2 extract, except for the TEAC assay, where the hydroalcoholic extract Aç2 and aqueous Aç3 values did not present significant differences. This fact was due to the scavenging capacity of the compounds that were extracted and their affinity with the radical. This assay demonstrated that radical ABTS$^{+\bullet}$ presented great affinity for hydrophilic systems [40], principally aqueous systems, and very poor affinity for ethanolic extraction. The antioxidant capacity of extracts is the expression of the different phenolic components, which behave through different mechanisms of interactions with oxidative species. Therefore, it is necessary to perform more than one antioxidant method to reflect both lipophilic and hydrophilic capacities. TEAC assay is based on the generation of the cationic radical ABTS$^{+\bullet}$ with blue–greenish coloration, which is applicable to both hydrophilic antioxidant systems as to the lipophilic ones, while the DPPH assay uses the radical DPPH$^{\bullet}$ dissolved in organic medium such as ethanol and, therefore, is more applicable to hydrophobic systems [41]. Thus, in agreement with Floegel et al. (2001), which compared both antioxidant methods by measuring antioxidant activities of different groups of fruit, vegetables, and beverages, TEAC results were higher than DPPH values [42]. On the other hand, FRAP assay revealed that Aç2 presented the highest antioxidant activity measured through SET. Schauss et al. (2006) also demonstrated that antioxidant activity of this fruit measured through ORAC method resulted in the highest reported scavenging activity values for a fruit or berry [43].

3.2. Characterization of the Zein Extract Solutions

Zein concentration of electrospinning solutions was one of the main parameters that determined the morphology of the fibers or capsules because they directly affected the viscosity and the conductivity values. Table 2 shows viscosity and conductivity results of zein solutions. As expected, viscosity significantly increased as the concentration of zein increased because this parameter is closely related to the polymeric chain entanglement and intercalation [44].

Table 2. Viscosity and conductivity of polymer solutions with and without açaí extract used during electrospinning process.

Zein concentration	Viscosity (cP)		Conductivity (mS cm^{-1})	
(%, w/v)	ZN	ZN-AÇ$_{CC}$	ZN	ZN-AÇ$_{CC}$
16	18.2 [a,x] ± 0.3	20.6 [a,y] ± 0.2	694 [b,x] ± 4	750 [b,y] ± 6
18	21.9 [b,x] ± 0.2	26.7 [b,y] ± 0.6	685 [a,x] ± 1	717 [a,y] ± 1
20	27.2 [c,x] ± 0.9	30.0 [c,y] ± 0.3	655 [a,x] ± 3	716 [a,y] ± 7

Letters a–c indicate significant differences among the different zein concentration samples. Letters x, y indicate significant differences between sample with and without açaí extract at the same zein concentration.

The incorporation of the açaí extract significantly increased the viscosity owing to the presence of more solutes in the solution, which could increase entanglements of zein protein molecules [45]. When the highest concentration of zein (20% w/v) was processed, the chains remained entangled enough to resist the electric charges that tend to break the jet during the electrospinning process, resulting in fiber formation. On the other hand, when using 16% (w/v) zein solution, the electric charge broke the chains, resulting in dispersed droplets whose evaporation originated spherical particles or capsules (see Supplementary Material Figure S1), thereby being the zein concentration selected.

The conductivity values of the solutions slightly decreased with increasing zein concentration and significantly increased when the extract was added. Possibly, this fact could be related to the presence of polyphenols solved in the hydroalcoholic extract that enhanced the conductivity [46]. When conductivity increased, the difference in electric charges between the Taylor cone and the collector plate increased, which promoted capsule development during the formation of droplets [44].

3.3. Morphological Studies of Açaí-Containing Zein Capsules

Once the zein concentration was fixed, samples S1–S6 were processed through an electrospinning process to obtain particles with great homogeneity. The size and homogeneity are highly dependent on electrospinning process parameters. The particle size distribution plots and Scanning Electronic Microscopy (SEM) micrographs of the samples at different electrospinning encapsulation conditions are shown in Figure 1. SEM micrographs of the electrosprayed zein–açaí capsules demonstrated particles with cavities and similar shape to those found in previous works [47,48].

Samples S1, S2, S4, and S6 were found to have a smaller average particle size: 882, 924, 899, and 896 nm, respectively, without significant differences between them. According to Duque et al. (2013), when increasing the injection flow, there is less solvent evaporation time, so there can be agglomerations of droplets and an increase in the diameter of particles [44]. This fact occurred with samples S1 and S3, where, by increasing the flow from 0.3 to 0.5 mL h^{-1}, the particle average diameter increased from 0.88 to 1.13 μm. On the other hand, the distance between the tip of the capillary and the collector plate also influenced on the homogeneity and morphology of the capsules. The sample with the highest homogeneous particle size, based on the smaller standard deviation, turned out to be sample S2. Hence, the parameters selected were 0.4 mL h^{-1} flow rate and 10 cm height.

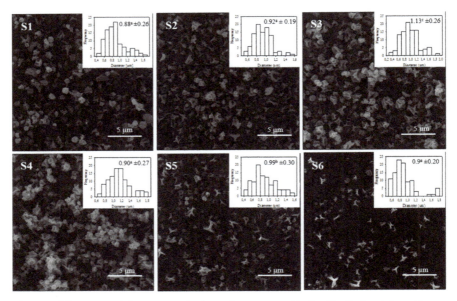

Figure 1. Histograms of electrosprayed zein–açaí structures with size distribution information as average and deviation (μm) and SEM micrographs of samples: S1, S2 and S3) 10 cm height and flow rates 0.3, 0.4, and 0.5 mL h^{-1}, respectively; and S4, S5, and S6) 12 cm height and flow rate 0.3, 0.4, and 0.5 mL h^{-1}, respectively. Values a, b, and c indicate significant differences between the samples, determined through a one-way analysis of variance (ANOVA) ($p < 0.05$). SD corresponds to the standard deviation.

3.4. Structural Characterization

The infrared spectra of electrosprayed zein particles (ZN$_e$), freeze-dried açaí extract (AÇ$_{EXT}$) and electrosprayed zein-containing açaí (ZN/AÇ$_{EXT}$) are displayed in Figure 2. Zein particles obtained through the electrospinning process presented characteristic peaks of zein protein at 3309 cm^{-1} which represented the N–H stretching of amide A, and peaks at 2966 and 2872 cm^{-1} derived from the stretching of the C–H aliphatic groups. The peak at 1655 cm^{-1} represented the stretching vibration of C=O group from amino acids and the bands at 1543 and 1450 cm^{-1} illustrated the flexion vibration of N–H and vibrational stretching of the C–N peptide bonds, respectively [49,50].

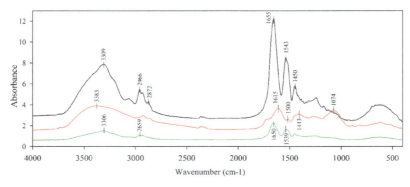

Figure 2. FTIR spectra of: electrosprayed zein particles, ZNe (black line); açaí extract, AÇ$_{EXT}$ (red line); and electrosprayed zein particles containing açaí, ZN/AÇ$_{EXT}$ (green line).

Lyophilized açaí extract spectra exhibited a similar pattern to other berries, such as maqui and murta [51,52]. A broad sign with peak at 3383 cm^{-1} could represent some hydroxyl groups O–H and aliphatic C–H from the polyphenolic compounds. The peak centered at 1615 cm^{-1} was assigned to the C=C vibrations from aromatic systems. The region between 1500 and 1340 cm^{-1} (centered at 1413 cm^{-1}) represented the deformation vibrations of phenolic O–H groups. A peak was distinguished in the region of 1150 and 1040 cm^{-1}, which was attributed to C–O stretching vibrations [53]. Although ZN/AÇ$_{EXT}$ presented similar bands to ZNe spectra, a certain displacement of these peaks was observed. This fact indicated the presence of a certain intermolecular interaction between zein protein and phenolics from açaí fruit extract. Zein protein, which contains mostly non-polar amino acids, favored chemical interactions with phenolic functional groups, increasing the protection of active compounds. Non-covalent hydrophobic interactions and hydrogen bonds were probably the main mechanisms of interaction between zein and polyphenols [54,55].

3.5. Loading Capacity (LC) and Encapsulation Efficiency (EE)

The encapsulation efficiency (EE) concept has given rise to different definitions in various works relating to compound encapsulation. In this study, EE was defined as the precise amount of active compounds that was actually protected in the capsule. On the other hand, the performance of the encapsulation process was distinguished as loading capacity (LC). Loading capacity (%LC) resulted in (98.6 ± 1.6)% value, since theoretical phenolic content of ZN/AÇ$_{EXT}$ capsules was 2104 mg gallic acid g^{-1}, and PT value after dissolution of ZN/AÇ$_{EXT}$ was 2075 mg gallic acid g^{-1} ZN/AÇ$_{EXT}$. This value justified the efficiency of this technology to encapsulate bioactive compounds without compromising its activity. Similar loading capacity values between 85% and 95% were observed to encapsulate other natural extracts by using the electrospinning technique, such as green tea extract in zein and carotenoids from tomato peel extract in gelatin [45,48].

Encapsulation efficiency value is essential to study the number of active compounds trapped in the capsule and the ability of the material to retain them [56]. EE value indicated (72.1 ± 1.7)% of phenolic content from açaí fruit extract was efficiently encapsulated. EE depends to a large extent on the affinity between the polymer matrix and active compounds. During the encapsulation process, the açaí fruit extract as core material was mixed with the substance of zein and the generated droplets were solidified by the evaporation of the ethanol and water [57]. Yao et al. (2016) also indicated that the EE would also be influenced by variations in the morphology that arise in the fibers or capsules due to the concentration of the solution and process conditions. In this study, the result indicated that more than 70% of the extract was efficiently trapped and distributed inside the capsule [58]. A similar value was found in the Flores et al. study of physical and storage properties of cranberry pulp encapsulated in whey protein by spray drying [59]. In addition, it is important to consider the methodology used to determine the efficiency, since EE results depend greatly on the methodology and, principally, the solvents used for the extraction of components.

3.6. Thermal Studies of Zein-Containing Açaí Extract Capsules

3.6.1. Thermal Stability Test

Figure 3 shows the thermogravimetric curves (TGA) (Figure 3A) and their respective derivatives (DTGA) (Figure 3B) of the samples. An initial stage of weight loss between 30 and 100 °C was observed for all the compounds, which indicated a loss of water and some volatile compounds. The thermogram of the electrosprayed zein particles presented a second degradation process between 270 and 450 °C, with a peak of maximum degradation at 332.8 °C. This second stage is attributed to the main degradation of the protein, causing changes in the structure due to the breakdown of low-energy intermolecular bonds that maintain their conformation [26,27,60]. Lyophilized açaí fruit extract presented an early degradation that started at approximately 100 °C, showing a maximum degradation at 162.5 °C. This extract was mainly composed of anthocyanins, flavonoids highly sensitive to temperature [16].

The anthocyanins of the extract were totally unprotected and exposed to degradation due to the increase in temperature. When the extract was encapsulated, the thermogram presented a similar degradation profile to zein. The incorporation of açaí fruit extract did not affect the protein stability. The degradation of açaí fruit was not exhibited because zein effectively protected the extract, delaying its degradation.

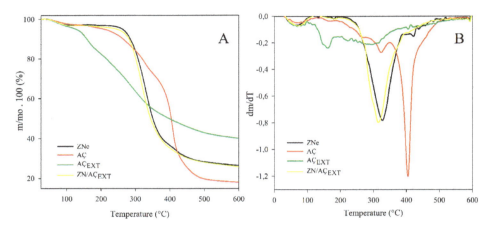

Figure 3. (**A**) Weight loss; and (**B**) derivative of the weight loss of electrosprayed zein (Zne), dried açaí fruit (AÇ), lyophilized açaí fruit extract (AÇ$_{EXT}$), electrosprayed zein capsules containing açaí (ZN/AÇ$_{EXT}$).

The main degradation of dehydrated fruit AÇ occurred at higher temperatures, approximately at 180–190 °C, and displayed two peaks of maximum degradation at 320.5 and 406 °C. This fact can be possibly explained because the dehydrated fruit contained a food matrix based on husk and pulp that could exert some protection to the active compounds, while the extract is a concentrated sample of antioxidants totally exposed to heating [12,61].

3.6.2. Thermal Protection of Açaí Phenolic Compounds Encapsulated in Zein

The thermal stability of the encapsulated extract was also evaluated by determining the loss of polyphenols when exposed to two thermal treatments of high-temperature processing: sterilization and baking. In addition, the stability of the phenolic content of dehydrated açaí fruit (AÇ) and lyophilized açaí extract (AÇ$_{EXT}$) was also analyzed. Figure 4 shows the loss of the phenolic content after each heat treatment.

In the case of the commercial dehydrated açaí fruit, a phenolic content reduction greater than 40% was displayed after both thermal processes without significant differences between both treatments. In the case of the encapsulated extract, phenolic content loss values were 5% and 20% approx. after sterilization and baking, respectively. Encapsulated açaí phenolic compounds presented a greater stability against both treatments compared to two other samples of açaí fruit, principally during a sterilization simulation process. Otherwise, during the baking process, the amount of phenolic loss correlated with the amount of phenols on the surface (shown in Section 3.5) which were the phenolic content available to degradation. This fact confirmed the protective effect of the encapsulation in zein by the electrospinning technique, which was clearly demonstrated when comparing with AÇ$_{EXT}$ phenolic loss. The sterilization process (121 °C) was a lesser influence than the baked (180 °C) over lyophilized açaí extract, exhibiting phenolic content reductions of 10 and 55%, respectively. Although sterilization did not cause a large degradation rate, in the case of baking, it turned out to be the sample that suffered the highest phenolic decrease, possibly due to higher temperature and longer exposure time. This fact is in agreement with AÇ$_{EXT}$ TGA thermogram (Figure 3B) that indicates AÇ$_{EXT}$ present an early degradation that starts at 100 °C and the maximum degradation temperature occurs at approximately 160 °C. Thus, the sterilization process clearly affects the phenolic content to a lesser extent than baking (180 °C), which occurs at a higher temperature than the maximum degradation. Other works have already shown that temperature is a determining factor in the degradation of polyphenols. Pacheco et al. studied the phytochemical composition and thermal stability of two commercial açaí species and concluded that the changes in antioxidant capacity during warming were highly related to the loss of anthocyanins because their polyphenols, such as phenolic acids and flavone glycosides, were not significantly altered [62]. The thermal degradation of anthocyanins can lead to a variety of species depending on the severity and nature of the heating. High temperature causes losses in the glycosidant sugar of the molecules and the opening of the ring, producing the so-called "colorless chalcones".

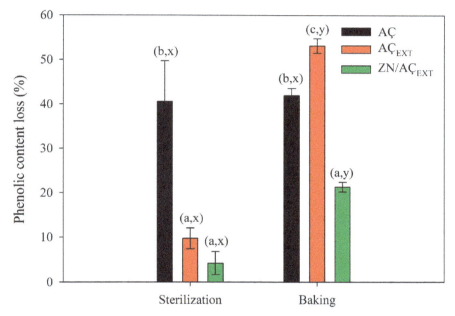

Figure 4. Loss of polyphenolic content (%) of the samples when subjected to thermal treatments. Values a, b, and c indicate significant differences between samples for the same thermal treatment. On the other hand, x and y indicate significant differences between thermal treatments for the same sample, determined through a one-way analysis of variance (ANOVA) ($p < 0.05$).

3.7. In Vitro Bioaccesibility Study

Phenolic content (PC) of AÇ, AÇ$_{EXT}$, and ZN/AÇ$_{EXT}$ were determined after each stage of in vitro digestion (gastric and intestinal stages) to evaluate phenolic content release within gastrointestinal tract (GI) and to assess bioaccessibility at the end of digestion process. Bioaccessibility can be defined as the fraction of a compound that is soluble in the gastrointestinal (GI) tract that is available for absorption, or the fraction of a compound released from its matrix in the GI tract [63]. Therefore, in this study, bioaccessibility was defined as phenolic content recovered from the intestinal phase after in vitro digestion. Table 3 shows phenolic content of gastric and intestinal stages per gram of each sample. Thus, significant differences to these values were because of their intrinsic phenolic content between samples. In general, polyphenols from açaí samples without encapsulation showed a moderate stability under gastric conditions since the phenolic content decreased between 20% and 35% with respect to intrinsic content (Table 3).

Table 3. Phenolic content as (mg gallic acid per gram) of açaí samples after in vitro digestive processes.

Sample	Gastric	Intestinal
AÇ	3044 [b,y] ± 32	988 [a,x] ± 108
AÇ$_{EXT}$	12,981 [b,z] ± 461	6462 [a,z] ± 402
ZN/AÇ$_{EXT}$	1486 [a,x] ± 148	2963 [b,y] ± 58

Letters a, b indicate significant differences among the different digestive stages of a sample. Letters x, y indicate significant differences between samples during the same digestive process.

As with during thermal analyses (Section 3.6.1), pulp matrix from açaí fruit generated some protection to açaí polyphenols during the gastric digestion phase. Other works have also shown flavonoid oligomers were degraded to smaller units at low pH values [64]. These results agreed with Gullón et al. (2015) who observed that the total phenolic recovery from pomegranate peel flour (35.8%) decreased after gastric digestion. Results after intestinal stage demonstrated phenolic content depended on açaí sample. Although AÇ polyphenols degraded to a lesser extent during the gastric phase, the dried fruit pulp showed the greatest PC decrease after intestinal digestion phase, remaining approx. 20% from initial phenolic content. On the other hand, although AÇ$_{EXT}$ suffered a higher reduction after gastric phase, the total PC loss was shorter than AÇ, with a PC reduction close to 60% with respect to non-digested samples. In general, these results suggested that several changes in phenolic compounds as a chemical structure modification, reduction of their solubility due to pH, and/or interaction with other compounds might have occurred during the duodenal stage [65,66]. On the other hand, encapsulated zein-containing açaí presented an interesting behavior because phenolic content values increased after both in vitro digestion processes. During the gastric phase, the phenolic components released from the zein capsules was approx. 60% with respect their intrinsic content, and this value increased after the intestinal stage. This fact is probably due to the breakage of the zein structures, thanks to the digestion of protein matrix, allowing the release of polyphenols. Gómez-Mascaraque et al. (2019) also revealed an increase to antioxidant capacity derived from catechin from zein and gelatin electrosprayed systems.

4. Conclusions

Polyphenolic content and antioxidant activity studies have demonstrated açaí to be the fruit with the highest content of active compounds. Studies of electrospinning encapsulation for the development of zein capsules containing hydroalcoholic açaí extract have also indicated the protective effect of these protein structures. This fact was certainly due to the positive chemical interactions observed through infrared spectroscopy between protein and active compounds from açaí extract. Zein was confirmed to be an adequate protein for the encapsulation of thermal sensitive active compounds by improving the thermal stability of polyphenols from açaí fruit when exposed to high-temperature treatments related to processed foods. After in vitro digestion processes, açaí polyphenols were also protected and diffused thanks to the breakage of zein protein electrosprayed capsules.

Supplementary Materials: The following are available online at http://www.mdpi.com/2076-3921/8/10/464/s1, Figure S1: Optical microscope images of electrosprayed samples.

Author Contributions: C.L.d.D. designed and supervised the research. L.G. optimized the conditions of zein solution and electrospinning equipment. C.P. executed experimental part related to characterization measurements and analysis of results. M.J.G. guided the execution of this work and C.A. carried out bioaccessibility studies. All the authors collaborated with the writing.

Acknowledgments: This work was supported by CONICYT through the Project Fondecyt Regular 1170624 and "Programa de Financiamiento Basal para Centros Científicos y Tecnológicos de Excelencia" (Project FB0807).

Conflicts of Interest: The authors declare no conflict of interest.

Abbreviations

TPC	Total Phenolic content
AÇ	dried açaí pulp
Aç1, Aç2, Aç3	açaí extracts under ethanol, ethanol 50% and water, respectively (3.3 mg açaí mL^{-1}).
AÇ$_{CC}$	concentrated açaí extract (0.4 g açaí mL^{-1})
AÇ$_{EXT}$	lyophilized açaí extract
ZN16, ZN18 and ZN20	zein solutions at 16, 18 and 20% (w/v)
ZN-AÇ$_{CC}$	zein solutions containing açaí concentrated extract
ZN/AÇ$_{EXT}$	electrosprayed zein capsules containing açaí extract
ZNe	electrosprayed zein capsules
TEAC	Trolox Equivalent Antioxidant Capacity
DPPH	2,2-diphenyl-1-picrylhydrazil
FRAP	Ferric Reducing Antioxidant Power
LC	loading capacity
EE	encapsulation efficiency

References

1. Farías, G.; Guzmán, L.; Delgado, C.; Maccioni, R. Nutraceuticals: A novel concept in prevention and treatment of alzheimer's disease and related disorders. *J. Alzheimer's Dis.* **2014**, *42*, 357–367. [CrossRef]
2. Rao, K. Safety assessment of nutraceuticals. *Biol. Eng. Med. Sci. Rep.* **2017**, *3*, 70–72. [CrossRef]
3. Li, A.N.; Li, S.; Zhang, Y.J.; Xu, X.R.; Chen, Y.M.; Li, H.B. Resources and Biological Activities of Natural Polyphenols. *Nutrients* **2014**, *6*, 6020–6047. [CrossRef] [PubMed]
4. Hogan, S.; Chung, H.; Zhang, L.; Li, J.; Lee, Y.; Dai, Y.; Zhou, K. Antiproliferative and antioxidant properties of anthocyanin-rich extract from açai. *Food Chem.* **2010**, *118*, 208–214. [CrossRef]
5. Kang, J.; Thakali, K.M.; Xie, C.; Kondo, M.; Tong, Y.; Ou, B.; Jensen, G.; Medina, M.B.; Schauss, A.G.; Wu, X. Bioactivities of açaí (*Euterpe precatoria* Mart.) fruit pulp, superior antioxidant and anti-inflammatory properties to *Euterpe oleracea* Mart. *Food Chem.* **2012**, *133*, 671–677. [CrossRef]
6. Soares, E.R.; Monteiro, E.B.; De Bem, G.F.; Inada, K.O.; Torres, A.G.; Perrone, D.; Soulage, C.O.; Monteiro, M.C.; Resende, A.C.; Moura-Nunes, N.; et al. Up-regulation of Nrf2-antioxidant signaling by Açaí (*Euterpe oleracea* Mart.) extract prevents oxidative stress in human endothelial cells. *J. Funct. Foods* **2017**, *37*, 107–115. [CrossRef]
7. Poulose, S.M.; Fisher, D.R.; Bielinski, D.F.; Gomes, S.M.; Rimando, A.M.; Schauss, A.G.; Shukitt-Hale, B. Restoration of stressor-induced calcium dysregulation and autophagy inhibition by polyphenol-rich açaí (*Euterpe* spp.) fruit pulp extracts in rodent brain cells in vitro. *Nutrition* **2014**, *30*, 853–862. [CrossRef]
8. Poulose, S.M.; Fisher, D.R.; Larson, J.; Bielinski, D.F.; Rimando, A.M.; Carey, A.N.; Schauss, A.G.; Shukitt-Hale, B. Attenuation of inflammatory stress signaling by açaí fruit pulp (*Euterpe oleracea* Mart.) extracts in BV-2 mouse microglial cells. *J. Agric. Food Chem.* **2012**, *60*, 1084–1093. [CrossRef]
9. Carey, A.N.; Miller, M.G.; Fisher, D.R.; Bielinski, D.F.; Gilman, C.K.; Poulose, S.M.; Shukitt-Hale, B. Dietary supplementation with the polyphenol-rich acai pulps (*Euterpe oleracea* Mart. And *Euterpe precatoria* Mart.) improves cognition in aged rats and attenuates inflammatory signaling in BV-2 microglial cells. *Nutr. Neurosci.* **2017**, *20*, 238–245. [CrossRef] [PubMed]
10. Xie, C.; Kang, J.; Burris, R.; Ferguson, M.E.; Schauss, A.G.; Nagarajan, S.; Wu, X. Açaí juice attenuates atherosclerosis in ApoE deficient mice through antioxidant and anti-inflammatory activities. *Atherosclerosis* **2011**, *216*, 327–333. [CrossRef] [PubMed]
11. Odendaal, A.Y.; Schauss, A.G. Chapter 18 -Potent Antioxidant and Anti-Inflammatory Flavonoids in the Nutrient-Rich Amazonian Palm Fruit, Açaí (*Euterpe* spp.). In *Polyphenols in Human Health and Disease*; Academic Press: San Diego, CA, USA, 2014; Volume 1, pp. 219–239.
12. Xie, C.; Kang, J.; Li, Z.; Schauss, A.G.; Badger, T.M.; Nagarajan, S.; Wu, T.; Wu, X. The açaí flavonoid velutin is a potent anti-inflammatory agent: Blockade of LPS-mediated TNF-α and IL-6 production through inhibiting NF-κB activation and MAPK pathway. *J. Nutr. Biochem.* **2012**, *23*, 1184–1191. [CrossRef] [PubMed]
13. Carvalho, A.V.; Ferreira da Silveira, T.; Mattietto, R.A.; Padilha de Oliveira, M.D.; Godoy, H.T. Chemical composition and antioxidant capacity of açai (*Euterpe oleracea*) genotypes and commercial pulps. *J. Sci. Food Agric.* **2017**, *97*, 1467–1474. [CrossRef] [PubMed]

14. Yamaguchi, K.K.D.L.; Pereira, L.F.R.; Lamarão, C.V.; Lima, E.S.; Da Veiga-Junior, V.F. Amazon acai: Chemistry and biological activities: A review. *Food Chem.* **2015**, *179*, 137–151. [CrossRef] [PubMed]
15. Schauss, A.G.; Wu, X.; Prior, R.L.; Ou, B.; Patel, D.; Huang, D.; Kababick, J.P. Phytochemical and Nutrient Composition of the Freeze-Dried Amazonian Palm Berry, *Euterpe oleraceae* Mart. (Acai). *J. Agric. Food Chem.* **2006**, *54*, 8598–8603. [CrossRef]
16. Ekici, L.; Simsek, Z.; Ozturk, I.; Sagdic, O.; Yetim, H. Effects of temperature, time, and pH on the stability of anthocyanin extracts: Prediction of total anthocyanin content using nonlinear models. *Food Anal. Methods* **2014**, *7*, 1328–2336. [CrossRef]
17. Costa, H.C.; Silva, D.O.; Vieira, L.G.M. Physical properties of açai-berry pulp and kinetics study of its anthocyanin thermal degradation. *J. Food Eng.* **2018**, *239*, 104–113. [CrossRef]
18. Espín, J.C.; García-Conesa, M.T.; Tomás-Barberán, F.A. Nutraceuticals: Facts and fiction. *Phytochemistry* **2007**, *68*, 2986–3008. [CrossRef]
19. Rein, M.J.; Renouf, M.; Cruz-Hernandez, C.; Actis-Goretta, L.; Thakkar, S.K.; Pinto, M.D.S. Bioavailability of bioactive food compounds: A challenging journey to bioefficacy. *Br. J. Clin. Pharmacol.* **2013**, *75*, 588–602. [CrossRef]
20. Munin, A.; Edwards-Lévy, F. Encapsulation of Natural Polyphenolic Compounds; a Review. *Pharmaceutics* **2011**, *3*, 793–829. [CrossRef]
21. Nedovic, V.; Kalusevic, A.; Manojlovic, V.; Levic, S.; Bugarski, B. An overview of encapsulation technologies for food applications. *Procedia Food Sci.* **2011**, *1*, 1806–1815. [CrossRef]
22. Bhushani, J.A.; Anandharamakrishnan, C. Electrospinning and electrospraying techniques: Potential food based applications. *Trends Food Sci. Technol.* **2014**, *38*, 21–33. [CrossRef]
23. Weiss, J.; Kanjanapongkul, K.; Wongsasulak, S.; Yoovidhya, T. Electrospun Fibers: Fabrication, Functionalities and Potential Food Industry Applications. In *Nanotechnology in the Food, Beverage and Nutraceutical Industries*; Woodhead Publishing Limited: Sawston, Cambridge, UK, 2012.
24. Jacobsen, C.; García, P.; Mendes, A.; Mateiu, R.; Chronakis, I. Use of electrohydrodynamic processing for encapsulation of sensitive bioactive compounds and applications in food. *Annu. Rev. Food Sci. Technol.* **2018**, *9*, 525–549. [CrossRef] [PubMed]
25. Paliwal, R.; Palakurthi, S. Zein in controlled drug delivery and tissue engineering. *J. Control. Release* **2014**, *189*, 108–122. [CrossRef] [PubMed]
26. Müller, V.; Piai, J.F.; Fajardo, A.R.; Fávaro, S.L.; Rubira, A.F.; Muniz, E.C. Preparation and Characterization of Zein and Zein-Chitosan Microspheres with Great Prospective of Application in Controlled Drug Release. *J. Nanomater.* **2011**, *2011*, 1–6. [CrossRef]
27. Patel, A.R.; Velikov, K.P. Zein as a source of functional colloidal nano- and microstructures. *Curr. Opin. Colloid Interface Sci.* **2014**, *19*, 450–458. [CrossRef]
28. De Dicastillo, C.L.; Bustos, F.; Valenzuela, X.; López-Carballo, G.; Vilariño, J.M.; Galotto, M.J. Chilean berry Ugni molinae Turcz. fruit and leaves extracts with interesting antioxidant, antimicrobial and tyrosinase inhibitory properties. *Food Res. Int.* **2017**, *102*, 119–128. [CrossRef] [PubMed]
29. Lamuela-Raventós, R.M. Folin-Ciocalteu Method for the Measurement of Total Phenolic Content and Antioxidant Capacity, Chapter 6. In *Measurement of Antioxidant Activity & Capacity*; Wiley Publishing: Hoboken, NJ, USA, 2017; pp. 107–115.
30. Prior, R.L.; Wu, X.; Schaich, K. Standardized Methods for the Determination of Antioxidant Capacity and Phenolics in Foods and Dietary Supplements. *J. Agric. Food Chem.* **2005**, *53*, 4290–4302. [CrossRef] [PubMed]
31. De Dicastillo, C.L.; Navarro, R.; Guarda, A.; Galotto, M.J. Development of Biocomposites with Antioxidant Activity Based on Red Onion Extract and Acetate Cellulose. *Antioxidants* **2015**, *4*, 533–547. [CrossRef] [PubMed]
32. Okada, Y.; Okada, M. Scavenging Effect of Water Soluble Proteins in Broad Beans on Free Radicals and Active Oxygen Species. *J. Agric. Food Chem.* **1998**, *46*, 401–406. [CrossRef]
33. Idham, Z.; Muhamad, I.; Sarmidi, M. Degradation kinetics and color stability of spray-dried encapsulated anthocyanins from Hibiscus sabdariffa. *J. Food Process Eng.* **2012**, *35*, 522–542. [CrossRef]
34. Minekus, M.; Alminger, M.; Alvito, P.; Balance, S.; Bohn, T.; Bourlieu, C.; Carrière, F.; Boutrou, R.; Corredig, M.; Dupont, D.; et al. A standardized static in vitro digestion method suitable for food—An international consensus. *Food Funct.* **2014**, *5*, 1113–1124. [CrossRef] [PubMed]
35. Alothman, M.; Bhat, R.; Karim, A. Antioxidant capacity and phenolic content of selected tropical fruits from Malaysia, extracted with different solvents. *Food Chem.* **2009**, *115*, 785–788. [CrossRef]

36. Boeing, J.S.; Barizão, É.O.; E Silva, B.C.; Montanher, P.F.; Almeida, V.D.C.; Visentainer, J.V. Evaluation of solvent effect on the extraction of phenolic compounds and antioxidant capacities from the berries: Application of principal component analysis. *Chem. Cent. J.* **2014**, *8*, 48. [CrossRef] [PubMed]

37. Kuskoski, E.M.; Asuero, A.G.; Troncoso, A.M.; Mancini-Filho, J.; Fett, R. Aplicación de diversos métodos químicos para determinar actividad antioxidante en pulpa de frutos. *Food Sci. Technol.* **2005**, *25*, 726–732. [CrossRef]

38. Rojano, B.; Zapata, I.; Alzate, A.; Mosquera, A.; Cortés, F.; Gamboa, L. Polifenoles y actividad antioxidante del fruto liofilizado de palma Naidi (*Açaí colombiano*) (*Euterpe oleracea Mart.*). *Rev. Fac. Nac. Agron. Medellín* **2011**, *64*, 6213–6220.

39. Fuentes, J.; Sandoval-Acuña, C.; Speisky, H.; López-Alarcón, C.; Gómez, M. First Web-Based Database on Total Phenolics and Oxygen Radical Absorbance Capacity (ORAC) of Fruits Produced and Consumed within the South Andes Region of South America. *J. Agric. Food Chem.* **2012**, *60*, 8851–8859.

40. Guan, X.; Jin, S.; Li, S.; Huang, K.; Liu, J. Antioxidant capacity of oat (*Avena Sativa* L.) bran oil extracted by subcritical butane extraction. *Molecules* **2018**, *23*, 1546. [CrossRef] [PubMed]

41. Thaipong, K.; Boonprakob, U.; Crosby, K.; Cisneros, L.; Hawkins, D. Comparison of ABTS, DPPH, FRAP, and ORAC assays for estimating antioxidant activity from guava fruit extracts. *J. Food Compos. Anal.* **2006**, *19*, 669–675. [CrossRef]

42. Floegel, A.; Kim, D.O.; Chung, S.J.; Koo, S.I.; Chun, O.K. Comparison of ABTS/DPPH assays to measure antioxidant capacity in popular antioxidant-rich US foods. *J. Food Compos. Anal.* **2011**, *24*, 1043–1048. [CrossRef]

43. Schauss, A.G.; Wu, X.; Prior, R.L.; Ou, B.; Huang, D.; Owens, J.; Agarwal, A.; Jensen, G.S.; Hart, A.N.; Shanbrom, E. Antioxidant Capacity and Other Bioactivities of the Freeze-Dried Amazonian Palm Berry, *Euterpe oleraceae* Mart. (Acai). *J. Agric. Food Chem.* **2006**, *54*, 8604–8610. [CrossRef]

44. Duque, L.; Rodriguez, L.; López, M. Electrospinning: The Nanofibers Age. *Rev. Iberoam. Polímeros* **2013**, *14*, 10–27.

45. Horuz, T. İnanç; Belibağlı, K.B.; Belibağli, K.B. Nanoencapsulation by electrospinning to improve stability and water solubility of carotenoids extracted from tomato peels. *Food Chem.* **2018**, *268*, 86–93. [CrossRef] [PubMed]

46. Darra, N.; Grimi, N.; Vorobiev, E.; Louka, N.; Maroun, R. Extraction of polyphenols from red grape pomace assisted by pulsed ohmic heating. *Food Bioprocess Technol.* **2013**, *6*, 1281–1289. [CrossRef]

47. Torres-Giner, S.; Gimenez, E.; Lagaron, J. Characterization of the morphology and thermal properties of Zein Prolamine nanostructures obtained by electrospinning. *Food Hydrocoll.* **2008**, *22*, 601–614. [CrossRef]

48. Gómez-Estaca, J.; Balaguer, M.; Gavara, R.; Hernández-Muñoz, P. Formation of zein nanoparticles by electrohydrodynamic atomization: Effect of the main processing variables and suitability for encapsulating the food coloring and active ingredient curcumin. *Food Hydrocoll.* **2012**, *28*, 82–91. [CrossRef]

49. Altan, A.; Aytac, Z.; Uyar, T. Carvacrol loaded electrospun fibrous films from zein and poly (lactic acid) for active food packaging. *Food Hydrocoll.* **2018**, *81*, 48–59. [CrossRef]

50. Arcan, I.; Yemenicioğlu, A. Development of flexible zein–wax composite and zein–fatty acid blend films for controlled release of lysozyme. *Food Res. Int.* **2013**, *51*, 208–216. [CrossRef]

51. Kumar, J.K.; Devi-Prasad, A.G. Identification and comparison of biomolecules in medicinal plants of Tephrosia tinctoria and Atylosia albicans by using FTIR. *Rom. J. Biophys.* **2011**, *21*, 63–71.

52. De Dicastillo, C.L.; Rodríguez, F.; Guarda, A.; Galotto, M.J. Antioxidant films based on cross-linked methyl cellulose and native Chilean berry for food packaging applications. *Carbohydr. Polym.* **2016**, *136*, 1052–1060. [CrossRef]

53. Lam, H.; Proctor, A.; Howard, L.; Cho, M. Rapid fruit extracts antioxidant capacity determination by fourier transform infrared spectroscopy. *Food Chem. Toxicol.* **2005**, *70*, 545–549. [CrossRef]

54. Gómez-Mascaraque, L.G.; Hernández-Rojas, M.; Tarancón, P.; Tenon, M.; Feuillère, N.; Ruiz, J.F.V.; Fiszman, S.; López-Rubio, A. Impact of microencapsulation within electrosprayed proteins on the formulation of green tea extract-enriched biscuits. *LWT* **2017**, *81*, 77–86. [CrossRef]

55. Jakobek, L. Interactions of polyphenols with carbohydrates, lipids and proteins. *Food Chem.* **2015**, *175*, 556–567. [CrossRef] [PubMed]

56. López, A.; Deladino, L.; Navarro, A.; Martino, M. Encapsulación de compuestos bioactivos con alginatos para la industria de alimentos. *Limentech Cienc. Y Tecnol. Aliment.* **2012**, *10*, 18–27.

57. Jaworek, A.; Sobczyk, A.; Krupa, A. Electrospray application to powder production and surface coating. *J. Aerosol Sci.* **2018**, *125*, 57–92. [CrossRef]
58. Yao, Z.-C.; Chang, M.-W.; Ahmad, Z.; Li, J.-S. Encapsulation of rose hip seed oil into fibrous zein films for ambient and on demand food preservation via coaxial electrospinning. *J. Food Eng.* **2016**, *191*, 115–123. [CrossRef]
59. Flores, F.P.; Singh, R.K.; Kong, F. Physical and storage properties of spray-dried blueberry pomace extract with whey protein isolate as wall material. *J. Food Eng.* **2014**, *137*, 1–6. [CrossRef]
60. Brahatheeswaran, D.; Mathew, A.; Aswathy, R.G.; Nagaoka, Y.; Venugopal, K.; Yoshida, Y.; Maekawa, T.; Sakthikumar, D. Hybrid fluorescent curcumin loaded zein electrospun nanofibrous scaffold for biomedical applications. *Biomed. Mater.* **2012**, *7*, 045001. [CrossRef]
61. Rawson, A.; Patras, A.; Tiwari, B.; Noci, F.; Koutchma, T.; Brunton, N.; Tiwari, B. Effect of thermal and non thermal processing technologies on the bioactive content of exotic fruits and their products: Review of recent advances. *Food Res. Int.* **2011**, *44*, 1875–1887. [CrossRef]
62. Pacheco-Palencia, L.A.; Duncan, C.E.; Talcott, S.T. Phytochemical composition and thermal stability of two commercial açai species, *Euterpe oleracea* and *Euterpe precatoria*. *Food Chem.* **2009**, *115*, 1199–1205. [CrossRef]
63. Cardoso, C.; Afonso, C.; Lourenço, H.; Costa, S.; Nunes, M.L. Bioaccessibility assessment methodologies and their consequences for the risk–benefit evaluation of food. *Trends Food Sci. Technol.* **2015**, *41*, 5–23. [CrossRef]
64. Tarko, T.; Duda-Chodak, A.; Zajac, N. Digestion and absorption of phenolic compounds assessed by in vitro simulation methods. A review. *Rocz. Państwowego Zakładu Hig.* **2013**, *64*, 79–84.
65. Gullón, B.; Pintado, M.E.; Fernández-López, J.; Pérez-Álvarez, J.A.; Viuda-Martos, M. In vitro gastrointestinal digestion of pomegranate peel (*Punica granatum*) flour obtained from co-products: Changes in the antioxidant potential and bioactive compounds stability. *J. Funct. Foods* **2015**, *19*, 617–628. [CrossRef]
66. Gómez-Mascaraque, L.G.; Tordera, F.; Fabra, M.J.; Martínez-Sanz, M.; López-Rubio, A. Coaxial electrospraying of biopolymers as a strategy to improve protection of bioactive food ingredients. *Innov. Food Sci. Emerg. Technol.* **2019**, *51*, 2–11. [CrossRef]

© 2019 by the authors. Licensee MDPI, Basel, Switzerland. This article is an open access article distributed under the terms and conditions of the Creative Commons Attribution (CC BY) license (http://creativecommons.org/licenses/by/4.0/).

Article

Influence of Bioactive Compounds Incorporated in a Nanoemulsion as Coating on Avocado Fruits (*Persea americana*) during Postharvest Storage: Antioxidant Activity, Physicochemical Changes and Structural Evaluation

Antonio de Jesus Cenobio-Galindo [1], Juan Ocampo-López [1], Abigail Reyes-Munguía [2], María Luisa Carrillo-Inungaray [2], Maria Cawood [3], Gabriela Medina-Pérez [1,4], Fabián Fernández-Luqueño [5] and Rafael Germán Campos-Montiel [1,*]

1. Instituto de Ciencias Agropecuarias, Universidad Autónoma del Estado de Hidalgo, Av. Rancho Universitario s/n Km. 1., Tulancingo C.P. 43600, Hidalgo, Mexico; anjec_hs@hotmail.com (A.d.J.C.-G.); jocampo@uaeh.edu.mx (J.O.-L.); gamepe@yahoo.com (G.M.-P.)
2. Unidad Académica Multidisciplinaria Zona Huasteca, Universidad Autónoma de San Luis Potosí, Romualdo del campo No. 501, Fracc. Rafael Curiel, C.P. Ciudad Valles, SLP C.P. 79060, Mexico; aby1974@hotmail.com (A.R.-M.); maluisa@uaslp.mx (M.L.C.-I.)
3. Department of Plant Sciences, University of the Free State, Bloemfontein 9301, South Africa; CawoodME@ufs.ac.za
4. Transdisciplinary Doctoral Program in Scientific and Technological Development for the Society, Cinvestav-Zacatenco, Mexico City C. P. 07360, Mexico
5. Sustainability of Natural Resources and Energy Program, Cinvestav-Saltillo, Coahuila de Zaragoza C. P. 25900, Mexico; fabian.fernandez@cinvestav.edu.mx
* Correspondence: ragcamposm@gmail.com; Tel.: +52-7717-172-000 (ext. 2422)

Received: 3 October 2019; Accepted: 17 October 2019; Published: 21 October 2019

Abstract: The objective of the present study was to determine the effect of the application of a nanoemulsion made of orange essential oil and *Opuntia oligacantha* extract on avocado quality during postharvest. The nanoemulsion was applied as a coating in whole fruits, and the following treatments were assessed: concentrated nanoemulsion (CN), 50% nanoemulsion (N50), 25% nanoemulsion (N25) and control (C). Weight loss, firmness, polyphenol oxidase (PPO) activity, total soluble solids, pH, external and internal colour, total phenols, total flavonoids, antioxidant activity by 2,2′-Azino-bis(3-ethylbenzothiazoline-6-sulfonic acid) (ABTS) and 2,2-diphenyl-1-picrylhydrazyl (DPPH), while the structural evaluation of the epicarp was assessed through histological cuts. Significant differences were found ($p < 0.05$) among the treatments in all the response variables. The best results were with the N50 and N25 treatments for firmness and weight loss, finding that the activity of the PPO was diminished, and a delay in the darkening was observed in the coated fruits. Furthermore, the nanoemulsion treatments maintained the total phenol and total flavonoid contents and potentiated antioxidant activity at 60 days. This histological study showed that the nanoemulsion has a delaying effect on the maturation of the epicarp. The results indicate that using this nanoemulsion as a coating is an effective alternative to improve the postharvest life of avocado.

Keywords: antioxidants; encapsulation; orange essential oil; xoconostle; maturation

1. Introduction

Avocado (*Persea americana*) is native to central and southern Mexico but consumed globally [1]. The main problem for avocado growers, marketers and consumers is the short shelf life of avocados.

Therefore, it is necessary to develop natural environmentally friendly products to improve the shelf life of avocados [2].

Xoconostle (*Opuntia oligacantha*) is a fruit from the central area of Mexico. Several reports have indicated that this fruit contains bioactive compounds such as phenolic compounds with antioxidant and antimicrobial properties [3]. However, these bioactive compounds are affected by pH, oxygen, and exposure to light or temperature, limiting their use in the food industry [4]. Cenobio-Galindo et al. [5] encapsulated *Opuntia oligacantha* extract and then incorporated it into starch films, finding that the encapsulated bioactive compounds were protected and available to act when released.

Orange essential oil (*Citrus sinensis*) is one of the most used essential oils in various industries, such as the food industry. It contains volatile monoterpenes with limonene identified as the major constituent, but its low solubility, volatility and instability during processing and storage makes it difficult to use. Therefore, it is necessary to incorporate this oil in a stable system that allows it to maintain its characteristics [6,7].

Nanoemulsions are encapsulation systems with characteristics of small particles ($r < 100$ nm), allowing excellent stability against separation and aggregation [8] and they also exhibit increased bioavailability of the encapsulated ingredients [9]. A potential advantage of these dispersions is that they have high compatibility with the components of foods and can even cross biological membranes without difficulty [8,10]. Nanoemulsions have been used mainly in the development of some foods and beverages to contribute to the stability of the foods that contain them [11]. Zambrano-Zaragoza et al. [12] studied the effect of a coating with α-tocopherol and cactus mucilage that were applied on apples and found that the nanoemulsion helped to maintain the firmness of the apples and decreased the activity of certain enzymes in the fruits. Oh et al. [13] developed a nanoemulsion with lemongrass oil and chitosan, finding an antimicrobial effect that contributes to colour retention and to maintaining the antioxidant activity of fruits. The objective of this study was to evaluate the effect of the application of a nanoemulsion based on orange essential oil and xoconostle extract on the physiological parameters, bioactive compounds and antioxidant activity of avocado fruits (*Persea americana*) during postharvest.

2. Materials and Methods

2.1. Materials

"Hass" avocado fruits were obtained from the production area of Uruapan, Michoacán, Mexico (19 ° 25′16 "N 102 ° 03′47" W). The xoconostle fruit variety *Opuntia oligacantha* var. Ulapa were from the municipality of Tetepango, Hidalgo, Mexico. The fruits that were used were in a state of physiological maturity.

The following reagents were used in this study: orange essential oil (Reasol, Mexico City, Mexico), soy lecithin (Reasol, Mexico City, Mexico), food-grade mineral oil (UltraSource, Kansas City, USA), catechol (Sigma-Aldrich, St. Louis, Missouri, USA), trichloroacetic acid (TCA) (Sigma-Aldrich, St. Louis, Missouri, USA), formaldehyde (analytical grade reagent), glacial acetic acid (analytical grade reagent), ethanol (analytical grade reagent), methanol (analytical grade reagent), paraplast (Sigma-Aldrich, St. Louis, Missouri, USA), Folin-Ciocalteu reagent (Sigma-Aldrich, St. Louis, Missouri, USA), nitrite sodium (analytical grade reagent), aluminium trichloride (Meyer, Mexico City, Mexico), sodium hydroxide (analytical grade reagent), 2,2-Azino-bis (3-ethylbenzthiazoline-6-sulfonic acid) (ABTS) (Sigma-Aldrich, St. Louis, Missouri, USA), potassium persulfate (analytical grade reagent), and 2,2-diphenyl-1-picrylhydrazyl (DPPH) (Sigma-Aldrich, St. Louis, Missouri, USA).

2.2. Preparation of the Nanoemulsion

The nanoemulsion (W/O) was prepared with 70% orange oil (according to Hashtjin and Abbasi [7] the composition of orange essential oil is limonene (94%), myrcene (2%), linalool (0.5%) and some others), 10% xoconostle extract (according to Cenobio-Galindo et al. [5]. The extract has various polyphenols, such as rutin, ferulic acid, quercetin, hydroxybenzoic acid, apigenin, caffeic acid, kaempferol) and 20% soy lecithin. This mixture was sonicated (Sonics Vibra-cell, Connecticut, USA) with a 6-mm-diameter probe for 20 intervals of 50 s of sonication with rest periods of 10 s using 80% amplitude with a frequency of 20 kHz. The obtained nanoemulsion was stored in a refrigerator at 6 °C and protected from light until analysis and use.

Determination of the Particle Size and Zeta Potential (ζ)

To verify that the dispersion made was a nanoemulsion, the droplet size was determined, in addition to the stability by means of the zeta potential. The droplet size and zeta potential (ζ) of the system were determined using the methodology developed by Zambrano-Zaragoza et al. [12]; the assessments were made in a Zetasizer Nano-ZS particle size analyzer (Malvern Instruments, Worcestershire, UK), equipped with laser light scattering at an angle of 90°, and the tests were performed in triplicate.

2.3. Application of the Nanoemulsion in Avocado "Hass"

Mature fruits homogeneous in weight and size were used and transported under refrigeration conditions (6 °C) to the laboratory located in the city of Tulancingo, Hidalgo, Mexico. The fruits were washed and sanitized with sodium hypochlorite (200 ppm) and were dried at room temperature. the nanoemulsion was applied by spraying the whole fruit. The fruits were coated with concentrated nanoemulsion (CN), nanoemulsion diluted 50% (N50), nanoemulsion 25% (N25) and a control (C) without the addition of nanoemulsion. Dilutions were made with food-grade mineral oil. The fruits were stored at 6 °C and were analyzed every 10 days until day 60.

2.4. Weight Loss

The mass was determined by weighing the fruits using a digital scale (OHAUS, Nueva Jersey, USA). The results are expressed as a percentage of weight loss and done in triplicate [14].

2.5. Firmness

The method developed by Maftoonazad and Ramaswamy [15] was used by means of a CT3 texture analyzer (Brookfield, Harlow, United Kingdom) adapted with a 5-mm conical strut probe. For the measurement, the epicarp was removed from both sides of the equatorial region. Ten repetitions were made per treatment and the results are expressed in Newtons (N).

2.6. pH and Total Soluble Solids

The pH was measured using a pH meter pH 211 microprocessor (Hanna, Rhode Island, USA). For the analysis, 10 g of mesocarp was homogenized with 90 mL of distilled water. The total soluble solids (TSS) was determined with a Pallete PR-101 refractometer (Atago, Washington, USA) and expressed in °Brix [16].

2.7. Determination of Polyphenol Oxidase Activity

The activity of polyphenol oxidase (PPO) was determined as reported by Vargas-Ortiz et al. [17]. For each treatment, 5 g of mesocarp was homogenized for 1 min in 15 mL of a 50 mM phosphate buffer (pH 6.5). The mixture centrifuged at 12000 rpm for 30 min at 4 °C and the supernatant used as the enzyme extract. Five hundred μL of 20 mM catechol was added as the substrate with 900 μL of 50 mM phosphate buffer (pH 6.5), and 100 μL of the enzyme extract. For the blank, 500 μL of 10% trichloroacetic acid (TCA) was added. The mixture was incubated for 20 min at 25 °C, and the reaction was stopped with the addition of 500 μL of 10% TCA. The absorbance was read at 410 nm using a spectrophotometer Jenway 6715 UV–Vis (USA). The PPO activity was reported as the increase in absorbance after the reaction time for 100 μL of extract.

2.8. Determination of Colour

The external and internal colour was determined using a CM-508d colourimeter (Minolta, Japan) to evaluate the parameters L* (lightness), a* (green to red) and b* (blue to yellow). Five measurements were made on each fruit, and 10 fruits were examined per treatment per day of analysis [15], for internal colour, fruits were cut longitudinally and the same measurements were made as for epicarp.

2.9. Bioactive Compounds and Antioxidant Activity

2.9.1. Extraction of Bioactive Compounds

The bioactive compounds of fruit were extracted according Vargas-Ortiz et al. [18]. Five grams of mesocarp was placed in a centrifuge tube, and 15 mL of ethanol/water solution (1:1) was added. The mixture was homogenized for 1 min at 4 °C and then was centrifuged at 12,000 rpm for 5 min at 4 °C. The supernatant was used for the determination of bioactive compounds and antioxidant activity.

2.9.2. Determination of Total Phenols

The total phenolic content was determined by using the Folin–Ciocalteu assay as described by Villa-Rodríguez et al. [19] with some modifications. One milliliter of the extract was mixed with 5 mL of diluted Folin–Ciocalteau reagent (1:10). After 6 min, 4 mL of Na_2CO_3 (20%) was added to the mixture, left for 2 h at room temperature and the absorbance against the reagent blank was determined at 760 nm with an UV-Visible spectrophotometer. Total phenolic content was expressed as gallic acid equivalents/100 g of fruit (wet base). All the assays were performed in triplicate.

2.9.3. Determination of Total Flavonoids

Total flavonoid content was measured by the aluminium chloride colorimetric assay using the method of Villa-Rodríguez et al. [19], with some modifications. An aliquot (1 mL) of extracts or standard solutions of quercetin was added to 4 mL of deionized water in a 10 mL flask. To the flask, 300 μL 5% $NaNO_2$ was added and after five minutes, 300 μL 10% $AlCl_3$. After another five minutes, 2 mL 1M NaOH was added and the volume was made up to 10 mL with deionized water. A blank was prepared in the same manner by using distilled water. The solution was mixed and absorbance was measured against the blank at 415 nm. The total flavonoid content was expressed as mg quercetin equivalents/100 g of fruit (wet base).

2.9.4. Determination of Antioxidant Activity by Inhibiting the DPPH Radical

A 2,2-diphenyl-1-picrylhydrazyl (DPPH) reagent was used to illustrate compounds with antioxidant activity [20]. A solution of DPPH 6.5×10^{-5} M in 80% methanol was prepared, maintaining stirring for 2 h in darkness. Then, 0.5 mL of sample was mixed with 2.5 mL of the DPPH solution, and the mixture was stirred. The absorbance of the mixture was immediately read at a wavelength of 515 nm. The mixture was left to react in darkness for 1 h. As a blank, 80% methanol was used. The results obtained are expressed in mg of ascorbic acid equivalents/100 g of fruit (wet base).

2.9.5. Determination of Antioxidant Activity by Inhibiting the ABTS Radical

Determination of antioxidant activity by inhibiting the radical 2,2-Azino-bis (3-ethylbenzthiazoline-6-sulfonic acid) (ABTS) was performed as described by Vargas-Ortiz et al. [18] with some modifications. A 10 mL solution of 7 mM ABTS was prepared and reacted with 10 mL of 2.45 mM $K_2S_2O_8$. The mixture was stirred for 16 h in a container in complete darkness. Afterwards the absorbance was measured in a spectrophotometer at 734 nm. The absorbance was adjusted with 20% ethanol to obtain a value of 0.7 ± 0.1. Two hundred μL of the sample was added to 2 mL of ABTS solution and allowed to react for 6 min and absorbance was measured at 734 nm. The results are expressed in mg of ascorbic acid equivalents /100 g of fruit (wet base).

2.10. Structural Evaluation of the Epicarp

Visually homogeneous fruits were selected and the fruit epicarp was cut into 1 cm² fragments of the equatorial part. Subsequently, according to the methodology developed by Hernández-Rivero et al. [21], the dissected material was fixed for 24 h in a mixture of formaldehyde, glacial acetic acid, 96% ethanol and distilled water at a ratio of 100:50:50:350. After this time, the tissue was washed with distilled water for 15 min, dehydrated, and infiltrated in an automatic tissue processor TP1020 (Leica, Germany) for 1 h in each of the following solutions: six sequential ethanol solutions at 60%, 70%, 80%, 90%, 96% and 100% and xylene. Then, the tissue was processed in two changes of paraplast solution for 2.5 h in each. The vegetal tissue was embedded in paraplast and cut transversely in 10 slices in a rotary microtome model 820 (Leica, Germany). The slices were mounted on slides, spread on a thermal plate at 25 °C for 24 h and stained with safranin-fast green to show the changes in the lignification of the cell walls, with the greater lignification staining red. The observations were made in an optical microscope model CX31RBSF (Olympus, Japan).

2.11. Statistical Analysis

For the statistical analysis, a completely randomized design was used. The results were analyzed with an analysis of variance and when significant differences were observed ($p < 0.05$) between the treatments, the mean comparison was performed by the Tukey method using the NCSS 2007 software (USA).

3. Results

3.1. Nanoemulsion

The results obtained for the emulsion showed a particle size of 94 ± 8 nm and a zeta potential (ζ) of −106 ± 5 mV, indicating excellent droplet size and stability against phase separation, in addition to translucent appearance, characteristics that are suitable for it to be considered a nanoemulsion [8].

3.2. Weight Loss

Figure 1a shows the result obtained for weight loss in avocado fruits. Significant differences ($p < 0.05$) between the treatments were observed. The control (C) fruits lost 9.81 ± 1.93% weight after 10 days as compared with the 2.31 ± 0.18%, 2.13 ± 0.99% and 2.26 ± 0.62% weight loss of fruits coated with nanoemulsion CN, N50 and N25, respectively; maintaining this behavior until the end of the analysis.

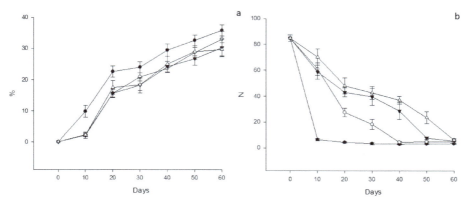

Figure 1. (a) Result for the percentage of weight loss in avocado. (b) Result for firmness in avocado. The results are expressed as means ± standard deviation. N = Newtons. ● = C, ▼ = N25, △ = N50, ○ = CN.

Antioxidants **2019**, *8*, 500

3.3. Firmness

Significant differences in firmness ($p < 0.05$) between the treatments were observed (Figure 1b) The firmness of the C group decreased from the beginning of the evaluation (6.33 ± 0.81 N to day 10), while the fruits covered with N25 and N50 showed the greatest firmness at day 30 (42.66 ± 4.88 and 39.66 ± 6.40 N, respectively), suggesting that the nanoemulsion is effective in maintaining firmness for longer compared to C.

3.4. pH and Total Soluble Solids

From Table 1 it is clear that significant differences in pH between treatments ($p < 0.05$) were found. At day 30, the C group showed a pH of 6.40 ± 0.03, and N25 had a pH of 6.15 ± 0.03, which was the most acid. The results for total soluble solids (Table 1) relate with the pH values, where C had higher total soluble solids than the rest of the treatments at day 30 (4.00 ± 0.01, 2.83 ± 0.05, 2.86 ± 0.05, and 3.23 ± 0.15 for C, CN, N50 and N25, respectively). These results are associated with the pH when indicating the lower acidity in the C group.

Table 1. Physicochemical changes in avocado during postharvest.

Days/Treatments	C	N25	N50	CN
	pH			
0	5.62 ± 0.06^{aA}	5.62 ± 0.04^{aA}	5.63 ± 0.04^{aA}	5.60 ± 0.09^{aA}
30	6.40 ± 0.03^{cB}	6.15 ± 0.03^{aB}	6.33 ± 0.02^{bB}	6.33 ± 0.04^{bB}
60	6.56 ± 0.05^{aC}	6.70 ± 0.07^{bC}	6.70 ± 0.01^{bC}	6.72 ± 0.01^{bC}
	TSS			
0	1.13 ± 0.05^{aA}	1.03 ± 0.05^{aA}	1.10 ± 0.10^{aA}	1.00 ± 0.01^{aA}
30	4.00 ± 0.01^{cB}	2.83 ± 0.04^{aB}	2.86 ± 0.05^{aB}	3.23 ± 0.15^{bB}
60	4.06 ± 0.11^{aB}	4.03 ± 0.05^{aC}	4.00 ± 0.07^{aC}	4.16 ± 0.05^{aC}
	PPO activity			
0	0.21 ± 0.01^{aA}	0.23 ± 0.01^{aA}	0.21 ± 0.01^{aA}	0.23 ± 0.01^{aA}
30	0.29 ± 0.01^{cB}	0.21 ± 0.01^{aA}	0.24 ± 0.00^{bB}	0.23 ± 0.01^{bA}
60	0.33 ± 0.01^{cC}	0.26 ± 0.01^{aB}	0.26 ± 0.02^{aB}	0.31 ± 0.01^{bB}

Different lowercase letters in the same row indicate significant differences ($p < 0.05$) between treatments at same analysis day. Different capital letters in the same column indicate significant differences ($p < 0.05$) between each treatment at different analysis day. Total soluble solids (TSS) are expressed in °Brix. Polyphenol oxidase (PPO) activity is expressed in increase in absorbance/0.1 mL of extract.

3.5. Determination of Polyphenol Oxidase Activity

PPO activity (Table 1) showed significant differences ($p < 0.05$) between the treatments. Greater activity in the C group at day 30 (0.29 ± 0.01) was observed, while N25 had the lowest PPO activity (0.21 ± 0.01), observing that the nanoemulsion is effective to keep inactive this enzyme causing the darkening of the mesocarp in fruits.

3.6. Determination of Colour

The external color of the avocado during ripening changed from green to black. In Table 2 the results of color are shown. No significant differences ($p > 0.05$) between the treatments were found in the L* parameter, although there is a decrease in value over time. The results of parameter a* (green color) displayed a significant decrease with all treatments after 30 days. The most significant effect was found in the controls (C), indicating a faster maturation which is associated with degradation of chlorophyll. The b* parameter showed significant differences ($p < 0.05$) and decreased in the all treatments after 30 days.

Antioxidants **2019**, 8, 500

Table 2. Color changes in the skin of the avocado during postharvest.

Days/Treatments	C	N25	N50	CN
		L*		
0	35.78 ± 1.70^{aA}	35.09 ± 2.18^{aA}	34.57 ± 1.52^{aA}	35.01 ± 1.47^{aA}
30	29.08 ± 1.81^{aB}	28.78 ± 1.66^{aB}	29.72 ± 1.05^{aB}	29.57 ± 1.13^{aB}
60	22.66 ± 0.67^{aC}	21.52 ± 0.85^{aC}	23.03 ± 0.29^{aC}	22.45 ± 0.29^{aC}
		a*		
0	-7.38 ± 0.84^{aA}	-7.24 ± 0.60^{aA}	-7.11 ± 0.73^{aA}	-6.90 ± 1.53^{aA}
30	2.05 ± 0.32^{cB}	-0.25 ± 1.15^{aB}	-0.94 ± 1.57^{aB}	1.40 ± 0.42^{bB}
60	4.51 ± 0.30^{aC}	4.11 ± 0.38^{aC}	4.99 ± 0.50^{aC}	4.43 ± 0.41^{aC}
		b*		
0	29.78 ± 2.65^{aA}	29.83 ± 2.24^{aA}	30.11 ± 1.81^{aA}	30.08 ± 2.31^{aA}
30	25.11 ± 0.57^{aB}	26.68 ± 0.56^{aB}	22.25 ± 0.39^{bB}	19.53 ± 1.59^{cB}
60	25.39 ± 0.64^{aB}	25.44 ± 1.07^{aB}	23.41 ± 1.53^{abB}	21.76 ± 0.43^{bB}

Different lowercase letters in the same row indicate significant differences ($p < 0.05$) between treatments at same analysis day. Different capital letters in the same column indicate significant differences ($p < 0.05$) between each treatment at different analysis day. L* = lightness, a* = green to red, b* = blue to yellow.

Table 3 shows the color results in the mesocarp of avocado fruits. In parameter L*, there are no significant differences ($p > 0.05$) between the treatments for each day of analysis, but a decrease in the brightness values respect to time is observed. Differences in the parameter a* ($p < 0.05$) are observed for treatments at day 30. C showed a greater increase in values, which is associated with the loss of green color, behavior similar to that observed for parameter b *, where C showed a loss of yellow color compared to the rest of the treatments at day 30.

Table 3. Color changes in the mesocarp of the avocado during postharvest.

Days/Treatments	C	N25	N50	CN
		L*		
0	77.42 ± 1.95^{aA}	77.32 ± 0.78^{aA}	76.59 ± 1.24^{aA}	75.99 ± 1.64^{aA}
30	69.73 ± 0.70^{aB}	71.43 ± 1.09^{aB}	71.31 ± 0.49^{aB}	69.60 ± 1.00^{aB}
60	59.94 ± 2.15^{aC}	59.94 ± 2.15^{aC}	62.32 ± 1.80^{aC}	60.12 ± 0.93^{aC}
		a*		
0	-7.67 ± 0.58^{aA}	-7.99 ± 0.63^{aA}	-7.95 ± 0.50^{aA}	-9.05 ± 0.77^{aA}
30	-2.38 ± 0.89^{bB}	-4.78 ± 0.47^{aB}	-4.26 ± 0.40^{aB}	-4.26 ± 0.40^{aB}
60	3.50 ± 0.29^{bC}	1.61 ± 0.23^{aC}	1.30 ± 0.34^{aC}	1.57 ± 0.19^{aC}
		b*		
0	45.50 ± 0.97^{aA}	45.70 ± 1.02^{aA}	44.65 ± 0.86^{aA}	47.98 ± 0.47^{aA}
30	41.96 ± 0.76^{bB}	44.67 ± 1.36^{aA}	44.75 ± 0.78^{aA}	45.57 ± 1.84^{aA}
60	33.77 ± 0.44^{bC}	35.73 ± 0.59^{aB}	35.95 ± 1.48^{abB}	33.85 ± 0.58^{bB}

Different lowercase letters in the same row indicate significant differences ($p < 0.05$) between treatments at same analysis day. Different capital letters in the same column indicate significant differences ($p < 0.05$) between each treatment at different analysis days. L* = lightness, a* = green to red, b* = blue to yellow.

3.7. Bioactive Compounds and Antioxidant Activity

3.7.1. Total Phenols

The results demonstrate significant differences ($p < 0.05$) in total phenolic compounds in the avocado mesocarp over a 60-day period as well as between controls and treatments with nanoemulsions (Table 4). After 30 days, there was a substantial drop in total phenolic concentration for all the treatments, but afterwards, a significant increase followed until day 60. The highest concentrations of total phenols at day 60 were found with N25 and N50 treatments (214.29 ± 7.78 and 240.15 ± 16.29 mg EAG/100 g).

145

Table 4. Bioactive compounds and antioxidant activity changes in the mesocarp during postharvest.

Days/Treatments	C	N25	N50	CN
		Total phenols		
0	247.33 ± 2.77^{aA}	238.14 ± 17.11^{aA}	244.46 ± 7.05^{aA}	239.00 ± 11.70^{aA}
30	122.07 ± 4.56^{bC}	160.85 ± 2.28^{aB}	127.24 ± 6.50^{bB}	120.34 ± 5.65^{bC}
60	152.23 ± 6.03^{cB}	214.29 ± 7.78^{bA}	240.15 ± 16.29^{aA}	164.87 ± 5.54^{cB}
		Total flavonoids		
0	37.07 ± 0.71^{aB}	36.25 ± 0.94^{aC}	36.46 ± 0.35^{aB}	36.04 ± 0.61^{aB}
30	32.34 ± 1.23^{cC}	44.27 ± 0.35^{aB}	36.04 ± 2.22^{bB}	46.74 ± 2.49^{aA}
60	42.63 ± 2.33^{bA}	48.18 ± 1.78^{aA}	47.77 ± 2.82^{abA}	44.27 ± 1.55^{bA}
		DPPH		
0	462.15 ± 25.61^{aA}	483.38 ± 15.13^{aA}	477.14 ± 32.29^{aA}	469.23 ± 25.59^{aA}
30	247.84 ± 10.91^{bB}	309.43 ± 17.04^{aC}	242.85 ± 20.33^{bB}	239.52 ± 7.52^{bC}
60	179.18 ± 17.22^{cC}	435.52 ± 13.69^{aB}	439.68 ± 18.47^{aA}	318.17 ± 13.75^{bB}
		ABTS		
0	228.44 ± 7.79^{aA}	225.60 ± 6.34^{aA}	223.28 ± 4.71^{aB}	221.73 ± 13.41^{aA}
30	112.48 ± 13.77^{bB}	211.40 ± 5.15^{aB}	112.48 ± 6.05^{bC}	128.75 ± 12.46^{bB}
60	68.32 ± 5.42^{dC}	233.87 ± 5.94^{bA}	310.57 ± 5.15^{aA}	195.13 ± 14.33^{cA}

Different lowercase letters in the same row indicate significant differences ($p < 0.05$) between treatments at same analysis day. The different capital letters in the same column indicate significant differences ($p < 0.05$) between each treatment at different analysis days. Total phenols are expressed in mg GAE/100 g, total flavonoids are expressed in mg QE/100 g, DPPH are expressed in mg AAE/100 g, and ABTS are expressed in mg AAE/100 g.

3.7.2. Total Flavonoid Content

Table 4 shows an increase in the total flavonoid concentration with respect to time. At day 60, the highest concentration of flavonoids was observed in the N25 and N50 treatment groups (48.18 ± 1.78 and 47.77 ± 2.82 mg EQ/100 g, respectively).

3.7.3. Determination of Antioxidant Activity

There were significant differences in the antioxidant activity when the DPPH radical was inhibited ($p < 0.05$) (Table 4). The treatments coated with the nanoemulsion presented an increase in antioxidant activity from day 30 to 60 that corresponded to the increase in bioactive compounds (phenols and flavonoids). The C treatment showed a different behavior, whereby there was a decrease in antioxidant activity with respect to time (Table 4). The results for the antioxidant activity in avocado mesocarp with the inhibition of the ABTS radical are shown in Table 4. These results are similar to those for the inhibition of the DPPH radical. Treatments coated with the nanoemulsion presented an increase in the antioxidant capacity from day 30 to day 60. The N25 treatment presented the best values of antioxidant activity at days 30 and 60 (211.40 ± 5.15 and 233.87 ± 5.94 mg EAA/100 g, respectively), while the C treatment presented the lowest values (112.48 ± 13.77 and 68.32 ± 5.42 mg EAA/100 g).

3.8. Structural Evaluation

In Figure 2, the changes in the pericarp of the avocados with respect to time are shown, demonstrating that the exocarp is constructed by epidermal cells in a vertical disposition. In the immature fruit (day 0) a thick cuticle covering the epidermal tissue is clearly visible. One of the most prominent changes of the treatments up to day 30 was the lignification of the cell wall as accessed through the red staining protocol by the action of safranin.

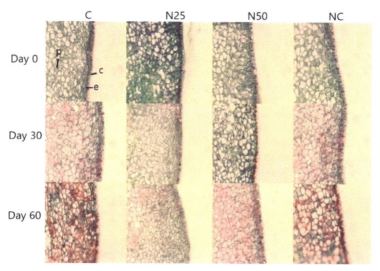

Figure 2. Structural evaluation of the avocado epicarp with respect to time. The columns indicate the same treatment with respect to time and the rows indicate the same day of analysis for different treatments, c = cuticle, e = epidermis, p = parenchyma.

4. Discussion

Postharvest weight loss is mainly attributed to perspiration caused by a deficit of steam pressure of the product in relation to its environment [22]. Sellamuthu et al. [23] found that weight loss decreased by applying thyme oil vapour and packaging of avocados from different cultivars in modified atmosphere and Russo et al. [24] observed that the firmness of the fruits decreased during storage from the fifth day in avocado fruits stored in different atmospheres (CO_2/O_2). Maftoonazad and Ramaswamy [15] applied a coating based on methylcellulose in avocado fruits, observing that coated treatments managed to maintain avocado firmness, which indicates that the use of a nanoemulsion as a coating is an effective alternative to prevent loss of firmness

Higher pH values can be associated with the conversion of organic acids and other complex molecules present in fruits into sugars, which is a source of energy reserve used in the metabolic processes during ripening [24,25]. Saucedo-Pompa et al. [26] coated avocado fruits with candelilla wax and ellagic acid and found that the fruits that were not coated had pH values closer to neutrality compared with the other treatments. Aguirre-Joya et al. [16] observed similar results in avocados coated with candelilla wax, pectin, aloe mucilage and polyphenols of *Larrea tridentate*, where the control showed a tendency to increase °Brix value as the storage time increased.

PPO is responsible for the darkening of the avocado, and it has been reported that PPO activity increases with a more accelerated ripening in stored fruits [23,27]. The results obtained in the present study are similar to those reported by Tesfay and Magwaza [2] in which the activity of PPO in the mesocarp of avocados decreased by applying edible moringa leaf extract coatings to avocado, showing that the maintenance of the integrity of the membrane in the coated fruit contributed to the reduction of darkening.

A decrease in luminosity in the epicarp of avocado has been related to the synthesis of some anthocyanins such as cyanidin 3-O-glucoside that confers dark colors [19]. The treatments that were coated with the nanoemulsion showed slower changes in colour, directly affecting the ripening attributes of the fruits [28]. Villa-Rodríguez et al. [19] evaluated the changes during the ripening of avocado fruits, reporting that there is a change in colour with respect to time that is related to the degradation (mainly chlorophyll) and synthesis of compounds. Correa-Pacheco et al. [29] determined color in mesocarp and epicarp of avocado fruits coated with chitosan-thyme essential oil nanoparticles, observing that there were no differences between the coated and uncoated treatments; however, the results of the present study show that the nanoemulsion was able to maintain the color of the mesocarp as opposed to the fruits that were not coated. Cenobio-Galindo et al. [5] mentioned that bioactive compounds from xoconostle can remain effectively included in certain coatings without losing their activity, finding that certain compounds, such as rutin, ferulic acid, quercetin and apigenin, were maintained even after being incorporated into films and provided antioxidant activity.

Tesfay et al. [30] mentioned that the production of polyphenols can be stimulated by some biotic or abiotic factors; therefore, the application of the nanoemulsion could act as stimulus to increase phenolic compounds. Taking into account the results obtained, it is possible that this behavior is related to the diluted nanoemulsion treatments forming a more stable semipermeable coating with respect to the CN, and the stress in the fruit can remain for a longer time, affecting the synthesis of compounds. Wang et al. [20] evaluated the concentration of total phenols in different cultivars of avocado, finding that the cultivar "Hass" had high concentrations in the seeds and peels. Similar results were found by Vinha et al. [25], in which the concentration of total phenolics in avocados from the Algarve region in Portugal were highest in the seed and skin.

Sellamuthu et al. [23] indicated that essential oils can act as signaling compounds, causing a slight stress situation in the fruit, resulting in the increase in certain compounds, such as flavonoids. The procyanidins (catechin and epicatechin) are the main flavonoids present in avocados and are considered potent antioxidants with beneficial health effects [19,20]. From the results of Vinha et al. [25], it was clear that the highest concentration of total flavonoids was detected in the pericarp and seeds. Their reported value in the mesocarp was lower at 21.9 ± 1.0 mg / 100 g than our findings, in which, even at day 0, a higher amount (37.07 ± 0.71 mg EQ/100 g) was noted.

The antioxidant activity present in the avocado is given by the bioactive compounds present in the fruit, which include phenolic compounds such as procyanidins, chlorophylls and carotenoids [15]. Wang et al. [20] determined the antioxidant activity in different cultivars of avocado by inhibiting the DPPH radical and found that the cultivar "Hass" had the greatest antioxidant activity. Villa-Rodríguez et al. [19] determined antioxidant activity in avocado with different extractions and found that lipophilic extracts showed a greater antioxidant activity than hydrophilic extracts, compared by the different methods analyzed. There are reports suggesting that essential oils acting as signaling compounds are also capable of increasing antioxidant activity; therefore, nanoemulsions can fulfil this function [31].

Higher lignification of the cell walls in fruits treated with C and CN, compared to the N25 and N50 treatments, is an indication of accelerated maturation of the fruit. The colour change among treatments is because the green colour prevails in cells that have a slower metabolism, and cells that showed a more lignified wall were stained red by the action of safranin [32]. The effects that retard the fruit maturation (weight loss, firmness and PPO activity) are attributed to the action of the nanoemulsion as a coating. In all the treatments, it was observed that the cuticle remained until the end of the analysis. Schroeder [33] mentioned that the elimination or destruction of this protective epidermal layer makes the surface of the fruit more susceptible to attack by fungal and bacterial infections or can cause physiological disorders as a result of drying out.

5. Conclusions

The incorporation of bioactive compounds with antioxidant activity through a nanoemulsion based on orange essential oil and xoconostle extract as a coating on avocado fruits had beneficial effects on postharvest storage of the fruit. The application of the nanoemulsion decreased the percentage of weight loss, retained firmness, decreased enzymatic activity, delayed its darkening and decreased effects on epidermal cells. These properties of the nanoemulsion translate into slower ripening of the avocados compared to the control. The nanoemulsion coating can be considered as an alternative treatment to increase the postharvest life of the avocado fruit.

Author Contributions: Investigation and writing—original draft preparation, A.d.J.C.-G.; methodology J.O.-L.; methodology A.R.-M.; validation M.L.C.-I.; writing—review and editing, M.C.; formal analysis, G.M.-P.; formal analysis, F.F.-L.; data curation and supervision R.G.C.-M.

Funding: Authors thanks CONACyT for the support granted (scholarship of A.d.J.C.G., number 295910).

Conflicts of Interest: The authors declare no conflict of interest.

References

1. Melgar, B.; Dias, M.I.; Ciric, A.; Sokovic, M.; Garcia-Castello, E.M.; Rodriguez-Lopez, A.D.; Barros, L.; Ferreira, I.C. Bioactive characterization of *Persea americana* Mill. by-products: A rich source of inherent antioxidants. *Ind. Crops Prod.* **2018**, *111*, 212–218. [CrossRef]
2. Tesfay, S.Z.; Magwaza, L.S. Evaluating the efficacy of moringa leaf extract, chitosan and carboxymethyl cellulose as edible coatings for enhancing quality and extending postharvest life of avocado (*Persea americana* Mill.) fruit. *Food Packag. Shelf Life* **2017**, *11*, 40–48. [CrossRef]
3. Espinosa-Muños, V.; Roldán-cruz, C.A.; Hernández-Fuentes, A.D.; Quintero-Lira, A.; Almaraz-Buendía, I.; Campos-Montiel, R.G. Ultrasonic-Assisted Extraction of Phenols, Flavonoids, and Biocompounds with Inhibitory Effect Against *Salmonella Typhimurium* and *Staphylococcus Aureus* from Cactus Pear. *J. Food Process. Eng.* **2017**, *40*, e12358. [CrossRef]
4. Pérez-Alonso, C.; Campos-Montiel, R.G.; Morales-Luna, E.; Reyes-Munguía, A.; Aguirre-Álvarez, G.; Pimentel-González, D.J. Estabilización de compuestos fenólicos de *Opuntia oligacantha* Först por microencapsulación con agave SAP (aguamiel). *Revista Mexicana de Ingeniería Química* **2015**, *14*, 579–588.
5. Cenobio-Galindo, A.J.; Pimentel-González, D.J.; Del Razo-Rodríguez, O.E.; Medina-Pérez, G.; Carrillo-Inungaray, M.L.; Reyes-Munguía, A.; Campos-Montiel, R.G. Antioxidant and antibacterial activities of a starch film with bioextracts microencapsulated from cactus fruits (*Opuntia oligacantha*). *Food Sci. Biotechnol.* **2019**, 1–9. [CrossRef]
6. Zhang, Z.; Vriesekoop, F.; Yuan, Q.; Liang, H. Effects of nisin on the antimicrobial activity of D-limonene and its nanoemulsion. *Food Chem.* **2014**, *150*, 307–312. [CrossRef]
7. Hashtjin, A.M.; Abbasi, S. Nano-emulsification of orange peel essential oil using sonication and native gums. *Food Hydrocoll.* **2015**, *44*, 40–48. [CrossRef]
8. McClements, D.J.; Rao, J. Food-grade nanoemulsions: Formulation, fabrication, properties, performance, biological fate, and potential toxicity. *Crit. Rev. Food Sci. Nutr.* **2011**, *51*, 285–330. [CrossRef]
9. Zhang, S.; Zhang, M.; Fang, Z.; Liu, Y. Preparation and characterization of blended cloves/cinnamon essential oil nanoemulsions. *LWT-Food Sci. Technol.* **2017**, *75*, 316–322. [CrossRef]
10. Donsì, F.; Cuomo, A.; Marchese, E.; Ferrari, G. Infusion of essential oils for food stabilization: Unraveling the role of nanoemulsion-based delivery systems on mass transfer and antimicrobial activity. *Innov. Food Sci. Emerg.* **2014**, *22*, 212–220. [CrossRef]
11. Cenobio-Galindo, A.J.; Campos-Montiel, R.G.; Jiménez-Alvarado, R.; Almaraz-Buendía, I.; Medina-Pérez, G.; Fernández-Luqueño, F. Development and Incorporation of Nanoemulsions in Food. *Int. J. Food Stud.* **2019**, *8*, 105–124.
12. Zambrano-Zaragoza, M.L.; Gutiérrez-Cortez, E.; Del Real, A.; González-Reza, R.M.; Galindo-Pérez, M.J.; Quintanar-Guerrero, D. Fresh-cut Red Delicious apples coating using tocopherol/mucilage nanoemulsion: Effect of coating on polyphenol oxidase and pectin methylesterase activities. *Food Res. Int.* **2014**, *62*, 974–983. [CrossRef]
13. Oh, Y.A.; Oh, Y.J.; Song, A.Y.; Won, J.S.; Song, K.B.; Min, S.C. Comparison of effectiveness of edible coatings using emulsions containing lemongrass oil of different size droplets on grape berry safety and preservation. *LWT-Food Sci. Technol.* **2017**, *75*, 742–750. [CrossRef]
14. Aguilar-Méndez, M.A.; Martín-Martínez, E.S.; Tomás, S.A.; Cruz-Orea, A.; Jaime-Fonseca, M.R. Gelatine–starch films: Physicochemical properties and their application in extending the post-harvest shelf life of avocado (*Persea americana*). *J. Sci. Food Agric.* **2008**, *88*, 185–193. [CrossRef]
15. Maftoonazad, N.; Ramaswamy, H.S. Postharvest shelf-life extension of avocados using methyl cellulose-based coating. *LWT-Food Sci. Technol.* **2005**, *38*, 617–624. [CrossRef]

16. Aguirre-Joya, J.A.; Ventura-Sobrevilla, J.; Martínez-Vazquez, G.; Ruelas-Chacón, X.; Rojas, R.; Rodríguez-Herrera, R.; Aguilar, C.N. Effects of a natural bioactive coating on the quality and shelf life prolongation at different storage conditions of avocado (*Persea americana* Mill.) cv. Hass. *Food Packag. Shelf Life* **2017**, *14*, 102–107. [CrossRef]
17. Vargas-Ortiz, M.; Rodríguez-Jimenes, G.; Salgado-Cervantes, M.; Pallet, D. Minimally Processed Avocado Through Flash Vacuum-Expansion: Its Effect in Major Physicochemical Aspects of the Puree and Stability on Storage. *J. Food Process. Preserv.* **2017**, *41*, e12988. [CrossRef]
18. Vargas-Ortiz, M.; Servent, A.; Rodríguez-Jimenes, G.; Pallet, D.; Salgado-Cervantes, M. Effect of thermal stage in the processing avocado by flash vacuum expansion: Effect on the antioxidant capacity and the qualitaty of the mash. *J. Food Process. Preserv.* **2017**, *41*, e13118. [CrossRef]
19. Villa-Rodríguez, J.A.; Molina-Corral, F.J.; Ayala-Zavala, J.F.; Olivas, G.I.; González-Aguilar, G.A. Effect of maturity stage on the content of fatty acids and antioxidant activity of 'Hass' avocado. *Food Res. Int.* **2011**, *44*, 1231–1237. [CrossRef]
20. Wang, W.; Bostic, T.R.; Gu, L. Antioxidant capacities, procyanidins and pigments in avocados of different strains and cultivars. *Food Chem.* **2010**, *122*, 1193–1198. [CrossRef]
21. Hernández-Rivero, R.; Arévalo-Galarza, M.; Valdovinos-Ponce, G.; González-Hernández, H.; Valdez-Carrasco, J.; Ramírez-Guzmán, M.E. Histología del daño en fruto y rama de aguacate 'Hass' por escamas armadas (Hemiptera: Diaspididae). *Revista Mexicana de Ciencias Agrícolas* **2013**, *4*, 739–751. [CrossRef]
22. Espinosa-Cruz, C.C.; Valle-Guadarrama, S.; Ybarra-Moncada, M.; Martínez-Damián, M.T. Comportamiento postcosecha de frutos de aguacate 'Hass' afectado por temperatura y atmósfera modificada con microperforado. *Revista Fitotecnia Mexicana* **2014**, *37*, 235–242.
23. Sellamuthu, P.S.; Mafune, M.; Sivakumar, D.; Soundy, P. Thyme oil vapour and modified atmosphere packaging reduce anthracnose incidence and maintain fruit quality in avocado. *J. Sci. Food Agric.* **2013**, *93*, 3024–3031. [CrossRef] [PubMed]
24. Russo, V.C.; Daiuto, E.R.; Vietes, R.L.; Smith, R.E. Postharvest Parameters of the "Fuerte" Avocado When Refrigerated in Different Modified Atmospheres. *J. Food Process. Preserv.* **2014**, *38*, 2006–2013. [CrossRef]
25. Vinha, A.F.; Moreira, J.; Barreira, S.V. Physicochemical parameters, phytochemical composition and antioxidant activity of the algarvian avocado (*Persea americana* Mill.). *J. Agric. Sci.* **2013**, *5*, 100. [CrossRef]
26. Saucedo-Pompa, S.; Rojas-Molina, R.; Aguilera-Carbó, A.F.; Saenz-Galindo, A.; de La Garza, H.; Jasso-Cantú, D.; Aguilar, C.N. Edible film based on candelilla wax to improve the shelf life and quality of avocado. *Food Res. Int.* **2009**, *42*, 511–515. [CrossRef]
27. Mpho, M.; Sivakumar, D.; Sellamuthu, P.S.; Bautista-Baños, S. Use of lemongrass oil and modified atmosphere packaging on control of anthracnose and quality maintenance in avocado cultivars. *J. Food Qual.* **2013**, *36*, 198–208. [CrossRef]
28. Cox, K.A.; McGhie, T.K.; White, A.; Woolf, A.B. Skin colour and pigment changes during ripening of 'Hass' avocado fruit. *Postharvest Biol. Technol.* **2004**, *31*, 287–294. [CrossRef]
29. Correa-Pacheco, Z.N.; Bautista-Baños, S.; Valle-Marquina, M.Á.; Hernández-López, M. The Effect of Nanostructured Chitosan and Chitosan-thyme Essential Oil Coatings on *Colletotrichum gloeosporioides* Growth in vitro and on cv Hass Avocado and Fruit Quality. *J. Phytopathol.* **2017**, *165*, 297–305. [CrossRef]
30. Tesfay, S.Z.; Bertling, I.; Bower, J.P. Effects of postharvest potassium silicate application on phenolics and other anti-oxidant systems aligned to avocado fruit quality. *Postharvest Biol. Technol.* **2011**, *60*, 92–99. [CrossRef]
31. Sellamuthu, P.S.; Sivakumar, D.; Soundy, P.; Korsten, L. Essential oil vapours suppress the development of anthracnose and enhance defence related and antioxidant enzyme activities in avocado fruit. *Postharvest Biol. Technol.* **2013**, *81*, 66–72. [CrossRef]
32. Deng, J.; Bi, Y.; Zhang, Z.; Xie, D.; Ge, Y.; Li, W.; Wang, J.; Wang, Y. Postharvest oxalic acid treatment induces resistance against pink rot by priming in muskmelon (*Cucumis melo* L.) fruit. *Postharvest Biol. Technol.* **2015**, *106*, 53–61. [CrossRef]
33. Schroeder, C.A. The structure of the skin or rind of the avocado. *Calif. Avocado Soc. Yearb.* **1950**, *34*, 169–176.

© 2019 by the authors. Licensee MDPI, Basel, Switzerland. This article is an open access article distributed under the terms and conditions of the Creative Commons Attribution (CC BY) license (http://creativecommons.org/licenses/by/4.0/).

Article

Enhancement of Minor Ginsenosides Contents and Antioxidant Capacity of American and Canadian Ginsengs (*Panax quinquefolius*) by Puffing

Min-Soo Kim [1], Sung-Joon Jeon [1], So Jung Youn [2], Hyungjae Lee [2], Young-Joon Park [3], Dae-Ok Kim [1], Byung-Yong Kim [1], Wooki Kim [1,*] and Moo-Yeol Baik [1,*]

1   Department of Food Science and Biotechnology, Kyung Hee University, Yongin 17104, Korea
2   Department of Food Engineering, Dankook University, Cheonan 31116, Korea
3   Department of Science in Korean Medicine, Kyung Hee University, Seoul 02447, Korea
*   Correspondence: kimw@khu.ac.kr (W.K.); mooyeol@khu.ac.kr (M.-Y.B.); Tel.: +82-31-201-3482 (W.K.); +82-31-201-2625 (M.-Y.B.)

Received: 4 October 2019; Accepted: 4 November 2019; Published: 5 November 2019

Abstract: The effects of puffing on ginsenosides content and antioxidant activities of American and Canadian ginsengs, *Panax quinquefolius*, were investigated. American and Canadian ginsengs puffed at different pressures were extracted using 70% ethanol. Puffing formed a porous structure, inducing the efficient elution of internal compounds that resulted in significant increases in extraction yields and crude saponin content. The content of minor ginsenosides (Rg2, Rg3, compound K) increased with increasing puffing pressure, whereas that of major ginsenosides (Rg1, Re, Rf, Rb1, Rc, Rd) decreased, possibly due to their deglycosylation and pyrolysis. Furthermore, 2,2'-azino-bis (3-ethylbenzothiazoline-6-sulphonic acid) (ABTS) radical scavenging activity, total phenolic content, total flavonoid content, amount of Maillard reaction products, and acidic polysaccharides content increased with increasing puffing pressure, but 2,2-diphenyl-1-picrylhydrazyl (DPPH) radical scavenging activity did not. There was no substantial difference in the results between puffed American and Canadian ginsengs. Consequently, these results suggest that puffing can be a promising novel technology for processing *P. quinquefolius* to achieve higher levels of minor ginsenosides and obtain value-added products.

Keywords: antioxidant activity; ginsenosides; *Panax quinquefolius*; puffing

---

1. Introduction

Panax quinquefolius L. (American and Canadian ginsengs) is the main ginseng cultivar in the United States and Canada [1]. *P. quinquefolius* administration is known to be effective in recovering fatigue, improving immunity, and controlling blood pressure and cholesterol level [2–4]. These effects are mediated by ginsenosides, ginseng-specific saponin components [5]. Ginsenosides have a glycoside structure, and their biological effectiveness varies depending on their structure. In addition to ginsenosides, ginseng is known to contain other active ingredients such as acidic polysaccharides and phenolic compounds [6].

The ginsenosides of ginseng include Rb1, Rb2, Rc, Rd, Re, Rf, Rg1, Rg2, Rg3, Rh, Compound K, F1, and F2 [7,8]. Ginsenosides are divided into protopanaxadiols (PPD) and protopanaxatriols (PPT) depending on the presence of a hydroxyl group at C3 or C6 and the presence or absence of sugar [9]. Among them, ginsenosides Rb1, Re, Rg1, Rc, and Rd account for more than 70% of the total amount of *P. quinquefolius* ginsenosides [10]. The genus *Panax* includes two main species, *Panax ginseng* (Korean ginseng) and *P. quinquefolius* [11]. There are distinctive differences in ginsenoside content between *P. ginseng* and *P. quinquefolius*. Ginsenoside Rg1, Rb1, and Rf contents are high in *P. ginseng*, while

P. quinquefolius contains low levels of them [12]. In general, *P. quinquefolius* is known to have higher amounts of ginsenoside Rb1 and Re [13]. It was reported that ginsenosides Rg2, Rg3, Rh1, and Rh2, specific components produced by thermal stimulation, showed preventive effects against cancer and inhibited cancer cell growth [14].

In Asia, there are two types of ginseng: white ginseng is produced by drying raw ginseng, and red ginseng is produced by steaming raw ginseng at 98–100°C for 2–3 h [15]. Red ginseng shows a larger number of pharmacological effects than white and raw ginseng [16]; in particular, antioxidant activity, antioxidant enzyme activity, and cell viability are high [15]. It was reported that the minor ginsenosides content of red ginseng is high [17], while these compounds are not found in white ginseng [18]. In 2015, ginseng was registered as an international food standard by CODEX (Codex Alimentarius International Food Standards). Changes in the content of ginsenosides and bioactive ingredients in ginsengs usually take a long time; therefore, red ginseng production and puffing may be useful to enhance the pharmacological properties of ginseng products. In the near future, processed ginseng products will be major items in the international functional food market.

To execute puffing, a unique food-processing method, pressure is suddenly lowered to atmospheric level from a high pressure in a high temperature regime. Puffing induces the gelatinization of starch and increases the volume of the material resulting from the evaporation of water [18]. Puffing occurs in two steps. After rapid heating at atmospheric pressure, the sudden evaporation of water causes a rapid decrease in pressure by transfer of superheated steam [19]. Puffing causes chemical and physical changes including starch gelatinization in cereal grains [20], ginsenoside profile changes in red ginseng [18], increase in antioxidant activity of white ginseng [21], denaturation and reorganization of proteins in ginseng [22], removal of water and formation of a porous structure in barley [23], and inactivation of enzymes that cause deterioration by lipid oxidation during storage of red ginseng [24]. *P. ginseng* has been mainly used in studies on the changes in ginsenoside content caused by puffing [18,25]. Puffing of herbal resources has been reported to the greatly increase antioxidant activities [18,21,24,25]. However, limited information is available on the effect of puffing on the ginsenosides and the bioactive components of *P. quinquefolius*. In this study, we thoroughly investigated the changes in ginsenosides content and antioxidant activity of American and Canadian ginsengs caused by puffing at different pressures.

2. Materials and Methods

2.1. Materials

Dried four-year-old commercial *P. quinquefolius* ginseng roots were purchased from a Wisconsin Ginseng Farm (Wisconsin, USA) and Rainey Ginseng Farms, LTD (Ontario, Canada) in 2014.

2.2. Chemicals

The compounds 2,2-diphenyl-1-picrylhydrazyl, 2,2′-azobis-(2-amidinopropane) dihydrochloride, 2,2′-azino-bis (3-ethylbenzothiazoline-6-sulphonic acid), Folin–Ciocalteu reagent, carbazole, galacturonic acid, gallic acid, and catechin were purchased from Sigma-Aldrich (St. Louis, MO, USA). Phosphate-buffered saline (PBS) was purchased from Welgene (Gyeongsan, Korea). Ascorbic acid was purchased from Reagent Duksan (Ansan-si, Republic of Korea). HPLC ginsenoside standards were purchased from Ambo Institute (Daejeon, Republic of Korea).

2.3. Puffing Process

To minimize the carbonization of ginseng at high temperature, dried ginseng and rice were mixed at a ratio of 1:4 (*w/w*). Dried ginseng (200 g) and rice (800 g) were placed in a rotary gun puffing machine (PPsori Co., Namyangju, Korea) and heated. When the gauge pressure reached 490 kPa, the medium pressure was released to 294 kPa, and then the rotary gun puffing machine was heated again. For subsequent measurements, when the pressure reached 686 kPa, 784 kPa, 882 kPa, and 980 kPa, the puffing machine door was opened to reach atmospheric pressure. Dried ginseng without puffing treatment was used as the control group.

2.4. Color Measurement

The color changes of puffed ginsengs were confirmed by a color difference meter (JC801, Color Techno System, Tokyo, Japan). Ginseng samples were finely ground with a blender and filtered through a 100-mesh sieve. The range of hunter L, a, and b value coordinates ranged from L = 0 (black) to 100 (white), a = −80 (greenness) to 100 (redness), and b = −80 (blueness) to 70 (yellowness). The colorimeter was calibrated with a standard plate of L = 98.26, a = 0.24, and b = −0.24 before measurement.

2.5. Extraction Yield

Five grams of puffed ginseng were ground and mixed with 125 mL of 70% ethanol (1:25, w/v), followed by stirring at room temperature for 30 min. The extracts were filtered through a funnel in a Kimble filtering flask using a Whatman No. 2 filter paper (Whatman, Maidstone, England). The extracts were dried in a hot-air dryer (HB-502M, Han Beak Scientific Co., Bucheon, Korea) at 105 °C, and the extraction yield was calculated using the following Equation (1):

$$Extraction\ yield\ (\%) = \frac{(W_2 - W_1)}{A} \times \frac{E}{E'} \times 100 \tag{1}$$

where

W_1 = Weight of empty aluminum dish (g)
W_2 = Weight of aluminum dish and solid (g)
A = Weight of dried ginseng (g)
E = Total volume of extract (mL)
E' = Used volume of extract (mL).

2.6. Crude Saponin Content

The crude saponin content of the extracts was determined by the diethyl ether method. Dried solids were mixed with distilled water to a total volume of 50 mL. Sample (5 mL), distilled water (20 mL), and diethyl ether (25 mL) were mixed in a separation funnel. The mixture was shaken well and allowed to stand for 30 min for the separation of the aqueous and the organic layers. The separated diethyl ether layer was discarded, and the remaining water layer was mixed with 25 mL of water-saturated n-butanol, followed by shaking and another 30 min of rest for further separation. The separated n-butanol layer was collected and washed with water-saturated n-butanol. This process was repeated three times. The collected n-butanol layer was concentrated using a rotary vacuum evaporator (N-11, EYELA, Tokyo, Japan) under reduced pressure. After concentration, the sample was dried in an oven at 105 °C for 2 h. The crude saponin content was calculated by weight changes using the following Equation (2):

$$Crude\ saponin\ content\left(\frac{mg}{g\ ginseng}\right) = \frac{(W_1 - W_2)}{W_3} \times \frac{A}{B} \tag{2}$$

where

W_1 = Weight of the dried sample and flask (mg)
W_2 = Weight of the flask (mg)
W_3 = Weight of total dried ginseng (g)
A = Weight of total concentration (g)
B = Weight of used concentration (g).

2.7. Ginsenoside Profile

Ginsenoside profile was analyzed by an HPLC system by dissolving the crude saponin concentrates in 5 mL of HPLC-grade methanol, followed by filtering with a Millipore filter (pore size 0.45 μm). The instrument was a 1260 Infinity II LC system (Agilent, Santa Clara, USA) equipped with a Kinetex C18 column (50 × 4.6 mm, Phenomenex, CA, USA) and a UV detector at 203 nm. The binary gradient-elution solvents consisted of distilled water (mobile phase A) and acetonitrile (mobile phase B). The flow rate of the mobile phase was 0.6 mL/min, the sample injection volume was 5 μL, and the analytical column temperature was maintained at 45 °C. The following gradient elution procedure was used: 0–7 min, 81% A, 19% B; 7–14 min, 71% A, 29% B; 14–25 min, 60% A, 40% B; 25–28 min, 44% A, 56% B; 28–30 min, 30% A, 70% B; 30–31.5 min, 10% A, 90% B; 31.5–34 min, 10% A, 90% B; 34–34.5 min, 81% A, 19% B; 34.5–40 min, 81% A, 19% B.

2.8. Antioxidant Activity

2.8.1. DPPH Radical Scavenging Activity

The 2,2-diphenyl-1-picrylhydrazyl (DPPH) radical scavenging activity of the extracts was measured with a modified version of the method of Blois (1958) [26] using 2,2-diphenyl-1-picrylhydrazyl and 80% methanol to prepare a 0.1 mM DPPH solution. The absorbance of the DPPH solution was adjusted to 0.650 ± 0.020 at 517 nm using 80% methanol. Extract (0.05 mL) was added to 2.95 mL of the adjusted DPPH solution, and the mixture was allowed to stand in a dark room at 23 °C for 30 min. Subsequently, the absorbance at 517 nm was measured. As standard and blank, ascorbic acid and distilled water were used, respectively. DPPH was calculated using the following Equation (3):

$$\% \; inhibition = \frac{A_{reference} - A_{sample}}{A_{reference}} \times 100 \tag{3}$$

where

$A_{reference}$ = absorbance of the blank
A_{sample} = absorbance of the sample
$A_{reference}$ = mixture of 0.1 mL of 80% MeOH and 2.9 mL of DPPH radical solution.

2.8.2. ABTS Radical Scavenging Activity

The 2,2′-azino-bis (3-ethylbenzothiazoline-6-sulphonic acid (ABTS) radical scavenging activity was determined according to the following procedure [27,28]. The solutions of 1.0 mM 2,2′-azobis-(2-amidinopropane) dihydrochloride and 2.5 mM 2,2′-azino-bis (3-ethylbenzothiazoline-6-sulphonic acid) were mixed in 100 mL PBS. The mixture was stirred at 70°C for 30 min and then cooled to room temperature. The solution was filtered using a 0.45 μm syringe filter and diluted with PBS to achieve an absorbance of 0.650 ± 0.020 at 734 nm. The radical solution was stored at 37 °C. Extract (20 μL) and radical solution (980 μL) were mixed and reacted at 37 °C for 10 min, and the absorbance was measured at 734 nm. As standard and blank, ascorbic acid and distilled water were used, respectively.

2.9. Total Phenolic Content

Total phenolic content (TPC) was determined using the following procedure with some modifications [29,30]. To 2.6 mL of deionized water, 200 μL of Folin and Ciocalteu's phenol reagent and 200 μL of extract were added and mixed. After 6 min, 2 mL of 7% Na_2CO_3 was added and mixed. The mixture was allowed to react at 25 °C for 90 min, and the absorbance was measured at 750 nm. Gallic acid was used as a standard, and deionized water was used as a blank.

2.10. Total Flavonoid Content

Total flavonoid content (TFC) was determined using the following method [31]. Distilled water (3.2 mL) and sample (0.5 mL) were mixed, and 0.15 mL of 5% $NaNO_2$ was added. After 5 min, 0.15 mL of 10% $AlCl_3$ was added. After 1 min, 1 mL of 1 M NaOH was added, and the absorbance was measured at 510 nm. Catechin was used as a standard, and distilled water was used as a blank.

2.11. Acidic Polysaccharides

Degradation of acidic polysaccharides was measured using a colorimetric method [32]. Samples (5 mL) were mixed with 3 mL of distilled water and 0.25 mL of 0.1% carbazole and placed in a water bath at 85 °C for 5 min. After cooling for 20 min at room temperature, the absorbance was measured at 525 nm. Galacturonic acid was used as a standard, and distilled water was used as a blank.

2.12. Maillard Reaction Products

To measure the level of the Maillard reaction products (MRPs), the extracts were diluted 50-fold, and the absorbance was measured at 420 nm using a spectrophotometer.

2.13. Statistical Analysis

All experiments were repeated in triplicate. Experimental data were analyzed by analysis of variance (ANOVA) and expressed as mean ± standard deviation. A Duncan's multiple range test was conducted to assess significant differences among experimental mean values ($p < 0.05, 0.01,$ or 0.001). All statistical computations and analyses were conducted with SAS software (version 8.2, SAS Institute, Inc., Cary, NC, USA).

3. Results and Discussion

3.1. Morphology

The appearance of both types of ginseng according to puffing pressure is shown in Figure 1. Originally, Canadian ginseng was bigger than American ginseng before puffing, and regardless of the puffing pressure, the volume of Canadian ginseng was larger than that of American ginseng. It was reported previously that browning was further enhanced at high pressure in puffed red ginseng [18]. In particular, the surface was frayed starting at 784 kPa and separated from the interior at 882 kPa, resulting in the formation of blisters or air pockets inside the ginseng.

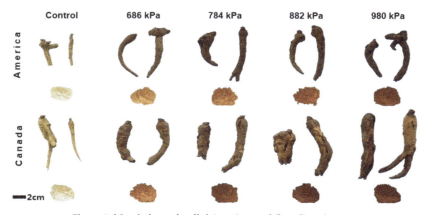

Figure 1. Morphology of puffed American and Canadian ginsengs.

Puffing also changed the color of both ginsengs (Table 1). Regardless of the ginseng type, the L value decreased as the puffing pressure increased, suggesting increased pigmentation with puffing. The a and b values were higher than those of the control groups but decreased as the puffing pressure increased. This indicated that puffing increased the redness and yellowness of both ginsengs. On the other hand, the highest redness and yellowness values were observed at the lowest puffing pressure tested. The L value was consistent with the trend of MRPs content, which increased with increasing puffing pressure (Table 2). The higher were the temperature and pressure, the more active was the Maillard reaction between amino acids and sugars in ginseng [23,33].

Table 1. Color, extraction yield, and crude saponin content of puffed American and Canadian ginsengs.

Sample (kPa)	L	a	b	Extraction Yield (%)	Crude Saponin Content (mg/g Dried Ginseng)
A-Control	76.86 ± 1.55 [a,*]	8.34 ± 0.15 [d]	16.99 ± 0.53 [d]	35.03 ± 8.49 [d]	72.44 ± 5.19 [e]
A-686	54.66 ± 10.14 [b]	12.91 ± 1.15 [ab]	27.64 ± 2.21 [ab]	46.86 ± 12.07 [ab]	141.20 ± 26.16 [a]
A-784	47.56 ± 14.73 [bc]	12.20 ± 1.00 [bc]	25.97 ± 3.90 [abc]	44.84 ± 5.74 [b]	126.56 ± 12.26 [abc]
A-882	44.87 ± 10.28 [bc]	11.73 ± 2.29 [bc]	24.55 ± 5.16 [bc]	40.44 ± 4.16 [c]	107.12 ± 17.39 [d]
A-980	39.44 ± 9.58 [c]	10.78 ± 2.51 [c]	21.52 ± 6.24 [c]	46.19 ± 3.20 [ab]	122.10 ± 14.78 [bcd]
C-Control	78.67 ± 1.92 [a]	8.41 ± 0.28 [d]	16.77 ± 1.04 [d]	38.57 ± 3.92 [cd]	104.33 ± 22.08 [d]
C-686	54.26 ± 4.08 [b]	14.27 ± 0.50 [a]	30.56 ± 1.02 [a]	46.95 ± 4.43 [ab]	129.44 ± 17.68 [ab]
C-784	53.53 ± 7.51 [b]	13.28 ± 1.05 [ab]	28.69 ± 2.36 [ab]	47.32 ± 3.54 [ab]	115.86 ± 13.42 [bcd]
C-882	48.04 ± 3.86 [b]	13.41 ± 1.30 [ab]	27.85 ± 2.55 [ab]	49.44 ± 5.26 [a]	109.61 ± 9.29 [cd]
C-980	39.82 ± 2.42 [c]	11.94 ± 1.24 [bc]	23.86 ± 2.68 [bc]	49.20 ± 5.04 [a]	126.33 ± 10.66 [abc]

A, American ginseng; C, Canadian ginseng. * Values with the same letter in the column are not significantly different ($p < 0.05$).

Table 2. Antioxidant activity (2,2-diphenyl-1-picrylhydrazyl (DPPH) and 2,2'-azino-bis (3-ethylbenzothiazoline-6-sulphonic acid (ABTS) radical scavenging activities), TPC, TFC, acidic polysaccharides, and Maillard reaction products (MRPs) of puffed American and Canadian ginsengs.

Sample (kPa)	DPPH (mg VCE/ g Dried Ginseng)	ABTS (mg VCE/ g Dried Ginseng)	TPC (mg GAE/ g Dried Ginseng)	TFC (mg CE/ g Dried Ginseng)	AP (mg GA/ g Dried Ginseng)	MRPs
A-Control	0.29 ± 0.11 [b*]	0.97 ± 0.15 [d]	1.11 ± 0.06 [f]	0.44 ± 0.05 [e]	1.78 ± 0.27 [c]	0.186 ± 0.014 [c]
A-686	1.88 ± 0.26 [a]	10.06 ± 4.88 [c]	9.24 ± 3.60 [e]	2.39 ± 1.17 [de]	2.60 ± 0.61 [bc]	0.295 ± 0.085 [bc]
A-784	1.97 ± 0.16 [a]	14.58 ± 5.23 [bc]	14.46 ± 5.05 [cde]	3.84 ± 1.66 [bcd]	2.87 ± 0.37 [b]	0.422 ± 0.129 [b]
A-882	1.89 ± 0.28 [a]	15.37 ± 3.99 [bc]	16.05 ± 4.80 [cd]	4.82 ± 2.15 [abc]	2.81 ± 0.10 [bc]	0.442 ± 0.033 [b]
A-980	1.67 ± 0.33 [a]	39.71 ± 1.02 [a]	24.23 ± 5.89 [a]	5.73 ± 2.11 [ab]	5.09 ± 1.21 [a]	0.694 ± 0.209 [a]
C-Control	0.41 ± 0.20 [b]	1.24 ± 0.19 [d]	1.65 ± 0.22 [f]	0.56 ± 0.06 [e]	1.79 ± 0.42 [c]	0.215 ± 0.043 [c]
C-686	2.04 ± 0.15 [a]	13.04 ± 3.70 [bc]	12.51 ± 4.69 [de]	3.54 ± 1.04 [cd]	3.11 ± 0.39 [b]	0.317 ±0.075 [bc]
C-784	1.94 ± 0.18 [a]	17.58 ± 7.58 [b]	12.87 ± 2.39 [cde]	4.20 ± 1.41 [abcd]	3.02 ± 0.15 [b]	0.400 ± 0.140 [b]
C-882	1.74 ± 0.22 [a]	17.28 ± 0.34 [b]	18.48 ± 3.02 [bc]	4.91 ± 1.73 [abc]	3.25 ± 0.45 [b]	0.616 ± 0.146 [a]
C-980	1.85 ± 0.23 [a]	39.64 ± 3.13 [a]	21.33 ± 6.21 [ab]	6.19 ± 1.34 [a]	4.69 ± 0.69 [a]	0.645 ± 0.204 [a]

A, American ginseng; C, Canadian ginseng; DPPH, DPPH radical scavenging activity; ABTS, ABTS radical scavenging activity; TPC, total phenolic content; TFC, total flavonoid content; AP, acidic polysaccharides; VCE: vitamin C equivalent; GAE: gallic acid equivalent; CE: catechin equivalent; GA, galacturonic acid; MRPs, Maillard reaction products. * Values with the same letter in the column are not significantly different ($p < 0.05$).

3.2. Extraction Yield and Crude Saponin Content

The extraction yields and crude saponin content of puffed American and Canadian ginsengs are shown in Table 1. Before puffing, in the control group, both extraction yield and crude saponin content of Canadian ginseng were higher than those of American ginseng. Puffed American and Canadian ginsengs revealed higher extraction yields and crude saponin contents than those of the control groups. Puffing caused cell wall breakdown and porous structure formation, resulting in easy elution of active components from ginseng [18]. Although puffing increased both extraction yield and crude saponin content in American and Canadian ginsengs, no increasing trends in yield and content with increasing puffing pressure were found. The puffed American and Canadian ginsengs showed the highest extraction yield at 784 kPa and 880 kPa, respectively. In contrast, both ginsengs showed the highest crude saponin content at 686 kPa. Consequently, puffing increased both extraction yield and crude saponin content of each ginseng type at different puffing pressures.

3.3. Changes in Ginsenosides

The HPLC chromatograms of puffed American and Canadian ginsengs are shown in Figure 2. In control groups of American and Canadian ginsengs, Rb1, Rb2, Rc, Rd, and Re were the major ginsenosides, all known to be characteristic of ginseng. In contrast, after puffing, the major ginsenoside content gradually decreased as puffing pressure increased in both American and Canadian ginsengs, resulting in the production and increase of minor ginsenosides such as Rg2, Rg3, Rh2, and compound K. Puffing of ginseng has been reported to decrease the content of major ginsenosides and increase that of minor ginsenosides, especially Rg3 in *P. ginseng* [18,25].

The changes in the ginsenoside profiles of puffed American and Canadian ginsengs are shown in Figure 3. The content of all major ginsenosides except Rg1 (Rb1, Rb2, Rc, Rd, Re) increased up to 686 kPa and then decreased as the puffing pressure increased in both ginsengs. Decomposition or thermal conversion of ginsenosides occur when heat is applied [34]. The increase in major ginsenoside content at 686 kPa may be attributed to non-soluble polymer components converting into soluble ones due to puffing and to the increased penetration of the extraction solvent due to the porous structure of the tissue [23]. Another possible explanation is the weakening of the binding force due to the destruction of cell walls and to changes in the molecular structure induced by puffing [18,35]. In this study, ginsenosides Rg2, Rg3, Compound K, and Rh2 were newly produced by deglycosylation and pyrolysis of major ginsenosides. Their concentrations increased with increasing puffing pressure. Especially, Compound K concentration greatly increased at 980 kPa compared to the control group in both puffed ginsengs. The contents of ginsenosides Rg2 and Rg3 also significantly increased in both puffed ginsengs. This result is slightly different from previous results on puffed ginseng [25] and puffed red ginseng [18], possibly because different ginseng varieties were used. *P. quinquefolius* has a different ginsenoside profile compared to *P. ginseng*, resulting in different ginsenoside products after puffing. Unlike *P. ginseng*, the levels of ginsenoside F1 and F2 of *P. quinquefolius* were relatively high and did not change with puffing, indicating that those ginsenosides are heat-resistant and not easily transformed by puffing.

3.4. Antioxidant Activity, TPC, and TFC

Antioxidant activities (DPPH and ABTS), TPC, and TFC of puffed American and Canadian ginsengs are shown in Table 2. In the control group, no significant differences ($p < 0.05$) in antioxidant activities were observed between American and Canadian ginsengs. Puffed American and Canadian ginsengs showed much higher antioxidant activities, TPC, and TFC compared to the control group. In DPPH, puffing increased the activity approximately 5–6 times compared to the control group, regardless of the puffing pressure, in both American and Canadian ginsengs. ABTS, TPC, and TFC but not DPPH increased with increasing puffing pressure. ABTS radical scavenging activity increased more than 10 times compared with the control group. Especially, at 980 kPa, ABTS of both ginsengs increased approximately 40 times over that of the control group. TPC and TFC in both American and Canadian ginsengs also showed the highest levels at 980 kPa, approximately 20 and 10 times higher than those of the control group, respectively. Water-soluble substances that affect the antioxidant activity exist in an insoluble state in combination with other substances. The increases of antioxidant activities, TPC, and TFC may be explained by the weakening of the molecular binding caused by heat treatment and elution of those materials into a water-soluble state [36] and the increase of certain substances such as maltol, a Maillard reaction product [15]. TPC [23] and TFC [37,38] have been reported to increase with increasing temperature and heat treatment time. Likewise, in this study, antioxidant activities, TPC, and TFC of American and Canadian ginsengs increased during puffing with increasing temperature and holding time.

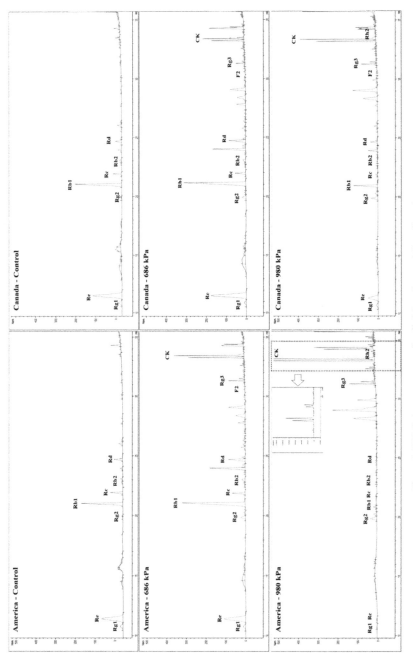

Figure 2. HPLC chromatograms of puffed American and Canadian ginsengs.

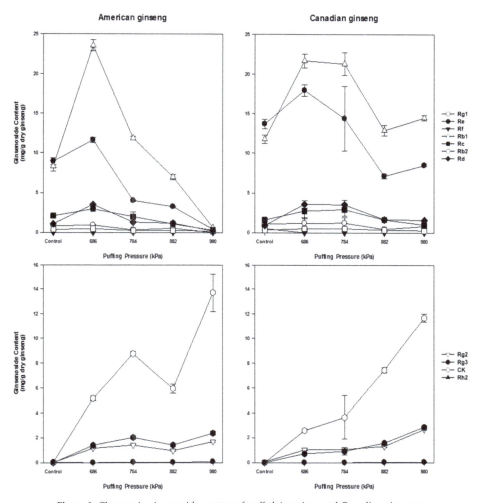

Figure 3. Changes in ginsenoside content of puffed American and Canadian ginsengs.

3.5. Acidic Polysaccharides and MRPs

The changes in acidic polysaccharides content of puffed American and Canadian ginsengs are shown in Table 2. Before puffing, American and Canadian ginsengs showed similar acidic polysaccharides contents. However, after puffing, the content of acidic polysaccharides of both ginsengs increased significantly ($p < 0.05$) compared to those of the non-puffed counterparts; the highest levels of acidic polysaccharides resulted from puffing at 980 kPa. No significant differences ($p < 0.05$) were observed by puffing at 686, 784, or 882 kPa in both American and Canadian ginseng. Ginseng contains 60–70% (dry basis) of carbohydrates, the major components of which are pectin and starch [39]. The acidic polysaccharides of ginseng consist of a pectin-like substance with a molecular weight of 34,600 Da, whose main component is galacturonic acid [22]. The majority of the polysaccharides from *P. quinquefolius* are in the neutral form, and their acidic form increases with thermal processing. The content of acidic polysaccharides of ginseng tends to increase with heat treatment [22,32]. Moreover, acidic polysaccharides are affected by temperature rather than heat treatment time [22]. Depending on the specific characteristics of puffing, the greatest concentration of acidic polysaccharides in this study was produced by puffing at 980 kPa in both American and Canadian ginsengs.

The MRPs of puffed American and Canadian ginsengs are shown in Table 2. The MRPs of both American and Canadian ginsengs before puffing were not significantly different ($p < 0.05$) and gradually increased with increasing puffing pressure. The increase in MRPs with thermal processing of American ginseng has been reported [15], and those changes were correlated with changes in antioxidant activity, TPC, and TFC [23,37,38].

3.6. Relationships between Antioxidant Activities, TPC, TFC, MRPs, and Acidic Polysaccharides

The results of the Pearson correlations among antioxidant activities, TPC, TFC, amount of MRPs, and acidic polysaccharides content of puffed American and Canadian ginsengs are shown in Table 3. In American ginseng, ABTS correlated well with acidic polysaccharides content, TPC, and amount of MRPs. TFC was in good correlation with TPC and amount of MRPs. TPC and amount of MRPs had good correlation with all other parameters except for DPPH. Moreover, acidic polysaccharides content had a good correlation with amount of MRPs, ABTS, and TPC. In contrast, the correlations for Canadian ginseng were slightly different from those of American ginseng; ABTS correlated well with acidic polysaccharides content, TPC, and TFC. The amount of MRPS had a good correlation with TPC and TFC. TPC and TFC h ad good correlation with all other parameters except for DPPH. Likewise, acidic polysaccharides content had a good correlation with ABTS and TFC.

Table 3. Pearson correlations between amount of MRPs, DPPH, ABTS, TPC, TFC, and acidic polysaccharides content of puffed American and Canadian ginsengs.

American Ginseng	MRPs	DPPH	ABTS	TPC	TFC	AP
MRPs						
DPPH	0.548					
ABTS	0.982 **	0.471				
TPC	0.981 **	0.693	0.942 *			
TFC	0.938 *	0.737	0.868	0.981 **		
AP	0.965 **	0.444	0.997 ***	0.918 *	0.831	
Canadian Ginseng						
MRPs						
DPPH	0.548					
ABTS	0.835	0.613				
TPC	0.937 *	0.793	0.88 **			
TFC	0.910 *	0.820	0.911 *	0.989 **		
AP	0.842	0.705	0.982 **	0.926 *	0.941 *	

MRPs, Maillard reaction products; DPPH, DPPH radical scavenging activity; ABTS, ABTS radical scavenging activity; TPC, total phenolic content; TFC, total flavonoid content; AP, acidic polysaccharides. * indicates $p < 0.05$; ** indicates $p < 0.01$; *** indicates $p < 0.001$.

Although the correlation patterns of American and Canadian ginsengs were not the same, two common correlations were observed. First, acidic polysaccharides content showed the strongest correlation with ABTS in both American and Canadian ginsengs. Second, DPPH did not correlate well with any other antioxidant parameter in either American or Canadian ginseng. These results suggest that compounds and mechanisms related to ABTS are very closely related to acidic polysaccharides content, and that DPPH has distinctive characteristics, different from the other antioxidant parameters evaluated in this study. It is interesting to note that this result is not casual, and this difference may come from the different sources of the samples, their different reactivity when subjected to puffing, and so on. Further analysis is needed for clarifying this result.

4. Conclusions

Puffing changed the structure of ginseng from dense to porous leading to easy elution of bioactive compounds from ginseng. This phenomenon increased the extraction yields and crude saponin content of both puffed American and Canadian ginsengs. It is notable that the levels of the major ginsenosides (Rb1, Rb2, Rc, Rd, Re) increased by puffing at 686 kPa, possibly as a result of the formation of a porous structure without thermal conversion of the ginsenosides. Deglycosylation and pyrolysis of major ginsenosides during the puffing process induced the production and increase of minor ginsenosides (Rg2, Rg3, Compound K, and Rh2). Especially, the level of compound K dramatically increased by puffing at 980 kPa in both American and Canadian ginsengs. Moreover, puffing greatly increased antioxidant activities (ABTS and DPPH), and the increase of MRPs level seemed to significantly affect the ABTS of puffed ginsengs. Puffed American and Canadian ginsengs showed higher antioxidant activities and higher minor ginsenoside levels than non-puffed ginsengs, a trend similar to that of puffed *P. ginseng*. There was no significant difference between American and Canadian ginsengs in either control or puffed samples. In the case of *P. ginseng*, red ginseng, and black ginseng, the preparation takes a very long time, which is not economical. On the other hand, the preparation of puffed ginseng takes relatively a very short time, and more effective changes in ginsenosides and bioactive compounds can be obtained. Therefore, puffing is an efficient and economic processing method appropriate for both American and Canadian ginsengs to enhance ginseng's antioxidant activity and produce minor ginsenosides, such as Rg3, Compound K, and Rh2.

Author Contributions: Conceptualization, H.L., W.K., and M.-Y.B.; funding acquisition, H.L., W.K., and M.-Y.B.; investigation, M.-S.K., S.-J.J., W.K., and M.-Y.B.; methodology, M.-S.K., S.-J.J., S.J.Y., Y.-J.P., D.-O.K., and B.-Y.K.; project administration, H.L., W.K., and M.-Y.B.; supervision, W.K., and M.-Y.B.; writing—original draft, M.-S.K., S.-J.J., W.K., and M.-Y.B.; writing—review and editing, H.L., Y.-J.P., D-O.K., B.-Y.K., W.K., and M.-Y.B.

Funding: This work was supported by the Rural Development Administration, Republic of Korea (PJ01314402).

Conflicts of Interest: The authors declare no conflicts of interest.

References

1. Ang-Lee, M.K.; Moss, J.; Yuan, C.-S. Herbal medicines and perioperative care. *Jama* **2001**, *286*, 208–216. [CrossRef] [PubMed]
2. Attele, A.S.; Wu, J.A.; Yuan, C.-S. Ginseng pharmacology: Multiple constituents and multiple actions. *Biochem. Pharmacol.* **1999**, *58*, 1685–1693. [CrossRef]
3. Wu, J.; Zhong, J.-J. Production of ginseng and its bioactive components in plant cell culture: Current technological and applied aspects. *J. Biotechnol.* **1999**, *68*, 89–99. [CrossRef]
4. Lee, S.-J.; Ko, W.-G.; Kim, J.-H.; Sung, J.-H.; Lee, S.-J.; Moon, C.-K.; Lee, B.-H. Induction of apoptosis by a novel intestinal metabolite of ginseng saponin via cytochrome c-mediated activation of caspase-3 protease. *Biochem. Pharmacol.* **2000**, *60*, 677–685. [CrossRef]
5. Park, J. Recent studies on the chemical constituents of Korean ginseng (Panax ginseng CA Meyer). *Korean J. Ginseng Sci.* **1996**, *20*, 389–415.
6. Park, C.-K.; Jeon, B.-S.; Yang, J.-W. The chemical components of Korean ginseng. *Food Ind. Nutr.* **2003**, *8*, 10–23.
7. Yang, X.-D.; Yang, Y.-Y.; Ouyang, D.-S.; Yang, G.-P. A review of biotransformation and pharmacology of ginsenoside compound K. *Fitoterapia* **2015**, *100*, 208–220. [CrossRef]
8. Huang, B.-M.; Chen, T.-B.; Xiao, S.-Y.; Zha, Q.-L.; Luo, P.; Wang, Y.-P.; Cui, X.-M.; Liu, L.; Zhou, H. A new approach for authentication of four ginseng herbs and their related products based on the simultaneous quantification of 19 ginseng saponins by UHPLC-TOF/MS coupled with OPLS-DA. *RSC Adv.* **2017**, *7*, 46839–46851. [CrossRef]
9. Nah, S.-Y. Ginseng ginsenoside pharmacology in the nervous system: Involvement in the regulation of ion channels and receptors. *Front. Physiol.* **2014**, *5*, 98. [CrossRef]
10. Chen, C.-F.; Chiou, W.-F.; Zhang, J.-T. Comparison of the pharmacological effects of *Panax ginseng* and *Panax quinquefolium*. *Acta Pharmacol. Sin.* **2008**, *29*, 1103. [CrossRef]

11. Hall, T.; Lu, Z.; Yat, P. Ginseng evaluation program part I: Standardized phase report. *J. Am. Bot. Counc. Herb.* **2001**, *52*, 26–45.
12. Yuan, C.-S.; Wang, C.-Z.; Wicks, S.M.; Qi, L.-W. Chemical and pharmacological studies of saponins with a focus on American ginseng. *J. Ginseng Res.* **2010**, *34*, 160. [CrossRef] [PubMed]
13. Assinewe, V.A.; Baum, B.R.; Gagnon, D.; Arnason, J.T. Phytochemistry of wild populations of Panax quinquefolius L. (North American ginseng). *J. Agric. Food Chem.* **2003**, *51*, 4549–4553. [CrossRef] [PubMed]
14. Helms, S. Cancer prevention and therapeutics: Panax ginseng. *Altern. Med. Rev.* **2004**, *9*, 259–274. [PubMed]
15. Kim, K.T.; Yoo, K.M.; Lee, J.W.; Eom, S.H.; Hwang, I.K.; Lee, C.Y. Protective effect of steamed American ginseng (Panax quinquefolius L.) on V79-4 cells induced by oxidative stress. *J. Ethnopharmacol.* **2007**, *111*, 443–450. [CrossRef] [PubMed]
16. Oh, C.-H.; Kang, P.-S.; Kim, J.-W.; Kwon, J.; Oh, S.-H. Water extracts of cultured mountain ginseng stimulate immune cells and inhibit cancer cell proliferation. *Food Sci. Biotechnol.* **2006**, *15*, 369–373.
17. Wang, C.-Z.; Li, B.; Wen, X.-D.; Zhang, Z.; Yu, C.; Calway, T.D.; He, T.-C.; Du, W.; Yuan, C.-S. Paraptosis and NF-κB activation are associated with protopanaxadiol-induced cancer chemoprevention. *BMC Complement. Altern. Med.* **2013**, *13*, 2. [CrossRef] [PubMed]
18. An, Y.-E.; Ahn, S.-C.; Yang, D.-C.; Park, S.-J.; Kim, B.-Y.; Baik, M.-Y. Chemical conversion of ginsenosides in puffed red ginseng. *LWT Food Sci. Technol.* **2011**, *44*, 370–374. [CrossRef]
19. Hoseney, R.C. *Principles of Cereal Science and Technology*; American Association of Cereal Chemists (AACC): Saint Paul, MN, USA, 1994.
20. Mishra, G.; Joshi, D.; Panda, B.K. Popping and puffing of cereal grains: A review. *J. Grain Process. Storage* **2014**, *1*, 34–46.
21. Kim, S.-T.; Jang, J.-H.; Kwon, J.-H.; Moon, K.-D. Changes in the chemical components of red and white ginseng after puffing. *Korean J. Food Preserv.* **2009**, *16*, 355–361.
22. Yoon, S.-R.; Lee, M.-H.; Park, J.-H.; Lee, I.-S.; Kwon, J.-H.; Lee, G.-D. Changes in physicochemical compounds with heating treatment of ginseng. *J. Korean Soc. Food Sci. Nutr.* **2005**, *34*, 1572–1578.
23. Kim, S.-B.; Do, J.-R.; Lee, Y.-W.; Gu, Y.-S.; Kim, C.-N.; Park, Y.-H. Nitrite-scavenging effects of roasted-barley extracts according to processing conditions. *Korean J. Food Sci. Technol.* **1990**, *22*, 748–752.
24. Ryu, G.; Lee, J. Development of extrusion process on red ginseng from raw ginseng and its products. *Final Rep. Ventur. Res. Minist. Health Welf. Seoul* **2003**, 9–10.
25. Kim, J.-H.; Ahn, S.-C.; Choi, S.-W.; Hur, N.-Y.; Kim, B.-Y.; Baik, M.-Y. Changes in effective components of ginseng by puffing. *Appl. Biol. Chem.* **2008**, *51*, 188–193.
26. Blois, M.S. Antioxidant determinations by the use of a stable free radical. *Nature* **1958**, *181*, 1199–1200. [CrossRef]
27. Van den Berg, R.; Haenen, G.R.; van den Berg, H.; Bast, A. Applicability of an improved Trolox equivalent antioxidant capacity (TEAC) assay for evaluation of antioxidant capacity measurements of mixtures. *Food Chem.* **1999**, *66*, 511–517. [CrossRef]
28. Hossain, A.; Moon, H.K.; Kim, J.-K. Antioxidant properties of Korean major persimmon (Diospyros kaki) leaves. *Food Sci. Biotechnol.* **2018**, *27*, 177–184. [CrossRef]
29. Singleton, V.L.; Rossi, J.A. Colorimetry of total phenolics with phosphomolybdic-phosphotungstic acid reagents. *Am. J. Enol. Vitic.* **1965**, *16*, 144–158.
30. Nam, T.G.; Kim, D.-O.; Eom, S.H. Effects of light sources on major flavonoids and antioxidant activity in common buckwheat sprouts. *Food Sci. Biotechnol.* **2018**, *27*, 169–176. [CrossRef]
31. Zhishen, J.; Mengcheng, T.; Jianming, W. The determination of flavonoid contents in mulberry and their scavenging effects on superoxide radicals. *Food Chem.* **1999**, *64*, 555–559. [CrossRef]
32. Do, J.; Lee, H.; Lee, S.; Jang, J.; Lee, S.; Sung, H. Colorimetric determination of acidic polysaccharide from Panax ginseng, its extraction condition and stability. *Korean J. Ginseng Sci.* **1993**, *17*, 139–144.
33. Kim, K.T.; Yoo, K.M. Effect of hot water boiling and autoclaving on physicochemical properties of American ginseng (*Panax quinquefolium* L.). *J. Ginseng Res.* **2009**, *33*, 40–47.
34. Hong, H.-D.; Kim, Y.-C.; Rho, J.-H.; Kim, K.-T.; Lee, Y.-C. Changes on physicochemical properties of Panax ginseng CA Meyer during repeated steaming process. *J. Ginseng Res.* **2007**, *31*, 222–229.
35. Hong, H.-D.; Kim, Y.-C.; Kim, S.-S.; Sim, G.-S.; Han, C.-K. Effect of puffing on quality characteristics of red ginseng tail root. *J. Ginseng Res.* **2007**, *31*, 147–153.

36. Choi, Y.; Lee, S.; Chun, J.; Lee, H.; Lee, J. Influence of heat treatment on the antioxidant activities and polyphenolic compounds of Shiitake (Lentinus edodes) mushroom. *Food Chem.* **2006**, *99*, 381–387. [CrossRef]
37. Dewanto, V.; Wu, X.; Adom, K.K.; Liu, R.H. Thermal processing enhances the nutritional value of tomatoes by increasing total antioxidant activity. *J. Agric. Food Chem.* **2002**, *50*, 3010–3014. [CrossRef]
38. Yang, S.-J.; Woo, K.-S.; Yoo, J.-S.; Kang, T.-S.; Noh, Y.-H.; Lee, J.-S.; Jeong, H.-S. Change of Korean ginseng components with high temperature and pressure treatment. *Korean J. Food Sci. Technol.* **2006**, *38*, 521–525.
39. Ko, S.; Choi, K.; Kim, H. Comparison of proximate composition, mineral nutrient, amino acid and free sugar contents of several Panax species. *Korean J. Ginseng Sci.* **1996**, *20*, 36–41.

© 2019 by the authors. Licensee MDPI, Basel, Switzerland. This article is an open access article distributed under the terms and conditions of the Creative Commons Attribution (CC BY) license (http://creativecommons.org/licenses/by/4.0/).

Article

Salicylic Acid and Melatonin Alleviate the Effects of Heat Stress on Essential Oil Composition and Antioxidant Enzyme Activity in *Mentha × piperita* and *Mentha arvensis* L.

Milad Haydari [1,†], Viviana Maresca [2,†], Daniela Rigano [3,†], Alireza Taleei [1], Ali Akbar Shahnejat-Bushehri [1], Javad Hadian [4], Sergio Sorbo [5], Marco Guida [2], Caterina Manna [6], Marina Piscopo [2], Rosaria Notariale [6], Francesca De Ruberto [2], Lina Fusaro [7] and Adriana Basile [2,*]

[1] Department of Agronomy and Plant Breeding, Collage of Agriculture and Natural Resources, University of Tehran, P.O. Box 31787-316, Karaj 77871-31587, Iran; milad.heydari@ut.ac.ir (M.H.); ataleei@ut.ac.ir (A.T.); ashah@ut.ac.ir (A.A.S.-B.)
[2] Department of Biology—University of Naples "Federico II", 80126 Naples, Italy; viviana.maresca@unina.it (V.M.); marco.guida@unina.it (M.G.); marina.piscopo@unina.it (M.P.); francesca.deruberto@gmail.com (F.D.R.)
[3] Department of Pharmacy, School of Medicine and Surgery, University of Naples Federico II, 80126 Naples, Italy; drigano@unina.it
[4] Medicinal Plants and Drug Research Institute, ShahidBeheshti University, G.C. Tehran 11369, Iran; j_hadian@sbu.ac.ir
[5] C.e.S.M.A. University of Naples "Federico II", 80126 Naples, Italy; sersorbo@unina.it
[6] Department of Precision Medicine, School of Medicine, University of Campania "Luigi Vanvitelli", via Luigi de Crecchio, 80138 Naples, Italy; caterina.manna@unicampania.it (C.M.); notarialer@gmail.com (R.N.)
[7] Department of Environmental Biology, Sapienza University of Rome, P.le Aldo Moro 5, 00185 Rome, Italy; lina.fusaro@uniroma1.it
* Correspondence: adbasile@unina.it; Tel.: +39-0812538508
† These authors contributed equally to this work.

Received: 16 October 2019; Accepted: 11 November 2019; Published: 13 November 2019

Abstract: The aim of this study was to evaluate changes in the chemical profile of essential oils and antioxidant enzymes activity (catalase CAT, superoxide dismutase SOD, Glutathione *S*-transferases GST, and Peroxidase POX) in *Mentha × piperita* L. (Mitcham variety) and *Mentha arvensis* L. (var. *piperascens*), in response to heat stress. In addition, we used salicylic acid (SA) and melatonin (M), two brassinosteroids that play an important role in regulating physiological processes, to assess their potential to mitigate heat stress. In both species, the heat stress caused a variation in the composition of the essential oils and in the antioxidant enzymatic activity. Furthermore both Salicylic acid (SA) and melatonin (M) alleviated the effect of heat stress.

Keywords: mentha; heat stress; antioxidant enzyme activity; salicylic acid; melatonin; essential oil

1. Introduction

The Lamiaceae family encompasses various genera, including aromatic herbs such as mint. Embracing half a dozen cultivated species, mint genus includes more than 30 species that are scattered worldwide, chiefly in temperate and tropical/subtropical regions. One of the distinctive features is that mint species possess essential oils [1].

Japanese mint or Cornmint (*Mentha arvensis* L. var. *piperascens* (Malinv. ex Holmes) Malinv. ex L.H.Bailey) is a fundamental natural source of monoterpenes, particularly L-menthol (up to 80% menthol), and it was already cultivated in ancient Japan as well as in China, India, and Brazil.

Mentha × piperita is an abortive hybrid of the species *M. aquatica* L. and *M. spicata*. Ecumenically, peppermint is one of the most commercial odorous scented herbs. The peppermint leaves have not only a peculiar, sweet, and strong odor, but also a redolent, warm, and spicy taste, with a cooling aftertaste. The supremacy of the essential oils of *Mentha × piperita* is due to the presence of menthone, isomenthone, and different isomers of menthol. Nowadays, extensive usage of peppermint oil in flavoring chewing gums, sugar confectioneries, ice creams, desserts, baked goods, tobacco, and alcoholic beverages is just one of the most prevalent applications of such oils. Furthermore, it is also commonly employed in the flavoring of pharmaceutical and oral preparations [2].

Menthol shows various biological activities, such as sedative, anesthetic, antiseptic, gastric, and antipruritic. It is also one of the few natural monocyclic monoterpene alcohols that have characteristics conducive to fragrances. As such, it has been used to flavor various goods such as candies, chewing gums, and toothpaste [3,4].

Heat stress has effects on metabolite synthesis in aromatic plants, changing phenolic and antioxidants concentrations [5]. Heat stress induces the generation of reactive oxygen species (ROS), such as superoxide radicals ($\bullet O_2^-$), hydrogen peroxide (H_2O_2), and hydroxyl radicals ($\bullet OH$), in plants, thereby creating a state of oxidative stress in them. This increased ROS level in plants causes oxidative damage to biomolecules such as lipids, proteins, and nucleic acids, thus altering the redox homeostasis [6,7]. To avoid potential damage by ROS, a balance between production and elimination of ROS at the intracellular level must be regulated. This equilibrium between production and detoxification of ROS is sustained by enzymatic and nonenzymatic antioxidants [8,9]. The enzymatic components comprise several antioxidant enzymes, such as superoxide dismutase (SOD), catalase (CAT), glutathione peroxidase (GPX), guaiacol peroxidase (POX), and peroxiredoxins.

Salicylic acid (SA) and melatonin (M) are two brassinosteroids playing an important role in regulating physiological processes. SA is a phenolic compound with antioxidant properties, involved in the regulation of physiological processes in plants [10]. SA can modulate plant responses to a wide range of oxidative stresses [11]. When applied exogenously at suitable concentrations, SA was found to enhance the efficiency of antioxidant system in plants [12]. M (N-acetyl-5-methoxytryptamine) is an indole hormone involved in multiple biological processes [13]. According to a lot of findings, M plays an important role in the regulation of plant growth and development [14] and provides a defense against abiotic stresses such as extreme temperature, excess copper, salinity, and drought [15–17]. A lot of studies have proven that M may act as a plant growth regulator in rooting, seed germination, and delay in leaf senescence and other morphogenetic features [18–20]. M has been observed to improve tolerance for multiple stresses including heat stress, and in particular, exogenous M treatments protect plants from temperature extremes [21].

The aim of this study was to evaluate the response of *M. arvensis* L. var. *piperascens* and *M. × piperita* to heat stress in relation to the production of essential oils in general and in particular of menthol, menthone, and isomenthone, which have considerable economic importance and play an important role in the industrial field. In particular, our goal was to investigate the potential of SA and M to mitigate the heat stress effects on the two plants, focusing on the variation of essential oils composition and antioxidant enzymes activity.

2. Materials and Methods

2.1. Plant Material, Culture, and Treatment

M. × piperita L. var. Mitcham and *M. arvensis* var. *piperascens* Malinv. ex L.H. Bailey were obtained from the "Safiabad agricultural and natural resources research and education center". Planting and cultivation conditions were carried out in the growth chambers. The method is described in detail in Heydari et al. [22].

After 40 culturing days, plants were sprayed with SA (2, 3, and 4 mM, reported as SA2, SA3, and SA4, respectively) and M (10 and 30 M, reported as M1 and M3), together with M and SA at the highest concentrations (M3SA4). Tap water was used for controls.

For each treatment, we selected 50 sample plants for subsequent experiments.

The abbreviation used to antioxidant enzyme activity and GC and GC-MS (Gas chromatography - Mass spectrometry) analysis are:

MpH1C = *M.* x *piperita* at the H1 temperature without treatment; MpH1M3 = *M.* x *piperita* at the H1 temperature treated with melatonin 3 mM; MpH1SA4 = *M.* x *piperita* at the H1 temperature treated with 4 mM salicylic acid; MpH1M3SA4 = *M.* x *piperita* at the H1 temperature treated with melatonin 3 mM and 4 mM salicylic acid; MpH2C = *M.* x *piperita* at the H2 temperature without treatment; MpH2M3 = *M.* x *piperita* at the H2 temperature treated with melatonin 3 mM; MpH2SA4 = *M.* x *piperita* at the H2 temperature treated with 4 mM salicylic acid; MpH2M3SA4 = *M.* x *piperita* at the H2 temperature treated with melatonin 3 mM and 4 mM salicylic acid; MpH3C = *M.* x *piperita* at the H3 temperature without treatment; MpH3M3 = *M.* x *piperita* at the H3 temperature treated with melatonin 3 mM; MpH3SA4 = *M.* x *piperita* at the H3 temperature treated with 4 mM salicylic acid; MpH3M3SA4 = *M.* x *piperita* at the H3 temperature treated with melatonin 3 mM and 4 mM salicylic acid; MaH1C = *M. arvensis* L. var. *piperascens* at the H1 temperature without treatment; MaH1M3 = *M. arvensis* L. var. *piperascens* at the H1 temperature treated with melatonin 3 mM; MaH1SA4 = *M. arvensis* L. var. *piperascens* at the H1 temperature treated with 4 mM salicylic acid; MaH1M3SA4= *M. arvensis* L. var. *piperascens* at the H1 temperature treated with melatonin 3 mM and 4 mM salicylic acid; MaH2C = *M. arvensis* L. var. *piperascens* at the H2 temperature without treatment; MaH2M3 = *M. arvensis* L. var. *piperascens* at the H2 temperature treated with melatonin 3 mM; MaH2SA4 = *M. arvensis* L. var. *piperascens* at the H2 temperature treated with 4 mM salicylic acid; MaH2M3SA4 = *M. arvensis* L. var. *piperascens* at the H2 temperature treated with melatonin 3 mM and 4 mM salicylic acid; MaH3C = *M. arvensis* L. var. *piperascens* at the H3 temperature without treatment; MaH3M3 = *M. arvensis* L. var. *piperascens* at the H3 temperature treated with melatonin 3 mM; MaH3SA4 = *M. arvensis* L. var. *piperascens* at the H3 temperature treated with 4 mM salicylic acid; MaH3M3SA4 = *M. arvensis* L. var. *piperascens* at the H3 temperature treated with melatonin 3 mM and 4 mM salicylic acid.

2.2. Relative Water Content (RWC)

Relative Water Content (RWC) was calculated according to Dhopte and Manuel [23]:

$$RWC = (FW-DW)/(TW-DW) \times 100$$

where FW is leaf fresh weight, DW is dry weight, and TW is leaf turgor mass of leaf samples [24] obtained measuring the leaf weight after 10–12 h in water saturating conditions.

2.3. Antioxidant Enzyme Activity

Protein extraction and the activity of antioxidant enzymes(SOD, CAT, GST ans PEROX) was carried out according to Maresca et al. [25].

2.4. Isolation of Essential Oils

Samples' shoots were air-dried in dark conditions at room temperature and were used for essential oils extraction. Each sample (50 g in three replications) was extracted using hydro-distillation for 3 h and Clevenger-type apparatus based on the standard procedure described by Russo et al. [26]. The essential oils were obtained with different yields (0.97 ± 0.02–3.26 ± 0.02%) on dry mass (w/w) and results were yellowish with a pleasant smell. The oils were dried with anhydrous sodium sulfate and stored under N2 at +4 °C in the dark for subsequent tests and analyses.

2.5. GC and GC-MS Analysis

Analytical gas chromatography was carried out on a Perkin-Elmer Sigma 115 gas chromatograph fitted with an Agilent HP-5 MS capillary column (30 m × 0.25 mm), 0.25 µm film thickness. The analysis was also performed by using a fused silica HP Innowax polyethylene glycol capillary column (50 m × 0.20 mm), 0.20 µm film thickness. Gas chromatography analysis was performed as done previously and described in detail by Rigano et al. [27]. Compounds identification and components relative percentages were carried out as described by Rigano et al. [27].

2.6. Statistical Analysis

For each species, the differences between treatments were analyzed by factorial ANOVA (Analysis of Variance) using the hormones and temperature as categorical predictors. The factorial ANOVA was followed by Student-Neuman-Keuls test for post hoc comparisons. Results were reported as mean ± standard deviation.

3. Results

3.1. Relative Water Content (RWC)

Significant decrease of Relative Water Contents (RWC%) in H2 and H3 suggested that plants could be under stress. *M. arvensis* L. var. *piperascens* lost more water than *M.* × *piperita* L. SA and M restored water content in a dose-dependant way (Figure 1). For this reason, for the next experiments, we only used the highest concentration for both treatments: melatonin 30 M (M3) and salicylic acid 4 mM (SA4).

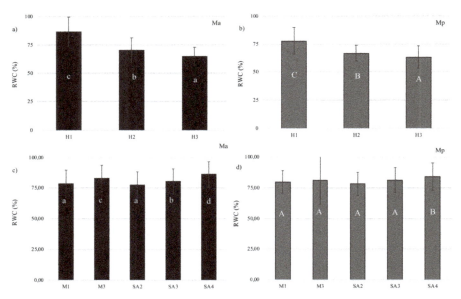

Figure 1. Relative Water Content for (**a**) and (**c**) *M. arvensis* L. var. *piperascens* (Ma) and for (**b**) and (**d**) *M.* × *piperita* (Mp). Values are presented as means ± standard deviation (n = 15); values not accompanied by the same letter are significantly different at $p < 0.05$, using the post-hoc Student–Newman–Keuls test. Lowercase letters (a–d) indicate significant differences between treatments for Ma; uppercase letters (A–C) indicate significant differences between treatments for Mp.

Antioxidants **2019**, *8*, 547

3.2. Antioxidant Enzyme Activity

As for the activity of antioxidant enzymes, heating determined a temperature-dependent increase, and treatments with SA4 and M3 determined a further increase, which proved to be extremely significant with the two hormones used simultaneously at their maximum concentrations.

In *M.* × *piperita*, the activity of all the measured antioxidant enzymes increased with increasing temperature both in the absence and presence of SA and M (M3, SA4). The only exception was the POX activity of samples C, which did not increase with increasing temperature but only under H3 conditions.

Moreover, in most cases the treatment with SA4 had a synergistic effect with the temperature compared to M3 on the activity of all the measured enzymes.

In general, for all the enzymatic activities measured, the samples treated with SA4M3 maintained a significantly higher enzyme activity compared to the C control samples and in the samples treated individually with SA and M3.

In *M. arvensis* L. var. *piperascens* the antioxidant enzymes activity in relation to temperature and treatment with M3, SA4, and M3SA4 followed the same trend shown in *M.* × *piperita* (Table 1).

Table 1. Enzyme activity in *M.* × *piperita* L. and *M. arvensis* L. var. *piperascens* for each treatment.

	Temperature	Hormons	CAT	GST	POX	SOD
Ma	H1	C	3.188 ± 0.89 [a]	1.02 ± 0.08 [a]	19.39 ± 1.25 [a]	6.88 ± 0.21 [a]
Ma		M3	34.76 ± 1.94 [a]	1.78 ± 0.08 [c]	126.42 ± 1.25 [c]	12.98 ± 1.03 [b]
Ma		SA4	42.56 ± 2.05 [a]	1.49 ± 0.06 [b]	98.12 ± 3.69 [c]	10.33 ± 1.05 [bc]
Ma		M3SA4	66.42 ± 2.45 [d]	2.54 ± 0.06 [d]	226.77 ± 5.11 [d]	18.45 ± 1.13 [bc]
Ma	H2	C	9.46 ± 1.83 [b]	1.52 ± 0.03 [d]	22.73± 1.04 [a]	9.98 ± 1.01 [c]
Ma		M3	63.86 ± 2.95 [c]	2.01 ± 0.08 [e]	245.13 ± 9.46 [d]	43.79 ± 1.13 [d]
Ma		SA4	98.73 ± 2.83 [b]	2.67 ± 0.08 [h]	108.42 ± 3.14 [f]	19.07 ± 1.16 [e]
Ma		M3SA4	134.52 ± 2.54 [e]	4.29 ± 0.08 [i]	269.67 ± 1.22 [g]	54.53 ± 1.51 [f]
Ma	H3	C	19.61 ± 1.67 [d]	2.28 ± 0.02 f	4.96 ±1.45 [b]	16.88 ± 0.71 [d]
Ma		M3	92.37± 2.04 [e]	2.67 ± 0.04 [g]	316.9 1± 7.70 [e]	49.07 ± 0.73 [g]
Ma		SA4	129.04 ± 3.75 [e]	3.74 ± 0.06 [l]	146.95 ± 4.22 [h]	49.12 ± 0.87 [g]
Ma		M3SA4	130.33 ± 2.16 f	4.61 ± 0.09 [m]	491.12 ± 11.77 [i]	58.77 ± 0.89 [h]
Mp	H1	C	9.53 ± 1.33 [a]	0.92 ±0.04 [a]	11.49 ± 1.25 [a]	5.64 ± 0.55 [a]
Mp		M3	10.51 ± 1.07 [a]	1.70 ± 0.03 [c]	61.50 ± 1.42 [c]	18.61 ± 1.62 [b]
Mp		SA4	12.54 ± 1.46 [a]	1.57 ± 0.07 [b]	66.26 ± 2.38 [c]	20.61 ± 0.83 [bc]
Mp		M3SA4	41.15 ± 2.03 [d]	1.93 ± 0.1 [d]	103.57 ±5.45 [d]	20.54 ± 0.53 [bc]
Mp	H2	C	19.42 ± 1.47 [b]	1.93 ± 0.03 [d]	15.19 ± 1.52 [a]	22.06 ± 0.90 [c]
Mp		M3	29.54 ± 1.19 [c]	2.28 ± 0.07 [e]	95.12 ± 1.20 [d]	28.75 ± 2.23 [d]
Mp		SA4	22.39 ± 1.54 [b]	3.12 ± 0.08 [h]	154.46 ± 4.56 [f]	33.48 ± 0.82 [e]
Mp		M3SA4	54.49 ± 2.50 [e]	3.74 ± 0.07 [i]	343.70 ± 6.38 [g]	51.21 ± 1.31 [f]
Mp	H3	C	42.75 ± 1.6 [d]	2.63 ± 0.04 f	36.79 ± 0.82 [b]	29.81 ± 1.19 [d]
Mp		M3	53.12 ± 2.04 [e]	2.99 ± 0.08 [g]	138.90 ± 7.00 [e]	56.40 ± 1.47 [g]
Mp		SA4	51.76 ± 2.26 [e]	6.13 ± 0.08 [l]	417.25 ± 13.36 [h]	54.50 ± 1.52 [g]
Mp		M3SA4	130.33 ± 2.16 f	6.52 ± 0.08 [m]	466.18 ± 16.72 [i]	66.61 ± 1.27 [h]

Values are presented as means ± standard deviation (n = 15); values not accompanied by the same letter[a–i,l,m], are significantly different at $p < 0.05$ using the post-hoc Student–Newman–Keuls test. For treatment details, see the Material and Methods Section 2.1.

3.3. Essential Oil Yield

Essential oil yields in *M.* × *piperita* and *M. arvensis* L. var. *piperascens* were not statistically different (Figure 2). Heat stress had a similar effect on both species, determining a significant reduction of essential oils as the temperature increased. In addition, both the treatments with SA4 and M3 determined an increase in the yield of essential oils in samples exposed to H3 heat stress conditions, even if there was no statistically significant difference between SA4 and M3.

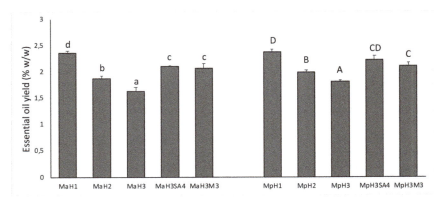

Figure 2. Essential oil yield in *M. arvensis* L. var. *piperascens* (Ma) and *M.* × *piperita* (Mp) shown in H1, H2, and H3 conditions, and the effect of melatonin (M3) and salicylic acid (SA4) at their highest concentrations on essential oil yield in H3 condition. Values are presented as means ± standard deviation (n = 15); values not accompanied by the same letter are significantly different at $p < 0.05$, using the post-hoc Student–Newman–Keuls test. Lowercase letters(a–d) indicate significant differences between treatments for Ma; uppercase letters(A–D) indicate significant differences between treatments for Mp. For treatments details see Material and Methods Section 2.1.

The oxygenated monoterpenes amount in *M. arvensis* L. var. *piperascens* increased by using SA4, M3, and the two of them used simultaneously in normal condition (H1). In H2 conditions, only SA4 increased the oxygenated monoterpenes, while in H3 conditions only M3 increased them. In *M.* × *piperita* SA, M, and the two hormones used simultaneously increased the oxygenated monoterpenes in H1, and the major effect was observed by using M3. In H2 the oxygenated monoterpenes increased only by using S4 and M3 together. Unfortunately, we could not observe an increase of the oxygenated monoterpenes in H3 (Figure 3).

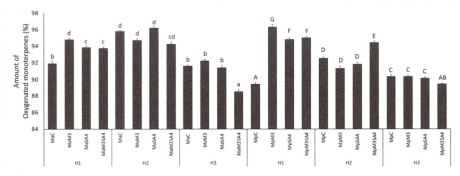

Figure 3. The amount of oxygenated monoterpenes in *M. arvensis* L. var. *piperascens* (Ma) and *M.* × *piperita* (Mp) under heat stress in H1, H2, and H3 and effects of melatonin (M3) and salicylic acid (SA4) on oxygenated monoterpenes. Values are presented as means ± standard deviation (n = 15); values not accompanied by the same letter are significantly different at $p < 0.05$, using the post-hoc Student-Newman-Keuls test. Lowercase letters(a–d) indicate significant differences between treatments for Ma; uppercase letters(A–G) indicate significant differences between treatments for Mp. For treatment details, see the Material and Methods Section 2.1.

A different trend was observed for monoterpene hydrocarbons on respect to oxygenated monoterpenes. In *M. arvensis* L. var. *piperascens* monoterpene hydrocarbons concentration was 4.3% for H1, 3.6% for H2, and 7.4% for H3 in control plants. Their amounts were decreased by using S4 (1.4%), M3 (1.3%), and both simultaneously (S4M3) (1.1%) in H1 conditions. In H2 they decreased by using M3 (0.8%), SA4 (0.8%), and the two of them simultaneously (1.0%), and the same applied for H3 by using SA4 (3.5%), M3 (2.5%) and the two of them simultaneously (4.8%). In *M. × piperita* the monoterpene hydrocarbons concentrations were in H1 4.0%, in H2 3.6% and in H3 5.2% in control plants. In H1 treatment by hormones decreased their amount (about 0.8%), and the same trend of reduction was observed for the other temperature conditions for all the treatments (data not shown).

Oxygenated sesquiterpenes were observed in very low concentrations in both the essential oils. Generally, in *M. arvensis* L. var. *piperascens* and in *M. × piperita* they were reduced or depleted by heat stress.

The dominant secondary metabolite in mint is menthol. Menthol dramatically decreased by heat stress, more than twice (in H1 56.6%, H2 39.0%, and H3 28.0%) in *M. arvensis* L. var. *piperascens* and about 4.5 fold (in H1 25%, H2 12.2%, and H3 5.6%) in *M. × piperita*. In *M. arvensis* L. var. *piperascens* only using SA4 and M3 simultaneously in H2 conditions the menthol concentration increased. In *M. × piperita* under H1 both SA4 and/or M3 increased the menthol concentration. In H3 by using SA4 and M3 simultaneously, the menthol concentration increased (14.3%) in comparison to control in H3 (Figure 4).

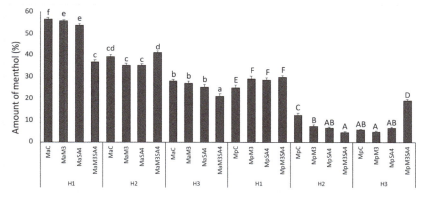

Figure 4. The amount of menthol in *M. arvensis* L. var. *piperascens* (Ma) and *M. × piperita* (Mp) under heat stress in H1, H2, and H3 and effect of melatonin (M3) and salicylic acid (SA4) on menthol. Values are presented as means ± standard deviation (n = 15); values not accompanied by the same letter are significantly different at $p < 0.05$, using the post-hoc Student-Newman-Keuls test. Lowercase letters(a–f) indicate significant differences between treatments for Ma; uppercase letters(A–F) indicate significant differences between treatments for Mp. For treatment details, see the Material and Methods Section 2.1.

Menthone is a precursor of menthol. The menthone concentration decreased in *M. arvensis* L. var. *piperascens* in H2 (7.6%) and H3 (12.0%), compared to H1 (14.5%). In *M. × piperita* the menthone concentration increased in H2 (15.9%) and decreased in H3 (9.5%), compared to H1 (13.0%). In normal condition (H1) in *M. arvensis* L. var. *piperascens* M3 decreased and SA4 increased menthone concentration. But in *M. × piperita* all the treatments increased menthone in H1 and H3 conditions. In H2 SA4 and M3 reduced menthone, but using both of them simultaneously increased menthone concentration (data not shown).

The menthofuran concentration in *M. arvensis* L. var. *piperascens* increased (in H1 7.8%, H2 34.0%, H3 13.0%) and in *M. × piperita* decreased (in H1 35.0%, H2 25.4%, H3 34.0%) under heat stress; the highest differences were observed in H2 condition. In *M. arvensis* L. var. *piperascens*, M3 increased, SA4 decreased, and the simultaneous use of both increased the menthofuran concentration, especially in the H1 condition. In H3 all treatments increased the menthofuran concentration. As regards *M. × piperita*, all treatments dramatically decreased the menthofuran concentration in H1. In H2, SA4 and M3 increased the menthofuran concentration. In H3, all the treatments decreased the menthofuran concentration, especially during simultaneous use (data not shown).

The pulegone concentrations increased for both the essential oils under heat stress condition. Particularly, the pulegone concentrations in *M. arvensis* L. var. *piperascens* were in H1 5.6%, in H2 8.0%, and in H3 11.0% and for *M.* × *piperita* in H1 15.0%, in H2 24.3% and in H3 28.1%. In *M. arvensis* L. var. *piperascens* SA4 and the simultaneous use of the two brassinosteroids caused an increase in the pulegone concentration in normal condition (H1). In H2, all treatments increased the pulegone amount. In H3, M3 and the simultaneous use of multiple treatments increased pulegone. In *M.* × *piperita*, all treatments decreased the pulegone concentration in H1. But as for the heat conditions, in H2 all treatments increased the pulegone amount, while in H3 only M3 increased the pulegone concentration (Figure 5).

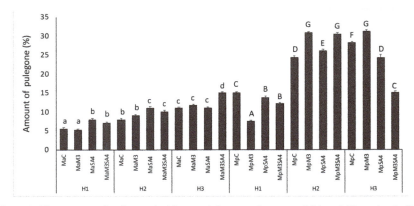

Figure 5. The amount of pulegone in *M. arvensis* L. var. *piperascens* (Ma) and *M.* × *piperita* (Mp) under heat stress in H1, H2, and H3 and effect of melatonin (M3) and salicylic acid (SA4) on pulegone. Values are presented as means ± standard deviation (n = 15); values not accompanied by the same letter are significantly different at $p < 0.05$, using the post-hoc Student-Newman-Keuls test. Lowercase letters(a–d) indicate significant differences between treatments for Ma; uppercase letters (A–G) indicate significant differences between treatments for Mp. For treatment details, see the Material and Methods Section 2.1.

In general, menthol has a significant negative correlation with menthofuran (r = −0.459 **) and pulegone (r = −0.912 **), that is to say that menthol is reduced and the pulegone increased under the heat stress. Also, menthofuran had a significant negative correlation with menthone (r = −0.527 **). The pulegone had a significant positive correlation with menthofuran (r = 0.345 *) (Table 2).

Table 2. Pearson correlation coefficients found among four important secondary metabolites (menthofuran, menthol, pulegone, and menthone) in the menthol pathway in *M. arvensis* L. var. *piperascens* and *M.* × *piperita* under the long-term extreme heat stress by using SA and M as a compensator of stress.

	Menthofuran	Menthol	Pulegone	Menthone
Menthofuran	1	−0.459 **	0.345 *	−0.527 **
Menthol		1	−0.912 **	−0.054
Pulegone			1	0.009
Menthone				1

** Correlation is significant at the 0.01 level (2-tailed).* Correlation is significant at the 0.05 level (2-tailed).

4. Discussion

In the mid-1980s, the relative water content (RWC) was introduced as the best reference point for the state of water in plants, expressing the balance between water absorption and consumption by transpiration [28]. The heat stress induces an increase of transpiration, with the effect of cooling and adapting the plant to heat. However a further transpiration dries up the tissues and the cells [29]. Generally, osmoregulation is one of the main mechanisms preserving turgor pressure in most plants against water loss; it causes the plant to continue to absorb water and maintain metabolic activity [30]. The RWC leaf may be the best biochemical growth/activity parameter that reveals the severity of stress [31]. Our data show a significant decrease in relative water content (RWC%) in both plants treated with different temperatures, suggesting that both plants were under stress. *M. arvensis* L. var. *piperascens* has lost more water than *M. × piperita* L.

Heat stress disturbs the stable physiological condition in plants and for this reason scientists are trying to find a way to relieve stress. Brassinosteroids such as SA and M have recently been studied in relation to this issue. Coban and Baydar [32] have shown that brassinosteroids reduce salt stress. He et al. [33] inhibited heat stress in bluegrass using SA. In particular, SA stimulates the production and/or an increase of secondary metabolites from polyphenols by acting as an elicitor [34,35]. SA activates phenylalanine ammonia lyase (PAL) [36] and plays a role in the regulation of physiological processes [37]. M also alleviates stress damage, and this has been reported in cucumber in the germination phase [38] and in *Arabidopsis* in which it compensates for heat stress [39]. Our data show that both plants undergo a strong heat stress reducing RWC in a temperature-dependent way. Treatment with SA and M in both plants significantly reduce heat stress effect on RWC, confirming the protective role of these two hormones against heat stress.

In a previous work, we have shown that heat stress determines a change in oxygenated monoterpenes, monoterpene hydrocarbons, oxygenated sesquiterpenes, sesquiterpene hydrocarbons, and other components in *M. × piperita* L. (Mitcham variety) and *M. arvensis* L. (var. *piperascens*) essential oils [22]. In this study, brassinosteroids treatments in both the oils subjected to heat stress determined a variation in the composition of the essential oils and in the antioxidant enzymatic activity.

Saharkhiz and Goudarzi [40], showed that application of 150 mgL^{-1} SA in *M. × piperita* L. significantly ($p < 0.05$) increased the oil content compared to control plants. In particular the treatment with different SA concentrations mostly increased menthone (15.8–18.1%) and menthol (46.3–47.4%) content.

In particular, monoterpenes (synthesized in Methylerythritol phosphate [MEP] pathway) and sesquiterpenes (synthesized in mevalonate [MVA] pathway) were the most important components. In the MEP pathway, menthol is synthesized in the cytoplasm and menthofuran is synthesized in the endoplasmic reticulum [41]. In particular, isopiperitenone from mitochondria is transferred to the cytoplasm and converts to pulegone. Pulegone can continue two branches of the MEP pathway: 1: It remains in the cytoplasm, is converted to menthone and finally to menthol; 2: it is transferred to the endoplasmic reticulum to be converted in menthofuran. So, in the MEP pathway, pulegone, menthone, menthofuran, and menthol have a crucial role and we should consider how they change under heat stress. In general, menthol and menthone have a significant negative correlation with menthofuran and pulegone, (menthol and menthone are reduced and the pulegone and menthofuran increased under the heat stress). Considering the (−)-Menthol biosynthesis pathway (Figure 6), we can hypothesize that under heat stress, pulegone reductase (PR) reduces its activity, leading to a decrease of the conversion of pulegone to menthone (that is the precursor of menthol). The increase of pulegone, due to the reduction of the activity of the enzyme that converts it in menthone, could also explain the increase of the mentofuran, which is synthesized from pulegone in the endoplasmic reticulum. In fact, pulegone had a significant positive correlation with menthofuran (Figure 6). In future studies, it will be necessary to verify the activity of the enzyme PR under heat stress and after treatment with the brassinosteroids.

As for the antioxidant activity, our studies have shown that SA and M have a positive effect on *M. arvensis* L. var. *piperascens* and *M.* × *piperita* L., increasing the activity of antioxidant enzymes in both species when used alone, but even more if applied simultaneously, demonstrating a synergistic effect.

On the other hand, an enhanced activity of CAT and SOD was observed in heat stressed plants of *Poa pratensis*, after the treatment with SA [33]. Xu et al. [42] reported that external M applications caused a significant increase in enzymatic antioxidants such as SOD, POX, CAT, and APX peroxidase and non enzymatic antioxidants such as ascorbic acid and vitamin E, resulting in decreased ROS levels and lipid peroxidation in cucumber under high temperature stress. Our data therefore not only confirms the effect of the two hormones on the activity of antioxidant enzymes and therefore the mitigating effect against the heat stress, but also show their ability to act in a synergistic way, which has not been demonstrated so far.

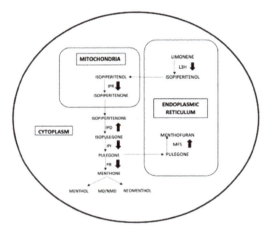

Figure 6. Menthol biosynthesis pathway. IPD: isopiperitenol dehydrogenase, IPR: isopiperitenone reductase, IPI: Isopentenyl diphosphate isomerase, PR: pulegone reductase, LH3:limonene 3-hydroxylase, MFS: menthofuran synthase, NMD: neomenthol reductase, MD: menthol dehydrogenases.

5. Conclusions

M and SA alleviate the effects of heat stress in *M. arvensis* L. var. *piperascens* and *M.* × *piperita* L. by changing the yield of essential oils and the activity of antioxidant enzymes.

It is possible that the activity of the brassinosteroids highlighted by us occurs through an action by the enzymatic game involved in the metabolism of the studied essential oils. Future studies will aim to highlight a possible modification of enzymatic activity and/or a different expression of the genes involved in their synthesis in response to the presence of M and SA.

Our results can be considered for future applications in the cosmetics, food, and pharmacological fields, given the extreme importance of menthol and menthone in these areas.

Author Contributions: Conceptualization, M.H., V.M. and D.R.; methodology, M.H., V.M., D.R., A.T., A.A.S.-B., J.H., S.S., M.G., C.M., M.P., R.N., F.D.R., L.F., A.B.; investigation, M.H., V.M., D.R. and A.B.; data curation, L.F. and R.N.; formal analysis, A.T., A.A.S.-B. and J.H.; writing—original draft preparation, A.B.; writing—review and editing, M.H., V.M., D.R. and A.B.; supervision, A.B.; project administration, A.B.

Funding: This research received no external funding.

Conflicts of Interest: The authors declare no conflict of interest.

References

1. Lawrence, B.M. *Mint: The Genus Mentha*; CRC Press: Boca Raton, FL, USA, 2006; ISBN 978-0-429-12587-4.

2. Kokkini, S.; Karousou, R.; Hanlidou, E. HERBS|Herbs of the Labiatae. In *Encyclopedia of Food Sciences and Nutrition*; Academic Press: Cambridge, MA, USA, 2003; pp. 3082–3090. ISBN 978-0-12-227055-0.

3. Taneja, S.C.; Chandra, S. 20—Mint. In *Handbook of Herbs and Spices (Second Edition)*; Peter, K.V., Ed.; Woodhead Publishing Series in Food Science, Technology and Nutrition; Woodhead Publishing: Sawston/Cambridge, UK, 2012; pp. 366–387. ISBN 978-0-85709-039-3.

4. Mander, L.; Liu, H.W. *Comprehensive Natural Products II: Chemistry and Biology*; Elsevier: Amsterdam, The Netherlands, 2010; ISBN 978-0-08-045382-8.

5. Fletcher, R.S.; Slimmon, T.; McAuley, C.Y.; Kott, L.S. Heat stress reduces the accumulation of rosmarinic acid and the total antioxidant capacity in spearmint (*Mentha spicata* L.). *J. Sci. Food Agric.* **2005**, *85*, 2429–2436. [CrossRef]

6. Smirnoff, N. The role of active oxygen in the response of plants to water deficit and desiccation. *New Phytol.* **1993**, *125*, 27–58. [CrossRef]

7. Gille, G.; Sigler, K. Oxidative stress and living cell. *Folia Microbiol.* **1995**, *40*, 131–152. [CrossRef] [PubMed]

8. Mittler, R. Oxidative stress, antioxidants and stress tolerance. *Trends Plant Sci.* **2002**, *7*, 405–410. [CrossRef]

9. Mittler, R.; Vanderauwera, S.; Gollery, M.; Van Breusegem, F. Reactive oxygen gene network of plants. *Trends Plant Sci.* **2004**, *9*, 490–498. [CrossRef]

10. Moghaddam, N.M.; Arvin, M.J.; Nezhad, G.R.K.; Maghsoudi, K. Effect of salicylic acid on growth and forage and grain yield of maize under drought stress in field conditions. *Seed Plant Prod. J.* **2011**, *272*, 41–55.

11. Shirasu, K.; Nakajima, H.; Rajasekhar, V.K.; Dixon, R.A.; Lamb, C. Salicylic acid potentiates an agonist-dependent gain control that amplifies pathogen signals in the activation of defense mechanisms. *Plant Cell* **1997**, *9*, 261–270.

12. Knörzer, O.C.; Lederer, B.; Durner, J.; Böger, P. Antioxidative defense activation in soybean cells. *Physiol. Plant.* **1999**, *107*, 294–302. [CrossRef]

13. Calvo, J.R.; González-Yanes, C.; Maldonado, M.D. The role of melatonin in the cells of the innate immunity: A review. *J. Pineal Res.* **2013**, *55*, 103–120. [CrossRef]

14. Manchester, L.C.; Coto-Montes, A.; Boga, J.A.; Andersen, L.P.H.; Zhou, Z.; Galano, A.; Vriend, J.; Tan, D.-X.; Reiter, R.J. Melatonin: An ancient molecule that makes oxygen metabolically tolerable. *J. Pineal Res.* **2015**, *59*, 403–419. [CrossRef]

15. Wang, P.; Sun, X.; Li, C.; Wei, Z.; Liang, D.; Ma, F. Long-term exogenous application of melatonin delays drought-induced leaf senescence in apple. *J. Pineal Res.* **2013**, *54*, 292–302. [CrossRef] [PubMed]

16. Zhang, N.; Sun, Q.; Zhang, H.; Cao, Y.; Weeda, S.; Ren, S.; Guo, Y.-D. Roles of melatonin in abiotic stress resistance in plants. *J. Exp. Bot.* **2015**, *66*, 647–656. [CrossRef] [PubMed]

17. Gao, W.; Zhang, Y.; Feng, Z.; Bai, Q.; He, J.; Wang, Y. Effects of Melatonin on Antioxidant Capacity in Naked Oat Seedlings under Drought Stress. *Molecules* **2018**, *23*, 1580. [CrossRef]

18. Nawaz, M.A.; Huang, Y.; Bie, Z.; Ahmed, W.; Reiter, R.J.; Niu, M.; Hameed, S. Melatonin: Current Status and Future Perspectives in Plant Science. *Front. Plant Sci.* **2016**, *6*, 1230. [CrossRef] [PubMed]

19. Arnao, M.B.; Hernández-Ruiz, J. Melatonin: Plant growth regulator and/or biostimulator during stress? *Trends Plant Sci.* **2014**, *19*, 789–797. [CrossRef]

20. Arnao, M.B.; Hernández-Ruiz, J. Melatonin and its relationship to plant hormones. *Ann. Bot.* **2018**, *121*, 195–207. [CrossRef]

21. Tan, D.-X.; Hardeland, R.; Manchester, L.C.; Paredes, S.D.; Korkmaz, A.; Sainz, R.M.; Mayo, J.C.; Fuentes-Broto, L.; Reiter, R.J. The changing biological roles of melatonin during evolution: From an antioxidant to signals of darkness, sexual selection and fitness. *Biol. Rev.* **2010**, *85*, 607–623. [CrossRef]

22. Heydari, M.; Zanfardino, A.; Taleei, A.; Bushehri, A.A.S.; Hadian, J.; Maresca, V.; Sorbo, S.; Napoli, M.D.; Varcamonti, M.; Basile, A.; et al. Effect of Heat Stress on Yield, Monoterpene Content and Antibacterial Activity of Essential Oils of *Mentha* x *piperita* var. Mitcham and *Mentha arvensis* var. piperascens. *Molecules* **2018**, *23*, 1903. [CrossRef]

23. Dhopte, A.M.; Livera-M, M. *Principles and Techniques for Plant Scientist [s]*; Agrobios: Jodhpur, India, 2002.

24. Arjenaki, F.G.; Morshedi, A.; Jabbari, R. Evaluation of Drought Stress on Relative Water Content, Chlorophyll Content and Mineral Elements of Wheat (*Triticum aestivum* L.) Varieties. *Int. J. Agric. Crop Sci.* **2012**, *4*, 726–729.

25. Maresca, V.; Fusaro, L.; Sorbo, S.; Siciliano, A.; Loppi, S.; Paoli, L.; Monaci, F.; Karam, E.A.; Piscopo, M.; Guida, M.; et al. Functional and structural biomarkers to monitor heavy metal pollution of one of the most contaminated freshwater sites in Southern Europe. *Ecotoxicol. Environ. Saf.* **2018**, *163*, 665–673. [CrossRef]
26. Russo, A.; Formisano, C.; Rigano, D.; Cardile, V.; Arnold, N.A.; Senatore, F. Comparative phytochemical profile and antiproliferative activity on human melanoma cells of essential oils of three lebanese Salvia species. *Ind. Crop. Prod.* **2016**, *83*, 492–499. [CrossRef]
27. Rigano, D.; Arnold, N.A.; Conforti, F.; Menichini, F.; Formisano, C.; Piozzi, F.; Senatore, F. Characterisation of the essential oil of *Nepeta glomerata* Montbret et Aucher ex Bentham from Lebanon and its biological activities. *Nat. Prod. Res.* **2011**, *25*, 614–626. [CrossRef] [PubMed]
28. Schonfeld, M.A.; Johnson, R.C.; Carver, B.F.; Mornhinweg, D.W. Water Relations in Winter Wheat as Drought Resistance Indicators. *Crop Sci.* **1988**, *28*, 526–531. [CrossRef]
29. Jiang, Y.; Huang, B. Drought and Heat Stress Injury to Two Cool-Season Turfgrasses in Relation to Antioxidant Metabolism and Lipid Peroxidation. *Crop Sci.* **2001**, *41*, 436–442. [CrossRef]
30. Gunasekera, D.; Berkowitz, G.A. Evaluation of contrasting cellular-level acclimation responses to leaf water deficits in three wheat genotypes. *Plant Sci.* **1992**, *86*, 1–12. [CrossRef]
31. Alizadeh, A. *Soil, Water, Plants Relationship*; Emam Reza University Press: Mashhad, Iran, 2002; Volume 3.
32. Çoban, Ö.; Göktürk Baydar, N. Brassinosteroid effects on some physical and biochemical properties and secondary metabolite accumulation in peppermint (*Mentha piperita* L.) under salt stress. *Ind. Crop. Prod.* **2016**, *86*, 251–258.
33. He, Y.; Liu, Y.; Cao, W.; Huai, M.; Xu, B.; Huang, B. Effects of Salicylic Acid on Heat Tolerance Associated with Antioxidant Metabolism in Kentucky Bluegrass. *Crop Sci.* **2005**, *45*, 988–995. [CrossRef]
34. Edreva, A.; Velikova, V.; Tsonev, T.; Dagnon, S.; Gürel, A.; Aktaş, L.; Gesheva, E. Stress-protective role of secondary metabolites: Diversity of functions and mechanisms. *Gen. Appl. Plant Physiol.* **2008**, *34*, 67–78.
35. Cohen, S.D.; Kennedy, J.A. Plant metabolism and the environment: Implications for managing phenolics. *Crit. Rev. Food Sci. Nutr.* **2010**, *50*, 620–643. [CrossRef]
36. Klessig, D.F.; Malamy, J. The salicylic acid signal in plants. *Plant Mol. Biol.* **1994**, *26*, 1439–1458. [CrossRef]
37. Raskin, I. Role of Salicylic Acid in Plants. *Annu. Rev. Plant Physiol. Plant Mol. Biol.* **1992**, *43*, 439–463. [CrossRef]
38. Zhang, H.-J.; Zhang, N.; Yang, R.-C.; Wang, L.; Sun, Q.-Q.; Li, D.-B.; Cao, Y.-Y.; Weeda, S.; Zhao, B.; Ren, S.; et al. Melatonin promotes seed germination under high salinity by regulating antioxidant systems, ABA and GA4 interaction in cucumber (*Cucumis sativus* L.). *J. Pineal Res.* **2014**, *57*, 269–279. [CrossRef] [PubMed]
39. Bajwa, V.S.; Shukla, M.R.; Sherif, S.M.; Murch, S.J.; Saxena, P.K. Role of melatonin in alleviating cold stress in *Arabidopsis thaliana*. *J. Pineal Res.* **2014**, *56*, 238–245. [CrossRef] [PubMed]
40. Saharkhiz, M.J.; Goudarzi, T. Foliar Application of Salicylic acid Changes Essential oil Content and Chemical Compositions of Peppermint (*Mentha piperita* L.). *J. Essent. Oil Bear. Plants* **2014**, *17*, 435–440. [CrossRef]
41. Croteau, R.B.; Davis, E.M.; Ringer, K.L.; Wildung, M.R. (−)-Menthol biosynthesis and molecular genetics. *Naturwissenschaften* **2005**, *92*, 562–577. [CrossRef]
42. Xu, X.; Sun, Y.; Guo, X.; Sun, B.; Zhang, J. Effects of exogenous melatonin on ascorbate metabolism system in cucumber seedlings under high temperature stress. *Yingyong Shengtai Xuebao* **2010**, *21*, 2580–2586.

© 2019 by the authors. Licensee MDPI, Basel, Switzerland. This article is an open access article distributed under the terms and conditions of the Creative Commons Attribution (CC BY) license (http://creativecommons.org/licenses/by/4.0/).

Article

Exhausted Woods from Tannin Extraction as an Unexplored Waste Biomass: Evaluation of the Antioxidant and Pollutant Adsorption Properties and Activating Effects of Hydrolytic Treatments

Lucia Panzella [1,*], Federica Moccia [1], Maria Toscanesi [1], Marco Trifuoggi [1], Samuele Giovando [2] and Alessandra Napolitano [1]

1. Department of Chemical Sciences, University of Naples "Federico II", Via Cintia 4, I-80126 Naples, Italy; federica.moccia@unina.it (F.M.); maria.toscanesi@unina.it (M.T.); marco.trifuoggi@unina.it (M.T.); alesnapo@unina.it (A.N.)
2. Centro Ricerche per la Chimica Fine Srl for Silvateam Spa, Via Torre 7, 12080 San Michele Mondovì, CN, Italy; sgiovando@silvateam.com
* Correspondence: panzella@unina.it; Tel.: +39-081-674131

Received: 5 February 2019; Accepted: 28 March 2019; Published: 1 April 2019

Abstract: Exhausted woods represent a byproduct of tannin industrial production processes and their possible exploitation as a source of antioxidant compounds has remained virtually unexplored. We herein report the characterization of the antioxidant and other properties of practical interest of exhausted chestnut wood and quebracho wood, together with those of a chestnut wood fiber, produced from steamed exhausted chestnut wood. 2,2-Diphenyl-1-picrylhydrazyl (DPPH) and ferric reducing/antioxidant power (FRAP) assays indicated good antioxidant properties for all the materials investigated, with exhausted chestnut wood, and, even more, chestnut wood fiber exhibiting the highest activity. High efficiency was observed also in the superoxide scavenging assay. An increase of the antioxidant potency was observed for both exhausted woods and chestnut wood fiber following activation by hydrolytic treatment, with an up to three-fold lowering of the EC_{50} values in the DPPH assay. On the other hand, exhausted quebracho wood was particularly effective as a nitrogen oxides (NO_x) scavenger. The three materials proved able to adsorb methylene blue chosen as a model of organic pollutant and to remove highly toxic heavy metal ions like cadmium from aqueous solutions, with increase of the activity following the hydrolytic activation. These results open new perspectives toward the exploitation of exhausted woods as antioxidants, e.g., for active packaging, or as components of filtering membranes for remediation of polluted waters.

Keywords: agri-food waste; exhausted wood; antioxidant; DPPH assay; FRAP assay; tannins; heavy metals; methylene blue; nitric oxides; acid hydrolysis

---

1. Introduction

In the light of their marked antioxidant properties, phenolic polymers from natural sources have been the focus of increasing interest as sustainable functional materials for the wide range of potential applications, for example as biocompatible materials for biomedical devices or as active components in functional food packaging [1–7]. Indeed, phenolic polymers display manifold advantages over their monomers, including higher stability properties thus offering easier handling and processing, as well as lower solubility reducing tendency to be released [4,8,9].

Among natural phenolic polymers, tannins occupy a prime position given the well-established antimicrobial and protein binding properties [1,10–14]. Tannins are traditionally ranked into the hydrolyzable and the condensed or non-hydrolyzable classes. *Castanea sativa* wood extract, commonly known as chestnut tannin, is composed mainly of hydrolyzable ellagitannins such as castalagin and its isomer vescalagin, whereas quebracho tannin, extracted from the hardwood of *Shinopsis balansae/lorentzii*, comprise mainly condensed tannins of which linear profisetinidins represent the major constituents (Figure 1) [1,15,16].

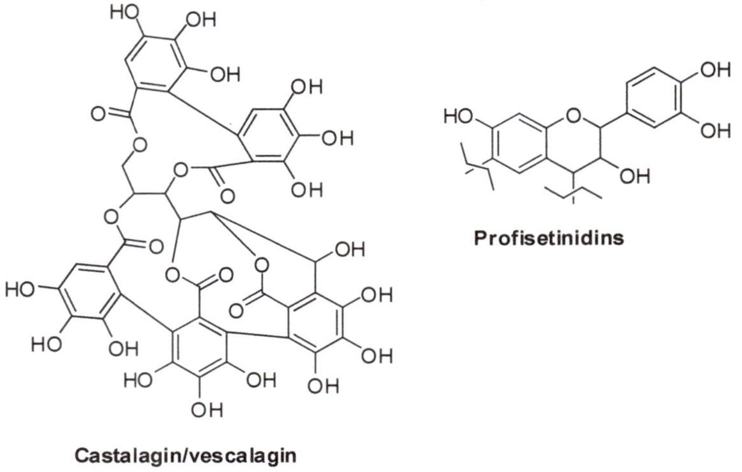

Profisetinidins

Castalagin/vescalagin

Figure 1. Representative structures of the main tannins occurring in chestnut and quebracho wood.

Both chestnut and quebracho tannins are commonly employed not only in the tanning industry, but also in the oenological, cosmetic, and pharmaceutical field as well as in products for animal feed [1,17–23]. Tannin extraction from wood is generally performed in hot water. The residual wood biomasses (exhausted woods), a by-product of the tannin extraction process, may be taken as renewable sources in their own right, since the wood has been treated only with hot water at high pressure, a treatment that is not expected to result in any significant change of the structural components. These exhausted woods are commonly used in the production of pellets for heating and energy production. The isolation of cellulose nanocrystals to be used e.g., as nanofillers for polymer composites from the exhausted acacia bark, obtained after the industrial process of extracting tannin, has also been reported [24]. However, the possible exploitation as antioxidant materials has remained virtually unexplored.

We report herein the characterization of the antioxidant properties of exhausted chestnut and quebracho wood, together with those of a chestnut wood fiber produced from steamed exhausted chestnut wood. The materials investigated are shown in Figure 2. For comparison, the corresponding fresh woods and tannins were investigated as well. Exhausted woods and chestnut wood fiber were also tested for their ability to remove environmental pollutants, either organic compounds or toxic heavy metals. Based on recent findings showing that acid hydrolytic treatment of natural phenolic polymers or wastes leads to materials with potent antioxidant efficiency [4,25,26], we also investigated the properties of the exhausted woods and chestnut wood fiber materials obtained by such treatment in comparison with those of the untreated materials, showing in most cases a significant enhancement of the activity. The results obtained open new perspectives for the exploitation of these by-products e.g., in active packaging or for remediation of polluted waters.

Figure 2. (**a**) Exhausted chestnut wood; (**b**) exhausted quebracho wood; (**c**) chestnut wood fiber.

2. Materials and Methods

2.1. General Experimental Methods

Chestnut and quebracho exhausted and fresh woods, chestnut wood fiber as well as chestnut and quebracho tannins were provided by Silvateam (Via Torre, S. Michele Mondovì, Cuneo, Italy). In particular, fresh chestnut and quebracho woods were obtained from hardwood of *Castanea sativa* and *Schinopsis lorentzii* respectively, and reduced in chips of ca. 1–3 cm. Tannins were obtained by soaking the wood chips in autoclaves with water, at 120 °C, under pressure; the extracts thus obtained were concentrated with a multiple-effect evaporator under vacuum until ca. 50% water was removed, and tannin powder finally obtained by spray-drying. Residual chips after the extraction were stored as exhausted woods. Chestnut wood fiber was obtained from chestnut exhausted wood after drying in oven overnight at 60 °C and milling to obtain <250 μm particles.

2,2-Diphenyl-1-picrylhydrazyl (DPPH), iron (III) chloride (97%), 2,4,6-tris(2-pirydyl)-s-triazine (TPTZ) (≥98%), (±)-6-hydroxy-2,5,7,8-tetramethylchromane-2-carboxylic acid (Trolox) (97%), ethylenediaminetetraacetic acid (EDTA) (>99%), nitroblue tetrazolium (NBT) chloride (98%), pyrogallol (≥98%), quercetin (≥95%), fluorescein, 2,2'-azobis(2-methylpropionamidine) dihydrochloride (AAPH) (97%), Folin & Ciocalteu's phenol reagent (2N), gallic acid (≥97.5%), activated carbon, sodium nitrite (≥97.0%), N-(1-naphthyl)ethylenediamine dihydrochloride (≥98%), sulfanilamide (≥99%), methylene blue (MB), cadmium carbonate (99%), and nitric acid (≥69% v/v, TraceSELECT® water solution) were obtained from Sigma-Aldrich (Milan, Italy) and used as obtained.

UV-Vis spectra were performed using a HewlettPackard 8453 Agilent spectrophotometer.

Fluorescence spectra were record on a HORIBA Jobin Yvon Inc. FluoroMax®-4 spectrofluorometer.

For metal removal experiments, 1% HNO_3, 0.1 M HCl and 0.01 M phosphate buffer (pH 7.0) were prepared using ultrapure deionized water with conductivity <0.06 μS/cm. All glassware used were carefully washed first with a 1% HNO_3 solution and then with ultrapure deionized water. Metal analysis was carried out on an inductively coupled plasma mass spectrometry (ICP-MS) instrument Aurora M90 model by Bruker.

2.2. Hydrolytic Treatment

The proper material (3 g) was treated with 70 mL of 6 M HCl under stirring at 100 °C overnight [4,26]. After cooling at room temperature, the mixture was centrifuged (8247× *g*, room temperature, 15 min) and the precipitate washed with water until neutrality and freeze dried. The recovery yields were 58% (chestnut wood fiber), 68% (exhausted chestnut wood) and 83% (quebracho exhausted wood).

2.3. DPPH Assay

To a 0.2 mM methanolic solution of DPPH wood or tannin powders were added (final dose 0.03–1.5 mg/mL) and after 10 min under stirring at room temperature the absorbance of the solution at 515 nm was measured [27,28]. Experiments were run in triplicate. Trolox was used as reference antioxidant.

2.4. Ferric Reducing/Antioxidant Power (FRAP) Assay

To 0.3 M acetate buffer (pH 3.6) containing 1.7 mM $FeCl_3$ and 0.83 mM TPTZ, wood or tannin powders (final dose 0.00625–0.3 mg/mL) were added and after 10 min under stirring at room temperature the absorbance of the solution at 593 nm was measured [29]. Results were expressed as Trolox equivalents. Experiments were run in triplicate.

2.5. Superoxide Scavenging Assay

Wood or tannin powders (final dose 0.0625 mg/mL) were added to 0.05 M ammonium hydrogen carbonate buffer (pH 9.3) containing 0.4 mM EDTA and 12 µM NBT, followed by a 20 mM pyrogallol solution in 0.05 mM HCl (final pyrogallol concentration 3.3 mM) [4,30]. The mixture was vigorously stirred for 5 min, after that absorbance at 596 nm was measured. Results were expressed as percentage of reduction of the absorbance at 596 nm of a control mixture run in the absence of sample. Experiments were run in triplicate.

2.6. Oxygen Radical Antioxidant Capacity (ORAC Assay)

Wood or tannin powders (final dose 0.075 mg/mL) were incubated in a 62.7 µM fluorescein solution in 75 mM phosphate buffer (pH 7.4) for 30 min, at 37 °C. A 153 mM AAPH solution was then added (final APPH concentration 19 mM) and after stirring at 37 °C for 45 min fluorescence was measured (λ_{ex} = 485 nm, λ_{em} = 511 nm) [31]. Results were expressed as relative fluorescence intensity with respect to a control mixture run in the absence of APPH. Experiments were run in triplicate.

2.7. MB Adsorption Assay

Adsorption experiments were performed at room temperature by adding wood samples (0.2 mg/mL) to a 5 or 25 mg/L aqueous solution of MB. The mixtures were taken under stirring and after 30 min the absorbance at 654 nm was measured [32]. Activated carbon was used as reference material. Experiments were run in triplicate.

2.8. Nitric oxides (NO_x) Scavenging Assay

A solution of sodium nitrite (1 M) in water was added to 10% sulfuric acid over 10 min [33]. The red orange gas that developed (0.2–0.6 mL) was withdrawn with a syringe and conveyed through a tip containing 5 mg of wood sample into 3 mL of Griess reagent (0.5% sulfanilamide and 0.05% N-(1-naphthyl)ethylenediamine dihydrochloride in 1.25% phosphoric acid) and the absorbance at 540 nm was measured. Experiments were run in triplicate.

2.9. Cd^{2+} Adsorption

A 1.5 mM stock solution of the heavy metal was prepared by dissolving 10 mg of cadmium carbonate in 39 mL of 0.1 M HCl. Prior to the adsorption experiments a 1.5 mg/mL suspension of each wood sample in 0.01 M phosphate buffer (pH 7.0) was obtained by homogenization in a Tenbroeck glass to glass homogenizer for 4 min. 0.7 mL of the wood suspensions and 0.1 mL of the metal solution were added to 10 mL of 0.01 M phosphate buffer at pH 7.0. After 2 h, the mixtures were filtered through a 0.45 µm nylon membrane, acidified by addition of 69% nitric acid (1:100 v/v), properly diluted with 1% nitric acid, and analyzed by ICP-MS [34]. A calibration curve was built with cadmium solutions at five different concentrations. For each binding experiment a blank experiment was planned in which the metal ion was added in the phosphate buffer and incubated for 2 h without addition of the wood sample. Experiments were run in triplicate.

2.10. Evaluation of the Solubility of Woods and Tannins in the Assay Media

Wood or tannin samples (3 mg) were added to methanol (20 mL), 0.3 M acetate buffer (pH 3.6) (20 mL), or water (15 mL), and taken under magnetic stirring. After 10 or 30 min the supernatants obtained after centrifugation (8247× g, room temperature, 15 min) were analyzed by UV-Vis spectrophotometry.

2.11. Determination of the Amount of Tannins in the Wood Samples

Wood or tannin samples (10 mg) were stirred in 1 mL of a 1:1 v/v acetone/water mixture containing 1% acetic acid [35]. After 60 min the supernatants obtained after centrifugation (3534× g, room temperature, 20 min) were analyzed by UV-Vis spectrophotometry after 1:500 v/v dilution in methanol. The amount of tannins in each wood sample was determined by comparison of the absorbance at 269 nm (chestnut-derived samples) or 280 nm (quebracho-derived samples) with that measured for chestnut or quebracho tannins, respectively.

2.12. Measurement of Total Phenolic Content (TPC)

Wood or tannin samples (10 mg) were stirred in DMSO (1 mL) for 1 h. After centrifugation (3534× g, room temperature, 20 min) 1–50 µL of the supernatant were added to 1.4 mL of water followed by 0.3 mL of a 75 g/L Na_2CO_3 solution and 0.1 mL of Folin & Ciocalteu's reagent. After 30 min incubation at 40 °C, absorbance at 765 nm was measured [36]. Gallic acid was used as reference compound. Experiments were run in triplicate.

3. Results and Discussion

3.1. Antioxidant Properties of Exhausted Woods

In a first series of experiments the antioxidant properties of chestnut wood fiber and exhausted chestnut and quebracho woods were investigated with respect to the corresponding fresh woods and tannins by widely used assays, i.e. DPPH, FRAP, superoxide scavenging, and ORAC assays following the "QUENCHER" method which allows one to measure the efficiency of electron transfer processes from a solid antioxidant [27–31].

3.1.1. DPPH and FRAP Assays

Table 1 reports the EC_{50} value, which is the dose of the material at which a 50% DPPH reduction is observed, determined for tannins and wood samples in the DPPH assay. For comparison data for the reference antioxidant Trolox are also reported.

Table 1. Antioxidant properties of tannins and wood samples.[1]

Sample	EC_{50} (mg/mL) [2] (DPPH Assay)	Trolox Equivalents (FRAP Assay)
Chestnut wood fiber	0.054 ± 0.003	0.17 ± 0.01
Exhausted chestnut wood	0.436 ± 0.003	0.06 ± 0.01
Fresh chestnut wood	0.128 ± 0.003	0.19 ± 0.01
Chestnut tannins	0.019 ± 0.002	1.2 ± 0. 1
Exhausted quebracho wood	1.14 ± 0.04	0.051 ± 0.005
Fresh quebracho wood	0.0990 ± 0.0007	0.08 ± 0.01
Quebracho tannins	0.026 ± 0.002	0.47 ± 0.03
Trolox	0.015 ± 0.001	-

[1] Reported are the mean ± SD values of at least three experiments. [2] EC_{50} is the dose of the material at which a 50% 2,2-diphenyl-1-picrylhydrazyl (DPPH) reduction is observed.

Among the waste materials, chestnut wood fiber displayed the most promising DPPH-reducing ability, with an EC_{50} value of 0.054 mg/mL that compares well with that of Trolox (3.6-fold higher). Also quebracho and particularly chestnut exhausted wood exhibited quite low EC_{50} values, much lower than those reported for other agro-food wastes, such as spent coffee grounds (EC_{50} = 5.00 mg/mL) [4]. As expected, higher antioxidant activities were exhibited by fresh wood samples, still containing tannins, and tannins themselves, characterized by EC_{50} values approaching that of Trolox.

The marked differences observed in the antioxidant activity may be interpreted considering the solubility of the materials in the assay medium. Spectra shown in Figure 3a clearly indicate release of UV absorbing species in the case of pure tannins, chestnut wood fiber, and fresh woods, whereas exhausted woods do not give rise to appreciable absorbance.

The results of the FRAP assay (Table 1) looked less encouraging than those obtained in the DPPH assay, with all the waste materials exhibiting an iron(III)-reducing activity far lower than Trolox (Trolox equivalents <<1). Only chestnut tannins showed a satisfactory antioxidant power in this assay, whereas the fresh wood samples as well as quebracho tannins performed much less. Here again the antioxidant activity parallels fairly well the solubility in the medium used for the FRAP assay (Figure 3b).

To obtain information about the compounds responsible for the antioxidant properties observed, the amounts of tannins and the TPC were determined for each wood sample (Table 2). As shown in Figure 4, a good linear correlation was found between Trolox equivalents determined in the FRAP assay and the tannin content (for both the hydrolyzable and non-hydrolyzable class), but not TPC ($R^2 = 0.59$), pointing to residual tannins as the main determinants of the iron-reducing properties. On the contrary, DPPH reducing ability was not apparently related to either TPC ($R^2 = 0.37$) or tannin content ($R^2 = 0.33$ and 0.50 for chestnut and quebracho-derived samples, respectively), suggesting that several factors (e.g., relative solubility in the assay medium) might be involved in the different activities observed.

The exhausted woods and the chestnut wood fiber that have been subjected to the hydrolytic treatment according to the protocol developed in previous studies [4,25,26] were then evaluated for their antioxidant activity by the DPPH and FRAP assays. Data shown in Figure 5a indicate for both exhausted quebracho and chestnut woods a decrease of EC_{50} values in the range of 25%–30% compared to the values obtained for the untreated materials. By contrast no significant increase of the activity was observed in the FRAP assay following the hydrolytic treatment, apart for chestnut wood fiber, which exhibited a two-fold increase in Trolox eqs (Figure 5b).

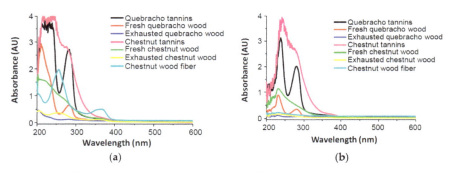

Figure 3. UV-Vis spectra of wood and tannin samples (0.15 mg/mL) in methanol (**a**) or 0.3 M acetate buffer (pH = 3.6) (**b**).

Table 2. Tannin content and total phenolic content (TPC) of wood samples.

Sample	Tannins (% w/w) [1]	TPC (mg gallic acid/g of Sample) [2]
Chestnut wood fiber	19	151 ± 17
Exhausted chestnut wood	11	51 ± 3
Fresh chestnut wood	21	153 ± 9
Chestnut tannins	100	457 ± 59
Exhausted quebracho wood	28	40 ± 1
Fresh quebracho wood	43	194 ± 8
Quebracho tannins	100	550 ± 99

[1] Reported are the mean values of at least three experiments (SD ≤ 5%). [2] Reported are the mean values ± SD of at least three experiments.

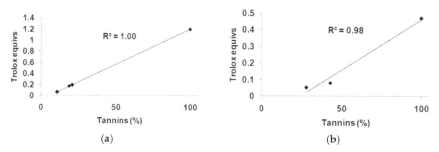

Figure 4. Correlation between tannin content and Fe^{3+}-reducing activity of tannin and wood samples. (**a**) Chestnut-derived samples; (**b**) Quebracho-derived samples.

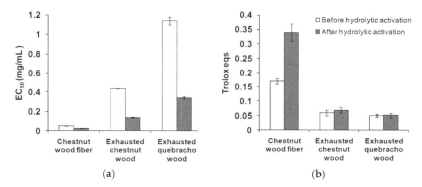

Figure 5. Antioxidant properties of exhausted wood samples before and after hydrolytic activation. (**a**) DPPH assay; (**b**) ferric reducing/antioxidant power (FRAP) assay. Reported are the mean ± SD values of at least three experiments.

3.1.2. Superoxide Scavenging Assay

A very high efficiency, compared to other agri-food waste products such as spent coffee grounds [4], was observed in the superoxide scavenging assay (Figure 6), with tannins and fresh woods being always the most active samples, although in the case of chestnut-derived materials both the exhausted woods exhibited an activity comparable to that of the native sample. Only a modest correlation ($R^2 = 0.90$ and 0.82 for chestnut and quebracho-derived samples, respectively) was found between the percentage of superoxide scavenging and the tannin content, whereas no correlation was found with TCP. Moreover, no significant improvement in the scavenging ability was detected in exhausted woods subjected to the hydrolytic treatment.

3.1.3. ORAC Assay

The relative fluorescence intensities determined in the ORAC assay are reported in Table 3. In this case major differences were apparent between tannins/fresh woods and exhausted samples, with the first ones almost totally inhibiting fluorescein oxidation and the second ones being completely inactive, with the only exception of chestnut wood fiber, which showed an activity comparable to that of fresh chestnut wood. Notably, in this case a good linear correlation ($R^2 = 0.95$) was found between the antioxidant activity of the wood samples and TPC; on the contrary no significant correlation was found with the amount of residual tannins. No effect of the hydrolytic treatment was apparent either in this assay.

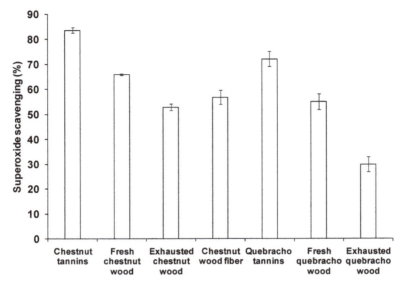

Figure 6. Superoxide scavenging activity of tannins and wood samples. Reported are the mean ± SD values of at least three experiments.

Table 3. Activity of tannins and wood samples in the oxygen radical antioxidant capacity (ORAC) assay.[1]

Sample	Relative Fluorescence Intensity (%)
-	0.2 [2]
Chestnut wood fiber	86
Exhausted chestnut wood	0.5
Fresh chestnut wood	89
Chestnut tannins	81
Exhausted quebracho wood	0.2
Fresh quebracho wood	90
Quebracho tannins	100

[1] Reported are the mean values of at least three experiments (SD ≤ 5%). [2] Mixture containing only fluorescein and APPH, without any antioxidant.

The results of the antioxidant assays further add to the potential of wood fiber as a reinforcement in polymers [37,38] or as an ecological-friendly medium for horticultural practice to increase the antioxidant activity of fruit and vegetables [39].

3.2. Pollutant Adsorption Properties of Exhausted Woods

In a further series of experiments the adsorption capacity of the exhausted woods toward various environmental pollutants was evaluated. These included MB, as a model organic dye, NO_x, that is nitric oxide (NO) and nitrogen dioxide (NO_2) which are reactive nitrogen species commonly present in cigarette smoke and exhaust gases able to induce oxidative and nitrosative stress in humans, and cadmium ions (Cd^{2+}), as a model of toxic heavy metals.

3.2.1. MB Adsorption Assay

Table 4 reports the percentages of MB adsorption by the wood samples.

Table 4. Pollutant adsorption properties of wood samples.[1]

Sample	MB Adsorption (%) [2]	NO$_x$ Scavenging (%) [3]
Chestnut wood fiber	100	32 ± 5
Exhausted chestnut wood	73 ± 3	23 ± 1
Fresh chestnut wood	79 ± 1	38 + 1
Exhausted quebracho wood	77 ± 1	79 ± 1
Fresh quebracho wood	77 ± 2	86 ± 5

[1] Reported are the mean ± SD values of at least three experiments. [2] Determined with methylene blue (MB) at 5 mg/L. [3] Determined with 0.6 mL of gas.

Both exhausted chestnut and quebracho woods proved to be very efficient at a dose of 0.2 mg/mL, and their adsorption properties were comparable to those exhibited by the fresh wood samples. Indeed, UV-Vis spectrophotometric analysis (Figure 7) indicated a lower solubility in the assay medium (water) of chestnut wood fiber and exhausted woods compared to the fresh samples, a feature that further adds to the potential of these materials for removal of organic compounds from waste waters.

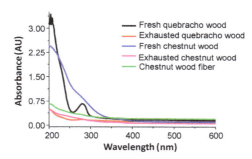

Figure 7. UV-Vis spectra of wood samples (0.2 mg/mL) in water.

A 20%–30% dye removal was still observed when wood samples were added to a more concentrated MB solution (25 mg/mL) (Figure 8). For comparison, under the same conditions a 70% MB removal was obtained with activated carbon, taken as a reference material. In contrast to what observed in the antioxidant assays, the MB adsorption ability of the exhausted woods did not change appreciably following to acid treatment.

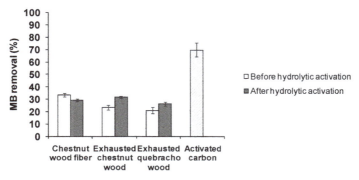

Figure 8. MB adsorption by chestnut wood fiber and exhausted woods before and after hydrolytic activation (MB starting concentration 25 mg/L). Reported are the mean ± SD values of at least three experiments. For comparison data relative to activated carbon are also reported.

3.2.2. NO$_x$ Scavenging Assay

All the samples examined led to >70% scavenging when 0.2 mL of NO$_x$ gases were passed through 5 mg of wood (data not shown). The percentages of NO$_x$ scavenging determined in the experiments run with 0.6 mL of gas are reported in Table 4, from which the superior ability of quebracho woods stands out compared to chestnut woods. These data would suggest a higher trapping efficiency of condensed tannins with respect to hydrolyzable tannins that could likely be ascribed to the reactivity of the resorcinol moieties of profisetinidins toward electrophiles like nitric oxides [40]. As observed above for MB adsorption, also in this case fresh woods were only slightly more active than exhausted samples. Notably, further to acid treatment a lower efficiency was observed for most of the tested woods (not shown), pointing to a role of the hydrolyzable cellulosic matrix in NO$_x$ scavenging.

3.2.3. Cd^{2+} Removal Assay

In a last series of experiments the ability of wood samples to remove heavy metals from aqueous solutions was investigated using Cd^{2+} as model ions. The assay was performed at pH 7.0, based on previous observations showing that natural phenolic polymers do not exhibit significant chelating properties at acidic pH [34]. In this case fresh woods were not evaluated because of the release of significant amounts of material under the testing conditions, likely due to solubilization of residual tannins and/or other low molecular weight components.

Based on the data reported in Figure 9, chestnut wood fiber was again the most active among the waste materials, followed by exhausted quebracho wood. An increase in Cd^{2+} removal was observed following hydrolytic activation for all the woods examined.

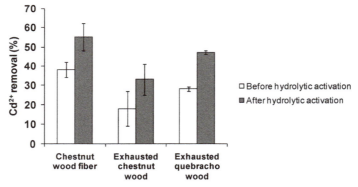

Figure 9. Cd^{2+} removal by exhausted woods before and after hydrolytic activation. Reported are the mean ± SD values of at least three experiments.

4. Conclusions

Exhausted woods represent a largely available waste product of the tannin extraction process, whose practical exploitation has not yet been duly considered. As reported in the present paper, these materials exhibit good antioxidant properties in the DPPH and superoxide scavenging assay, and are also able to efficiently adsorb pollutants such as toxic gases, organic dyes and heavy metals. In particular, chestnut woods, containing mainly hydrolyzable tannins, were particularly active as antioxidants, whereas quebracho woods, characterized by the presence of condensed tannins, were found to be more effective as adsorbent materials, especially toward NO$_x$ fumes. Of particular interest is the enhancement of the antioxidant properties of exhausted woods and chestnut wood fiber following hydrolytic treatment, a methodology that had already been applied to materials containing lignins like spent coffee grounds [4] and the black sesame seed pigment [25], but also ellagitannins like pomegranate wastes [26]. An interpretation of such effects at a molecular level invoke aromatization and dehydration processes enhancing the hydrogen donor ability of the OH functionalities stemming from delocalization of the resulting phenoxyl radical over a more conjugated molecular backbone [4,8]. On the other hand, from consideration of the moderate weight loss further to the hydrolytic treatment it cannot be argued that the observed activation effects may simply be ascribed to removal of cellulosic or other inactive components of the materials.

Overall, these results would point to exhausted woods as novel potential functional additives to be used for example in active packaging, as food stabilizers against oxidative deterioration, or in filtering devices as pollutant removal agents.

Author Contributions: Conceptualization and writing—original draft preparation, L.P.; methodology and data curation, L.P. and M.T. (Marco Trifuoggi); investigation, F.M. and M.T. (Maria Toscanesi); resources and supervision, S.G.; writing—review and editing, A.N.

Funding: This research received no external funding.

Conflicts of Interest: The authors declare no conflict of interest.

References

1. Panzella, L.; Napolitano, A. Natural phenol polymers: Recent advances in food and health applications. *Antioxidants* **2017**, *6*, 30. [CrossRef] [PubMed]
2. Piccinino, D.; Capecchi, E.; Botta, L.; Bizzarri, B.M.; Bollella, P.; Antiochia, R.; Saladino, R. Layer-by-layer preparation of microcapsules and nanocapsules of mixed polyphenols with high antioxidant and UV-shielding properties. *Biomacromolecules* **2018**, *19*, 3883–3893. [CrossRef] [PubMed]
3. Kurisawa, M.; Chung, J.E.; Uyama, H.; Kobayashi, S. Oxidative coupling of epigallocatechin gallate amplifies antioxidant activity and inhibits xanthine oxidase activity. *Chem. Commun.* **2004**, 294–295. [CrossRef] [PubMed]
4. Panzella, L.; Cerruti, P.; Ambrogi, V.; Agustin-Salazar, S.; D'Errico, G.; Carfagna, C.; Goya, L.; Ramos, S.; Martín, M.A.; Napolitano, A.; et al. A superior all-natural antioxidant biomaterial from spent coffee grounds for polymer stabilization, cell protection, and food lipid preservation. *ACS Sustain. Chem. Eng.* **2016**, *4*, 1169–1179. [CrossRef]
5. Panzella, L.; Pérez-Burillo, S.; Pastoriza, S.; Martín, M.Á.; Cerruti, P.; Goya, L.; Ramos, S.; Rufián-Henares, J.Á.; Napolitano, A.; d'Ischia, M. High antioxidant action and prebiotic activity of hydrolyzed spent coffee grounds (HSCG) in a simulated digestion-fermentation model: Toward the development of a novel food supplement. *J. Agric. Food Chem.* **2017**, *65*, 6452–6459. [CrossRef] [PubMed]
6. Olejar, K.J.; Ray, S.; Ricci, A.; Kilmartin, P.A. Superior antioxidant polymer films created through the incorporation of grape tannins in ethyl cellulose. *Cellulose* **2014**, *21*, 4545–4556. [CrossRef]
7. Memoli, S.; Napolitano, A.; d'Ischia, M.; Misuraca, G.; Palumbo, A.; Prota, G. Diffusible melanin-related metabolites are potent inhibitors of lipid peroxidation. *Biochim. Biophys. Acta* **1997**, *1346*, 61–68. [CrossRef]
8. Panzella, L.; D'Errico, G.; Vitiello, G.; Perfetti, M.; Alfieri, M.L.; Napolitano, A.; d'Ischia, M. Disentangling structure-dependent antioxidant mechanisms in phenolic polymers by multiparametric EPR analysis. *Chem. Commun.* **2018**, *54*, 9426–9429. [CrossRef]
9. Ambrogi, V.; Panzella, L.; Persico, P.; Cerruti, P.; Lonz, C.A.; Carfagna, C.; Verotta, L.; Caneva, E.; Napolitano, A.; d'Ischia, M. An antioxidant bioinspired phenolic polymer for efficient stabilization of polyethylene. *Biomacromolecules* **2014**, *15*, 302–310. [CrossRef]
10. Ekambaram, S.P.; Perumal, S.S.; Balakrishnan, A. Scope of hydrolysable tannins as possible antimicrobial agent. *Phytother. Res.* **2016**, *30*, 1035–1045. [CrossRef]
11. Freire, M.; Cofrades, S.; Perez-Jimenez, J.; Gomez-Estaca, J.; Jimenez-Colmenero, F.; Bou, R. Emulsion gels containing n-3 fatty acids and condensed tannins designed as functional fat replacers. *Food Res. Int.* **2018**, *113*, 465–473. [CrossRef]
12. Cheaib, D.; El Darra, N.; Rajha, H.N.; El-Ghazzawi, I.; Mouneimne, Y.; Jammoul, A.; Maroun, R.G.; Louka, N. Study of the selectivity and bioactivity of polyphenols using infrared assisted extraction from apricot pomace compared to conventional methods. *Antioxidants* **2018**, *7*, 174. [CrossRef]
13. Smeriglio, A.; Barreca, D.; Bellocco, E.; Trombetta, D. Proanthocyanidins and hydrolysable tannins: Occurrence, dietary intake and pharmacological effects. *Br. J. Pharmacol.* **2017**, *174*, 1244–1262. [CrossRef]

14. Hagerman, A.E. Fifty years of polyphenol–protein complexes. In *Recent Advances in Polyphenol Research*; Cheynier, V., Sarni-Manchado, P., Quideau, S., Eds.; Wiley-Blackwell: Hoboken, NJ, USA, 2012; Volume 3, pp. 71–97.

15. Comandini, P.; Lerma-García, M.J.; Simó-Alfonso, E.F.; Toschi, T.G. Tannin analysis of chestnut bark samples (*Castanea sativa* Mill.) by HPLC-DAD-MS. *Food Chem.* **2014**, *157*, 290–295. [CrossRef]

16. Reid, D.G.; Bonnet, S.L.; Kemp, G.; van der Westhuizen, J.H. Analysis of commercial proanthocyanidins. Part 4: Solid state ^{13}C NMR as a tool for in situ analysis of proanthocyanidin tannins, in heartwood and bark of quebracho and acacia, and related species. *Phytochemistry* **2013**, *94*, 243–248. [CrossRef]

17. Redondo, L.M.; Chacana, P.A.; Dominguez, J.E.; Fernandez Miyakawa, M.E. Perspectives in the use of tannins as alternative to antimicrobial growth promoter factors in poultry. *Front. Microbiol.* **2014**, *27*, 118. [CrossRef]

18. Cardullo, N.; Muccilli, V.; Saletti, R.; Giovando, S.; Tringali, C. A mass spectrometry and ^{1}H NMR study of hypoglycemic and antioxidant principles from a *Castanea sativa* tannin employed in oenology. *Food Chem.* **2018**, *268*, 585–593. [CrossRef]

19. Molino, S.; Fernandez-Miyakawa, M.; Giovando, S.; Rufian-Henares, J.A. Study of antioxidant capacity and metabolization of quebracho and chestnut tannins through in vitro gastrointestinal digestion-fermentation. *J. Funct. Foods* **2018**, *49*, 188–195. [CrossRef]

20. Poaty, B.; Dumarcay, S.; Gerardin, P.; Perrin, D. Modification of grape seed and wood tannins to lipophilic antioxidant derivatives. *Ind. Crops Prod.* **2010**, *31*, 509–515. [CrossRef]

21. Aires, A.; Carvalho, R.; Saavedra, M.J. Valorization of solid wastes from chestnut industry processing: Extraction and optimization of polyphenols, tannins and ellagitannins and its potential for adhesives, cosmetic and pharmaceutical industry. *Waste Manag.* **2016**, *48*, 457–464. [CrossRef] [PubMed]

22. Versari, A.; du Toit, W.; Parpinello, G.P. Oenological tannins: A review. *Aust. J. Grape Wine* **2013**, *19*, 1–10. [CrossRef]

23. Campo, M.; Pinelli, P.; Romani, A. Hydrolyzable tannins from sweet chestnut fractions obtained by a sustainable and eco-friendly industrial process. *Nat. Prod. Commun.* **2016**, *11*, 409–415. [CrossRef]

24. Taflick, T.; Schwendler, L.A.; Rosa, S.M.L.; Bica, C.I.D.; Nachtigall, S.M.B. Cellulose nanocrystals from acacia bark-Influence of solvent extraction. *Int. J. Biol. Macromol.* **2017**, *101*, 553–561. [CrossRef] [PubMed]

25. Panzella, L.; Eidenberger, T.; Napolitano, A.; d'Ischia, M. Black sesame pigment: DPPH assay-guided purification, antioxidant/antinitrosating properties, and identification of a degradative structural marker. *J. Agric. Food Chem.* **2012**, *60*, 8895–8901. [CrossRef] [PubMed]

26. Verotta, L.; Panzella, L.; Antenucci, S.; Calvenzani, V.; Tomay, F.; Petroni, K.; Caneva, E.; Napolitano, A. Fermented pomegranate wastes as sustainable source of ellagic acid: Antioxidant properties, anti-inflammatory action, and controlled release under simulated digestion conditions. *Food Chem.* **2018**, *246*, 129–136. [CrossRef] [PubMed]

27. Gökmen, V.; Serpen, A.; Fogliano, V. Direct measurement of the total antioxidant capacity of foods: The 'QUENCHER' approach. *Trends Food Sci. Technol.* **2009**, *20*, 278–288. [CrossRef]

28. Goupy, P.; Dufour, C.; Loonis, M.; Dangles, O. Quantitative kinetic analysis of hydrogen transfer reactions from dietary polyphenols to the DPPH radical. *J. Agric. Food Chem.* **2003**, *51*, 615–622. [CrossRef]

29. Benzie, I.F.F.; Strain, J.J. The Ferric Reducing Ability of Plasma (FRAP) as a measure of "antioxidant power": The FRAP assay. *Anal. Biochem.* **1996**, *239*, 70–76. [CrossRef] [PubMed]

30. Xu, C.; Liu, S.; Liu, Z.; Song, F.; Liu, S. Superoxide generated by pyrogallol reduces highly water-soluble tetrazolium salt to produce a soluble formazan: A simple assay for measuring superoxide anion radical scavenging activities of biological and abiological samples. *Anal. Chim. Acta* **2013**, *793*, 53–60. [CrossRef]

31. Drinkwater, J.M.; Tsao, R.; Liu, R.; Defelice, C.; Wolyn, D.J. Effects of cooking on rutin and glutathione concentrations and antioxidant activity of green asparagus (*Asparagus officinalis*) spears. *J. Funct. Foods* **2015**, *12*, 342–353. [CrossRef]

32. Panzella, L.; Melone, L.; Pezzella, A.; Rossi, B.; Pastori, N.; Perfetti, M.; D'Errico, G.; Punta, C.; d'Ischia, M. Surface-functionalization of nanostructured cellulose aerogels by solid state eumelanin coating. *Biomacromolecules* **2016**, *17*, 564–571. [CrossRef]

33. Panzella, L.; Manini, P.; Crescenzi, O.; Napolitano, A.; d'Ischia, M. Nitrite-induced nitration pathways of retinoic acid, 5,6-epoxyretinoic acid, and their esters under mildly acidic conditions: Toward a reappraisal of retinoids as scavengers of reactive nitrogen species. *Chem. Res. Toxicol.* **2003**, *16*, 502–511. [CrossRef]

34. Manini, P.; Panzella, L.; Eidenberger, T.; Giarra, A.; Cerruti, P.; Trifuoggi, M.; Napolitano, A. Efficient binding of heavy metals by black sesame pigment: Toward innovative dietary strategies to prevent bioaccumulation. *J. Agric. Food Chem.* **2016**, *64*, 890–897. [CrossRef]
35. Bosso, A.; Guaita, M.; Petrozziello, M. Influence of solvents on the composition of condensed tannins in grape pomace seed extracts. *Food Chem.* **2016**, *207*, 162–169. [CrossRef] [PubMed]
36. Seifzadeha, N.; Saharia, M.A.; Barzegara, M.; Gavlighia, H.A.; Calanib, L.; Del Rio, D.; Galaverna, G. Evaluation of polyphenolic compounds in membrane concentrated pistachio hull extract. *Food Chem.* **2019**, *277*, 398–406. [CrossRef] [PubMed]
37. Peng, Y.; Liu, R.; Cao, J. Characterization of surface chemistry and crystallization behavior of polypropylene composites reinforced with wood flour, cellulose, and lignin during accelerated weathering. *Appl. Surf. Sci.* **2015**, *332*, 253–259. [CrossRef]
38. Neagu, R.C.; Cuénoud, M.; Berthold, F.; Bourban, P.E.; Gamstedt, E.K.; Lindström, M.; Månson, J.A.E. The potential of wood fibers as reinforcement in cellular biopolymers. *J. Cell. Plast.* **2012**, *48*, 71–103. [CrossRef]
39. Gajewski, M.; Kowalczyk, K.; Bajer, M. Influence of ecological friendly mediums on chemical composition of greenhouse-grown eggplants. *Ecol. Chem. Eng. A* **2010**, *17*, 1103–1109.
40. Panzella, L.; Manini, P.; Napolitano, A.; d'Ischia, M. The acid-promoted reaction of the green tea polyphenol epigallocatechin gallate with nitrite ions. *Chem. Res. Toxicol.* **2005**, *18*, 722–729. [CrossRef]

© 2019 by the authors. Licensee MDPI, Basel, Switzerland. This article is an open access article distributed under the terms and conditions of the Creative Commons Attribution (CC BY) license (http://creativecommons.org/licenses/by/4.0/).

Article

Synthesis of Lipophilic Esters of Tyrosol, Homovanillyl Alcohol and Hydroxytyrosol

Roberta Bernini *, Isabella Carastro, Francesca Santoni and Mariangela Clemente

Department of Agriculture and Forest Sciences (DAFNE), University of Tuscia, Via S. Camillo de Lellis, 01100 Viterbo, Italy; isabella109@alice.it (I.C.); santonifrancesca@yahoo.it (F.S.); marian.clem@unitus.it (M.C.)
* Correspondence: roberta.bernini@unitus.it; Tel.: +39-0761-357-452

Received: 26 April 2019; Accepted: 10 June 2019; Published: 14 June 2019

Abstract: Low-molecular weight phenols such as tyrosol, homovanillyl alcohol and hydroxytyrosol are valuable compounds that exhibit a high number of health-promoting effects such as antioxidant, anti-inflammatory and anticancer activity. Despite these remarkable properties, their applications such as dietary supplements and stabilizers of foods and cosmetics in non-aqueous media are limited for the hydrophilic character. With the aim to overcome this limitation, the paper describes a simple and low-cost procedure for the synthesis of lipophilic esters of tyrosol, homovanillyl alcohol and hydroxytyrosol. The reactions were carried out under mild and green chemistry conditions, at room temperature, solubilizing the phenolic compounds in dimethyl carbonate, an eco-friendly solvent, and adding a little excess of the appropriate C2–C18 acyl chloride. The final products were isolated in good yields. Finally, according to the "circular economy" strategy, the procedure was applied to hydroxytyrosol-enriched extracts obtained by *Olea europaea* by-products to prepare a panel of lipophilic extracts that are useful for applications where solubility in lipid media is required.

Keywords: tyrosol; homovanillyl alcohol; hydroxytyrosol; dimethyl carbonate; lipophilic alkyl esters; hydroxytyrosol-enriched extracts; *Olea europaea*; green chemistry; circular economy

1. Introduction

Tyrosol **1** (4-hydroxyphenethyl alcohol), homovanillyl alcohol **2** (3-hydroxy-4-methoxyphenethyl alcohol) and hydroxytyrosol **3** (3,4-dihydroxyphenethyl alcohol) are low-molecular weight phenols (Figure 1) found mainly in olive mill wastewater, the by-products of the olive oil production [1,2], from which they can be extracted using environmentally and economically sustainable technologies [3,4] and reused according to the circular economy strategy [5].

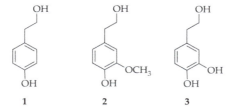

Figure 1. Low-molecular weight phenols found in olive mill wastewater.

They are valuable compounds, displaying a high number of health-promoting effects such as antioxidant, anti-inflammatory and anticancer activity [6,7]. In particular, tyrosol **1** suppresses allergic inflammatory disorders [8], and prevents apoptosis in irradiated keratinocytes [9]. In addition, it is useful to synthetize novel antioxidants [10,11]. Homovanillyl alcohol **2** is a good antioxidant agent [12] and exhibits anti-inflammatory effects on human gastric adenocarcinoma cells [13]. In accordance with the theoretical predictions on *ortho*-diphenols [14], hydroxytyrosol 3 is the most powerful antioxidant found in olive oil by-products. In addition, it increases high-density lipoprotein (HDL)-cholesterol, inhibits inflammation, improves endothelial function, decreases tumor growth and metastasis, and protects the central nervous system [15–18]. In our experiment, hydroxytyrosol was synthetized by the 2-iodoxybenzoic acid (IBX) oxidation of tyrosol by safe and eco-friendly procedures [19,20] and used to prepare both hydroxytyrosol-derived compounds [21–23] and poly(vinyl alcohol)-based films with antioxidant activity [24–26].

Despite the remarkable biological properties of these compounds, their applicability as active ingredients in lipophilic foods and cosmetic products requiring solubility in non-aqueous media is limited for the hydrophilic character. With the aim to overcome this limitation, common for almost all phenolic compounds, recently, a growing interest has been devoted to synthesizing lipophilic derivatives. They could be prepared incorporating one or more halogen atoms [27–29] or by introducing different chain lengths [30–40] into the molecular skeleton, avoiding the derivatization of the phenolic moiety to which the biological effects, and in particular the antioxidant activity, are attributed.

Among them, tyrosyl, homovanillyl and hydroxytyrosyl esters with different chain lengths proved promising for their beneficial effects in non-aqueous media related to their lipophilicity and bioavailability. In particular, antioxidant activity and this relationship with chain length were the objects of many studies. In emulsified systems, the antioxidant activity of alkyl esters is directly related to the chain length: it increases up to a point (C8–C10) and then dramatically decreases [33], revisiting the polar paradox theory [41]. This effect, named "the cut-off effect", was explained with the surfactant property of lipophilic derivatives. C8–C10 hydroxytyrosyl esters are both the best antioxidants and surfactants [33]. Other studies have shown that esters with medium length chains have comparable or higher antioxidant activity than esters with long chains [42]. Tyrosyl esters were tested as antimicrobial and antileishmanial agents [35] and, recently, their absorption and stability in the intestinal lumen were investigated [43]. This study has demonstrated that lipophilic tyrosol esters were well absorbed by the intestinal lumen, confirming the relevant role of the alkyl chain [43]. Lipophilic hydroxytyrosol esters are good candidates as active ingredients of formulations for the treatment of cutaneous inflammations, permeating through the human stratum corneum and viable epidermis membrane and releasing hydroxytyrosol, which is the antioxidant and anti-inflammatory agent [44].

Based on the literature data, and the need to prepare lipophilic phenolic derivatives for extending their use in food and cosmetic sectors, we planned a simple and low-cost procedure to obtain tyrosyl, homovanillyl and hydroxytyrosyl alkyl esters. Finally, the procedure was applied to hydroxytyrosol-enriched extracts obtained by *Olea europaea* by-products to prepare a panel of lipophilic extracts that are useful for applications where solubility in lipid media is required.

2. Materials and Methods

2.1. Chemicals

Solvents, reagents, standards, silica gel 60 F254 plates, and silica gel 60 were purchased from Sigma-Aldrich (Milan, Italy). Hydroxytyrosol was prepared by the IBX oxidation of tyrosol as already reported [19]. A patented procedure based on membrane technology was applied to obtain hydroxytyrosol-enriched extracts from *Olea europaea* by-products [45]. The vegetal material was extracted with water and then subject to microfiltration (MF), nanofiltration (NF), and reverse osmosis (RO). Microfiltration was carried out with tubular ceramic membranes in titanium oxide; nanofiltration and reverse osmosis were carried out using spiral wound module membranes of poly(ether sulfone) [45]. The collected fraction was concentrated using a heat pump evaporator (Vacuum Evaporators—Scraper Series, C&G Depurazione Industriale Srl, Firenze, Italy) [3,4]. High Performance Liquid Chromatographic (HPLC) analysis of the resulting extracts was carried out at $\lambda = 280$ nm using a diode array detector (DAD) to identify the polyphenolic compounds found in

Antioxidants **2019**, *8*, 174

the sample. The quantitative data were referred to pure standards (tyrosol, hydroxytyrosol, oleuropein, and caffeic acid).

2.2. Instruments

NMR spectra (^{1}H NMR, ^{13}C NMR) were recorded using a 400-MHz nuclear resonance spectrometer Advance-III Bruker (Munich, Germany). Each sample (20 mg) was dissolved in chloroform-d3 or methanol-d4 (0.5 mL). The chemical shifts were expressed in parts per million (δ scale) and referenced to either the residual protons or carbon of the solvent.

HPLC analysis of the hydroxytyrosol-enriched extracts and the corresponding lipophilic fractions were carried out using a HP 1200 liquid chromatograph (Agilent Technologies, Palo Alto, CA, USA). The detector was a diode array and the column was a LiChrosorb RP-18 (250 × 4.60 mm, 5 μm i.d.; Merck, Darmstadt, Germany). A flow rate of 0.8 mL min^{-1} was used for 88 min working from 100% of solvent A (H_2O at pH = 3.2) to 100% of solvent B (CH_3CN).

2.3. Synthesis of Esters: General Procedure

The appropriate phenethyl alcohol (0.5 mmol) was solubilized in dimethyl carbonate (1.5 mL) at 25 °C. Then the acyl chloride was added (0.6 mmol) and the mixture was kept under stirring for 24 h. The reaction was monitored by thin-layer chromatography on silica gel plates using mixtures of dichloromethane and methanol (9.8/0.2, 9.5/0.5 or 9.0/1.0) as eluents. At the end, the solvent was distilled under reduced pressure and the residue was solubilized with ethyl acetate (10 mL); then a saturated solution of NaCl was added (5.0 mL). After the extraction with ethyl acetate (3 × 10 mL), the combined organic phases were washed with NaCl s.s. (10 mL), dried over Na_2SO_4, and filtered. The solvent was distilled under reduced pressure and the residue was purified by a silica gel chromatographic column using mixtures of dichloromethane and methanol (9.8:0.2, 9.5:0.5 or 9.0/1.0) as eluents. All compounds were characterized by NMR analysis. Tyrosyl and homovanillyl alcohol were obtained in yields ranging from 90 to 98%, and hydroxytyrosyl esters from 60 to 68%, as detailed in Table 1.

Table 1. Esterification reactions of tyrosol **1**, homovanillyl alcohol **2** and hydroxytyrosol **3** (yields calculated after chromatographic purification).

Substrate	Product	Yield (%)	Substrate	Product	Yield (%)	Substrate	Product	Yield (%)
1	4	94	2	15	92	3	26	66
1	5	98	2	16	92	3	27	62
1	6	96	2	17	95	3	28	68
1	7	98	2	18	94	3	29	62
1	8	96	2	19	92	3	30	62
1	9	96	2	20	98	3	31	64
1	10	97	2	21	95	3	32	60
1	11	95	2	22	98	3	33	62
1	12	94	2	23	92	3	34	60
1	13	96	2	24	90	3	35	64
1	14	90	2	25	90	3	36	60

4-Hydroxyphenethyl acetate (Tyrosyl acetate) **4**. Colorless oil. NMR spectra are in accordance with those reported in the literature [19,32,35].

4-Hydroxyphenethyl butanoate (Tyrosyl butyrate) **5**. Colorless oil. NMR spectra are in accordance with those reported in the literature [32].

4-Hydroxyphenethyl hexanoate (Tyrosyl hexanoate) **6**. Colorless oil. NMR spectra are in accordance with those reported in the literature [19,33].

4-Hydroxyphenethyl octanoate (Tyrosyl caprylate) **7**. Yellow oil. NMR spectra are in accordance with those reported in the literature [33,35].

4-Hydroxyphenethyl decanoate (Tyrosyl capriate) **8**. Yellow oil. NMR spectra are in accordance with those reported in the literature [33,35].

4-Hydroxyphenethyl decanoate (Tyrosyl laurate) **9**. Yellow oil. NMR spectra are in accordance with those reported in the literature [32,35].

4-Hydroxyphenethyl tetradecanoate (Tyrosyl myristate) **10**. Yellow oil. NMR spectra are in accordance with those reported in the literature [32,35].

4-Hydroxyphenethyl palmitate (Tyrosyl palmitate) **11**. Colorless oil. NMR spectra are in accordance with those reported in the literature [19,32,35].

4-Hydroxyphenethyl stearate (Tyrosol stearate) **12**. Colorless oil. NMR spectra are in accordance with those reported in the literature e [32,35].

4-Hydroxyphenethyl oleate (Tyrosol oleate) **13**. Yellow oil. NMR spectra are in accordance with those reported in the literature e [32,35].

4-Hydroxyphenethyl linoleate (Tyrosol linoleate) **14**. Yellow oil. NMR spectra are in accordance with those reported in the literature e [19,32,35].

4-Hydroxy-3-methoxyphenethyl acetate (Homovanillyl acetate) **15**. Colorless oil. NMR spectra are in accordance with those reported in the literature [12,19].

4-Hydroxy-3-methoxyphenethyl butanoate (Homovanillyl butyrate) **16**. Colorless oil. NMR spectra are in accordance with those reported in the literature [12].

4-Hydroxy-3-methoxyphenethyl hexanoate (Homovanillyl hexanoate) **17**. Yellow oil. NMR spectra are in accordance with those reported in the literature [19].

4-Hydroxy-3-methoxyphenethyl octanoate (Homovanillyl caprylate) **18**. Yellow oil. ^{1}H-NMR (400 MHz, CDCl$_3$) δ: 6.88–6.86 (1H, m, Ph-H), 6.74–6.72 (2H, m, Ph-H), 5.57 (1H, bs, OH), 4.27 (2H, t, J = 6.0 Hz, OCH$_2$), 3.90 (3H, s, OCH$_3$), 2.88 (2H, t, J = 6.0 Hz, Ph-CH$_2$), 2.31 (2H, t, J = 8.0 Hz, COCH$_2$), 1.60 (2H, m, CH$_2$), 1.28 (8H, m, 4CH$_2$), 0.90 (3H, m, CH$_3$). ^{13}C-NMR (100 MHz, CDCl$_3$) δ: 173.4, 145.9, 143.8, 129.2, 121.1, 113.8, 110.9, 64.5, 55.4, 34.3, 33.9, 31.1, 28.6, 28.5, 24.5, 22.1, 13.5.

4-Hydroxy-3-methoxyphenethyl decanoate (Homovanillyl capriate) **19**. Yellow oil. NMR spectra are in accordance with those reported in the literature [12].

4-Hydroxy-3-methoxyphenethyl dodecanoate (Homovanillyl laurate) **20**. Yellow oil. ^{1}H-NMR (400 MHz, CDCl$_3$) δ: 6.97–6.95 (1H, m, Ph-H), 6.87–6.72 (2H, m, Ph-H), 5.40 (1H, bs, OH), 4.27 (2H, t, J = 8.0 Hz, OCH$_2$), 3.89 (3H, s, OCH$_3$), 2.88 (2H, t, J = 8.0 Hz, Ph-CH$_2$), 2.31 (2H, t, J = 8.0 Hz, COCH$_2$), 1.64–1.60 (2H, m, CH$_2$), 1.28 (16H, m, 8CH$_2$), 0.92–0.89 (3H, m, CH$_3$). ^{13}C-NMR (100 MHz, CDCl$_3$) δ: 173.4, 145.9, 143.8, 129.1, 121.1, 113.9, 110.9, 64.5, 55.4, 34.3, 33.9, 31.4, 29.1, 29.0, 28.9, 28.8, 28.7, 28.5, 24.4, 22.1, 13.6.

4-Hydroxy-3-methoxyphenethyl tetradecanoate (Homovanillyl myristate) **21**. Yellow oil. ^{1}H-NMR (400 MHz, CDCl3) δ: 6.87–6.85 (1H, m, Ph-H), 6.74–6.72 (2H, m, Ph-H), 5.60 (1H, bs, OH), 4.27 (2H, t, J = 8.0 Hz, OCH$_2$), 3.89 (3H, s, OCH$_3$), 2.88 (2H, t, J = 8.0 Hz, Ph-CH$_2$), 2.31 (2H, t, J = 8.0 Hz, COCH$_2$), 1.64–1.60 (2H, m, CH$_2$), 1.28 (20H, m, 10CH$_2$), 0.92–0.89 (3H, m, CH$_3$). ^{13}C-NMR (100 MHz, CDCl3) δ: 173.4, 145.9, 143.8, 129.1, 121.1, 113.9, 110.9, 64.5, 55.4, 34.3, 33.9, 31.4, 29.1, 29.0, 28.9, 28.8, 28.7, 28.6, 28.5, 28.4, 24.5, 22.2, 13.6.

4-Hydroxy-3-methoxyphenethyl palmitate (Homovanillyl palmitate) **22**. Colorless oil. NMR spectra are in accordance with those reported in the literature [19,30].

4-Hydroxy-3-methoxyphenethyl stearate (Homovanillyl stearate) **23**. Yellow oil. NMR spectra are in accordance with those reported in the literature [12].

4-Hydroxy-3-methoxyphenethyl oleate (Homovanillyl oleate) **24**. Yellow oil. NMR spectra are in accordance with those reported in the literature [19].

4-Hydroxy-3-methoxyphenethyl linoleate (Homovanillyl linoleate) **25**. Yellow oil. NMR spectra are in accordance with those reported in the literature [19].

3,4-Dihydroxyphenethyl acetate (Hydroxytyrosyl acetate) **26**. Yellow oil. NMR spectra are in accordance with those reported in the literature [12,19,46].

3,4-Dihydroxyphenethyl butanoate (Hydroxytyrosyl butyrate) **27**. Colorless oil. NMR spectra are in accordance with those reported in the literature [4,12,44].

3,4-Dihydroxyphenethyl hexanoate (Hydroxytyrosyl hexanoate) **28**. Yellow oil. NMR spectra are in accordance with those reported in the literature [19].

3,4-Dihydroxyphenethyl octanoate (Hydroxytyrosyl caprylate) **29**. Colorless oil. NMR spectra are in accordance with those reported in the literature [34].

3,4-Dihydroxyphenethyl decanoate (Hydroxytyrosyl capriate) **30**. Yellow oil. NMR spectra are in accordance with those reported in the literature [12,44].

3,4-Dihydroxyphenethyl dodecanoate (Hydroxytyrosyl laurate) **31**. Yellow oil. NMR spectra are in accordance with those reported in the literature [31].

3,4-Dihydroxyphenethyl tetradecanoate (Hydroxytyrosyl myristate) **32**. Yellow oil. NMR spectra are in accordance with those reported in the literature [30].

3,4-Dihydroxyphenethyl palmitate (Hydroxytyrosyl palmitate) **33**. Yellow oil. NMR spectra are in accordance with the literature [19,31,44].

3,4-Dihydroxyphenethyl stearate (Hydroxytyrosyl stearate) **34**. Yellow oil. NMR spectra are in accordance with those reported in the literature [12,31,44].

3,4-Dihydroxyphenethyl oleate (Hydroxytyrosyl oleate) **35**. Colorless oil. NMR spectra are in accordance with those reported in the literature [19,31,44].

3,4-Dihydroxyphenethyl linoleate (Hydroxytyrosyl linoleate) **36**. Colorless oil. NMR spectra are in accordance with those reported in the literature [19,31,44].

2.4. Esterification of Hydroxytyrosol Present in the Extracts: General Procedure

A total of 25 mg (0.16 mmol) of hydroxytyrosol-enriched extracts were dissolved in dimethyl carbonate (5.0 mL), and 0.19 mmol (40–200 µL) of the appropriate acyl chloride was introduced. The reaction was kept at 25°C under magnetic stirring for 24 h; then, dimethyl carbonate was distilled under reduced pressure by using a rotary evaporator (Laborota 4000, Heidolph, Munich, Germany). The mixture was solubilized with ethyl acetate and washed with a saturated solution of NaCl; then, the combined organic phases were dried over Na_2SO_4. After filtration, the solution was recovered, and the solvent was evaporated under reduced pressure. The content of hydroxytyrosol and the corresponding alkyl ester present in each sample was determined by HPLC–DAD analysis at $\lambda = 280$ nm. The yields of hydroxytyrosyl esters range from 60 to 66%.

3. Results and Discussion

Tyrosol **1**, homovanillyl alcohol **2** or hydroxytyrosol **3** (Scheme 1) was solubilized in dimethyl carbonate (DMC), an eco-friendly solvent [47], and then a little excess of the appropriate acyl chloride (1.2 equiv.) was added. The reactions were stirred at room temperature for 24 h. After the work-up and column chromatographic purification, the corresponding esters were isolated in good yields (Table 1). The experimental results confirmed that the esterification reactions proceeded chemoselectively on the alcoholic group due the higher nucleophilicity compared to the phenolic moiety, emphasized by DMC, as already observed by us [19,47]. Even if hydroxytyrosol esters were isolated in lower yields compared to the enzymatic procedures reported in the literature [30–35], the simplicity of the operations and low cost of the reagents makes the described procedure attractive.

Most of the isolated esters exhibit a strong antioxidant activity in lipid media as oils and emulsions [33,35,41]. Tyrosol caprylate **7**, capriate **8**, and laurate **9** show remarkable antimicrobial activity against *Leishmania major, Leishmania infantum, Staphylococcus aureus, Staphylococcus xylosus, Bacillus cereus* and *Brevibacterium flavum* [35]. Hydroxytyrosol acetate **26**, found in olive oil [46], exhibits antioxidant activity in oil and emulsions [48]; hydroxytyrosol oleate **35**, recently found in olive oil by-products [49], is effective as an anti-inflammatory agent. Both derivatives **26** and **35** show antiproliferative activity on human cervical cells (HeLa) [50]. Hydroxytyrosol butanoate **27**, decanoate **30**, palmitate **33**, stearate **34**, oleate **35** and linoleate **36** are promising therapeutic agents for topical use in consideration to their cutaneous permeability [44].

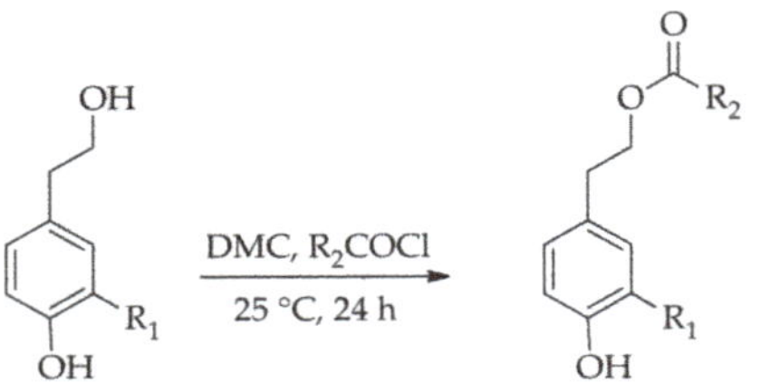

Antioxidants 2019, 8, 174

Scheme 1. Esterification reactions of phenolic compounds **1**, **2** and **3**.

Finally, hydroxytyrosol-enriched extracts were esterified under the same experimental conditions using the C2–C18 acyl chlorides. These samples were obtained by a selective extraction of *Olea europaea* by-products using a sustainable process based on membrane technologies [46]. The extracts contain 60.53 ± 0.41 mg/g of hydroxytyrosol (6.0 % w/w) on a total polyphenols content of 98.14 ± 2.43 mg/g [3,4]. After each esterification reaction, the hydroxytyrosyl ester found in the mixture was characterized and quantified by HPLC–DAD analysis. According to already observed using pure hydroxtytyrosol, the yields of the esterification reactions range from 60 to 66%.

Recently, we evaluated the antiproliferative activity of lipophilic fractions containing hydroxytyrosyl butanoate, octanoate and oleate on the human colon cancer cell line HCT8-β8, a model of colorectal cancer [4]. The experimental data has shown that all fractions exhibited antiproliferative activity. The relevant effect of hydroxytyrosol oleate was related to the high lipophilicity and bioavailability of the compound for the presence of the unsaturated C18 chain [4].

4. Conclusions

Lipophilic tyrosyl, homovanillyl and hydroxytyrosyl esters were prepared using a green and mild procedure. The products were obtained in good yields, solubilizing **1**, **2** or **3** in dimethyl carbonate, an eco-friendly solvent, and adding a little excess of the appropriate C2–C18 acyl chloride. The procedure was then applied to hydroxytyrosol-enriched extracts obtained by olive oil by-products to afford a panel of lipophilic fractions containing hydroxytyrosyl esters of different chain lengths to use for applications where the solubility in lipid media is required.

Author Contributions: Experimental design: R.B.; acquisition and analysis of data: I.C.; F.S.; M.C.; interpretation of data: M.C.; R.B.; paper drafting, editing and revision: R.B.

Funding: The authors gratefully acknowledge MIUR (Ministry for Education, University and Research) for financial support (Law 232/216, Department of Excellence).

Conflicts of Interest: The authors declare no conflict of interest.

References

1. Bendini, A.; Cerretani, L.; Carrasco-Pancorbo, A.; Gomez-Caravaca, A.M.; Segura-Carretero, A.; Fernandez-Gutierrez, A.; Lercker, G. Phenolic molecules in virgin olive oils: A survey of their sensory properties, health effects, antioxidant activiy and analytical methods. An overview of the last decade. *Molecules* **2007**, *12*, 1679–1719. [CrossRef] [PubMed]
2. Della Greca, M.; Previtera, L.; Temussi, F.; Zarrelli, A. Low-molecular weight components of olive oil mill wastewaters. *Phytochem. Anal.* **2004**, *15*, 184–188. [CrossRef] [PubMed]
3. Romani, A.; Pinelli, P.; Ieri, F.; Bernini, R. Sustainability, innovation and green chemistry in the production and valorization of phenolic extracts from Olea europaea L. *Sustainability* **2016**, *8*, 1002. [CrossRef]
4. Bernini, R.; Carastro, I.; Palmini, G.; Tanini, A.; Zonefrati, R.; Pinelli, P.; Brandi, M.L.; Romani, A. Lipophilization of hydroxytyrosol-enriched fractions from *Olea europaea* L. by-products and evaluation of the in vitro effects on a model of colorectal cancer cells. *J. Agric. Food Chem.* **2017**, *65*, 6506–6512. [CrossRef] [PubMed]
5. Stahel, W.R. The circular economy. *Nature* **2016**, *531*, 435–438. [CrossRef] [PubMed]
6. Serino, A.; Salazar, G. Protective role of polyphenols against vascular inflammation, aging and cardiovascular disease. *Nutrients* **2019**, *11*, 53. [CrossRef]
7. Borzì, A.M.; Biondi, A.; Basile, F.; Luca, S.; Vicari, E.S.D.; Vacante, M. Olive oil effects on colorectal cancer. *Nutrients* **2018**, *11*, 33. [CrossRef]
8. In-Gyu Je, I.-G.; Kim, D.-S.; Kim, S.-W.; Lee, S.; Lee, H.-S.; Park, E.K.; Khang, D.; Kim, S.-H. Tyrosol suppresses allergic inflammation by inhibiting the activation of phosphoinositide 3-kinase in mast cells. *PLoS ONE* **2015**. [CrossRef]
9. Saluccia, S.; Burattini, S.; Battistelli, M.; Buontempo, F.; Canonico, B.; Martelli, A.M.; Papa, S.; Falcieri, E. Tyrosol prevents apoptosis in irradiated keratinocytes. *J. Dermatol. Sci.* **2015**, *80*, 61–68. [CrossRef]
10. Barontini, M.; Bernini, R.; Carastro, R.; Gentili, P.; Romani, A. Synthesis and DPPH radical scavenging activity of novel compounds obtained from tyrosol and cinnamic acid derivatives. *New J. Chem.* **2014**, *38*, 809–816. [CrossRef]
11. Bernini, R.; Crisante, F.; D'Acunzo, F.; Gentili, P.; Ussia, E. Oxidative cleavage of 1-aryl-isochroman derivatives by the Trametes villosa laccase/1-hydroxybenzotroazole system. *New J. Chem.* **2016**, *40*, 3314–3322. [CrossRef]
12. Grasso, S.; Siracusa, L.; Spatafora, C.; Renis, M.; Tringali, C. Hydroxytyrosol lipophilic analogues: enzymatic synthesis, radical scavenging activity and DNA oxidative damage protection. *Bioorg. Chem.* **2007**, *35*, 137–152. [CrossRef] [PubMed]
13. Serreli, G.; Deiana, M. Biological relevance of extra virgin olive oil polyphenols metabolites. *Antioxidants* **2018**, *7*, 170. [CrossRef] [PubMed]
14. Goupy, P.; Dufour, C.; Loonis, M.; Dangles, O. Quantitative kinetic analysis of hydrogen transfer reactions from dietary polyphenols to the DPPH radical. *J. Agric. Food Chem.* **2003**, *51*, 615–622. [CrossRef] [PubMed]
15. Bernini, R.; Merendino, N.; Romani, A.; Velotti, F. Naturally occurring hydroxytyrosol: Synthesis and anticancer potential. *Curr. Med. Chem.* **2013**, *20*, 655–670. [CrossRef] [PubMed]
16. Bernini, R.; Gilardini Montani, M.S.; Merendino, N.; Romani, A. Velotti, F. Hydroxytyrosol-derived compounds: a basis for the creation of new pharmacological agents for cancer prevention and therapy. *J. Med. Chem.* **2015**, *58*, 9089–9107. [CrossRef] [PubMed]
17. Fuccelli, R.; Fabiani, R.; Rosignoli, P. Hydroxytyrosol exerts anti-inflammatory and anti-oxidant activities in a mouse model of systemic inflammation. *Molecules* **2018**, *23*, 3212. [CrossRef] [PubMed]
18. de Pablos, R.M.; Espinosa, O.; Ornedo-Hortega, R.; Cano, M.; Arguelles, S. Hydroxytyrosol protects from aging process via AMPK and autophagy; a review of its effects on cancer, metabolic syndrome, osteoporosis, immune-mediated and neurodegenerative diseases. *Pharmacol. Res.* **2019**, *143*, 58–72. [CrossRef]
19. Bernini, R.; Mincione, E.; Barontini, M.; Crisante, F. Convenient synthesis of hydroxytyrosol and its lipophilic derivatives from tyrosol or homovanillyl alcohol. *J. Agric. Food Chem.* **2008**, *56*, 8897–8904. [CrossRef] [PubMed]
20. Bernini, R.; Mincione, E.; Crisante, F.; Barontini, M.; Fabrizi, G. A novel use of the recyclable polymer-supported IBX: an efficient chemoselective and regioselective oxidation of phenolic compounds. The case of hydroxytyrosol derivatives. *Tetrahedron Lett.* **2009**, *50*, 1307–1310. [CrossRef]

21. Bernini, R.; Cacchi, S.; Fabrizi, G.; Filisti, E. 2-Arylhydroxytyrosol derivatives via Suzuki-Miyaura cross-coupling. *Org. Lett.* **2008**, *10*, 3457–3460. [CrossRef]
22. Bernini, R.; Crisante, F.; Merendino, N.; Molinari, R.; Soldatelli, M.C.; Velotti, F. Synthesis of a novel ester of hydroxytyrosol and alpha-lipoic acid exhibiting an antiproliferative effect on human colon cancer HT-29 cells. *Eur. J. Med. Chem.* **2011**, *46*, 439–446. [CrossRef]
23. Bernini, R.; Crisante, F.; Fabrizi, G.; Gentili, P. Convenient synthesis of 1-aryl-dihydroxyisochromans exhibiting antioxidant activity. *Curr. Org. Chem.* **2012**, *16*, 1051–1057. [CrossRef]
24. Fortunati, E.; Luzi, F.; Dugo, L.; Fanali, C.; Tripodo, G.; Santi, L.; Kenny, J.M.; Torre, L.; Bernini, R. Effect of hydroxytyrosol methyl carbonate on the thermal, migration and antioxidant properties of PVA based films for active food packaging. *Polym. Int.* **2016**, *65*, 872–882. [CrossRef]
25. Fortunati, E.; Luzi, F.; Dugo, L.; Fanali, C.; Tripodo, G.; Santi, L.; Kenny, J.M.; Torre, L.; Bernini, R. Hydroxytyrosol as active ingredient in poly(vinyl alcohol) films for food packaging applications. *J. Renew. Mat.* **2017**, *5*, 81–95. [CrossRef]
26. Luzi, F.; Fortunati, E.; Di Michele, A.; Pannucci, E.; Botticella, E.; Santi, L.; Kenny, J.M.; Torre, L.; Bernini, R. Nanostructured starch combined with hydroxytyrosol in poly(vinyl) alcohol based ternary films as active packaging system. *Carbohyd. Polym.* **2018**, *193*, 239–248. [CrossRef]
27. Bovicelli, P.; Bernini, R.; Antonioletti, R.; Mincione, E. Selective halogenation of flavanones. *Tetrahedron Lett.* **2002**, *43*, 5563–5567. [CrossRef]
28. Bernini, R.; Pasqualetti, M.; Provenzano, G.; Tempesta, S. Ecofriendly synthesis of halogenated flavonoids and evaluation of their antifungal activity. *New J. Chem.* **2015**, *39*, 2980–2987. [CrossRef]
29. Bernini, R.; Barontini, M.; Cis, V.; Carastro, I.; Tofani, D.; Chiodo, R.A.; Lupattelli, P.; Incerpi, S. Synthesis and evaluation of the antioxidant activity of lipophilic phenethyl trifluoroacetate esters by in vitro ABTS, DPPH and in cell-culture DCF assays. *Molecules* **2018**, *23*, 208. [CrossRef]
30. Torres de Pinedo, A.; Penalver, P.; Rondon, D.; Morales, J.C. Efficient lipase-catalyzed synthesis of new lipid antioxidants based on a catechol structure. *Tetrahedron* **2005**, *61*, 7654–7660. [CrossRef]
31. Trujillo, N.; Mateos, R.; Collantes de Teran, L.; Espartero, J.L.; Cert, R.; Jover, M.; Alcudia, F.; Bautista, J.; Cert, A.; Parrado, J. Lipophilic hydroxytyrosyl esters. Antioxidant activity in lipid matrices and biological systems. *J. Agric. Food Chem.* **2006**, *54*, 3779–3785. [CrossRef]
32. Mateos, R.; Trujillo, M.; Pereira-Caro, G.; Madrona, A.; Cert, A.; Espartero, J. New lipophilic tyrosyl esters. Comparative antioxidant evaluation with hydroxytyrosyl esters. *J. Agric. Food Chem.* **2008**, *56*, 10960–10966. [CrossRef]
33. Lucas, R.; Comelles, F.; Alcántara, D.; Maldonado, O.S.; Curcuroze, M.; Parra, J.L.; Morales, J.C. Surface-active properties of lipophilic antioxidants tyrosol and hydroxytyrosol fatty acid esters: a potential explanation for the nonlinear hypothesis of the antioxidant activity in oil-in-water emulsions. *J. Agric. Food Chem.* **2010**, *58*, 8021–8026. [CrossRef]
34. Tofani, D.; Balducci, V.; Gasperi, T.; Incerpi, S.; Gambacorta, A. Fatty acid hydroxytyrosyl esters: structure/antioxidant activity relationship by ABTS and in cell-culture DCF assays. *J. Agric. Food Chem.* **2010**, *58*, 5292–5299. [CrossRef]
35. Aissa, I.; Sghair, R.M.; Bouaziz, M.; Laouini, D.; Sayadi, S. Synthesis of lipophilic tyrosyl esters derivatives and assessment of their antimicrobial and antileishmanial activities. *Lip. Health Dis.* **2012**, *11*, 1–8. [CrossRef]
36. Bernini, R.; Crisante, F.; Barontini, M.; Tofani, D.; Balducci, V.; Gambacorta, A. Synthesis and structure/antioxidant activity relationship of novel catecholic antioxidant structural analogues to hydroxytyrosol and its lipophilic esters. *J. Agric. Food Chem.* **2012**, *60*, 7408–7416. [CrossRef]
37. Pande, G.; Akoh, C. Enzymatic synthesis of tyrosol-based phenolipids: characterization and effect of alkyl chain unsaturation on the antioxidant activities in bulk oil and oil-in-water emulsion. *J. Am. Oil. Chem. Soc.* **2016**, *93*, 329–337. [CrossRef]
38. Sun, Y.; Zhou, D.; Shahidi, F. Antioxidant properties of tyrosol and hydroxytyrosol saturated fatty acid. *Food Chem.* **2018**, *245*, 1262–1268. [CrossRef]
39. Zhou, D.Y.; Sun, Y.X.; Shahidi, F. Preparation and antioxidant activity of tyrosoland hydroxytyrosol esters. *J. Funct. Foods* **2017**, *37*, 66–73. [CrossRef]
40. Madrona, A.; Pereira-Caro, G.; Mateos, R.; Rodrıguez, G.; Trujillo, M.; Fernandez-Bolanos, J.; Espartero, J.L. Synthesis of hydroxytyrosyl alkyl ethers from olive oil waste waters. *Molecules* **2009**, *14*, 1762–1772. [CrossRef]

41. Shahidi, F.; Zhong, Y. Revisiting the polar paradox theory: A critical overview. *J. Agric. Food Chem.* **2011**, *59*, 3499–3504. [CrossRef]
42. Akanbi, T.O.; Barrow, C.J. Lipase-produced hydroxytyrosyl eicosapentaenoate is an excellent antioxidant for the stabilization of omega-3 bulk oils, emulsions and microcapsules. *Molecules* **2018**, *23*, 275. [CrossRef]
43. Yin, F.W.; Hu, X.P.; Zhou, D.Y.; Ma, X.C.; Tian, X.G.; Huo, X.K.; Rakariyatham, F.; Shahidid, F.; Zhua, B.-W. Evaluation of the stability of tyrosol esters during in vitro gastrointestinal digestion. *Food Funct.* **2018**, *9*, 3610–3616. [CrossRef]
44. Procopio, A.; Celia, C.; Nardi, M.; Oliverio, M.; Paolino, D.; Sindona, G. Lipophilic hydroxytyrosol esters: fatty acid conjugates for potential topical administration. *J. Nat. Prod.* **2011**, *74*, 2377–2381. [CrossRef]
45. Pizzichini, D.; Russo, C.; Vitagliano, M.; Pizzichini, M.; Romani, A.; Ieri, F.; Pinelli, P.; Vignolini, P. Process for producing concentrated and refined actives from tissues and byproducts of *Olea europaea* with membrane technologies. Eur. Pat. Appl. 2338500 A1, 2011.
46. Brenes, M.; Garcia, A.; Garcia, P.; Rios, J.J.; Garrido, A. Phenolic compounds in Spanish olive oils. *J. Agric. Food Chem.* **1999**, *47*, 3335–3340. [CrossRef]
47. Bernini, R.; Mincione, E.; Crisante, F.; Barontini, M.; Fabrizi, G.; Gentili, P. Chemoselective and efficient carboxymethylation of the alcoholic chain of phenols by dimethyl carbonate (DMC). *Tetrahedron Lett.* **2007**, *48*, 7000–7003. [CrossRef]
48. Gordon, M.H.; Paiva-Martins, F.; Almeida, M. Antioxidant activity of hydroxytyrosol acetate compared with other olive oil polyphenols. *J. Agric. Food Chem.* **2001**, *49*, 2480–2485. [CrossRef]
49. Plastina, P.; Benincasa, C.; Perri, E.; Fazio, A.; Augimeri, G.; Poland, M.; Witkamp, R.; Meijerink, J. Indentification of hydroxytyrosyl oleate, a derivative of hydroxytyrosol with anti-inflammatory properties, in olive oil by-products. *Food Chem.* **2019**, *279*, 105–113. [CrossRef]
50. Bouallagui, Z.; Bouaziz, M.; Lassoued, S.; Engasser, J.M.; Ghoul, M.; Sayadi, S. Hydroxytyrosol acyl esters: biosynthesis and activities. *Appl. Biochem. Biotech.* **2011**, *163*, 592–599. [CrossRef]

© 2019 by the authors. Licensee MDPI, Basel, Switzerland. This article is an open access article distributed under the terms and conditions of the Creative Commons Attribution (CC BY) license (http://creativecommons.org/licenses/by/4.0/).

Review

Antioxidative and Anti-Inflammatory Properties of Cannabidiol

Sinemyiz Atalay, Iwona Jarocka-Karpowicz and Elzbieta Skrzydlewska *

Department of Analytical Chemistry, Medical University of Białystok, 15-089 Białystok, Poland; sinemyiz.atalay@umb.edu.pl (S.A.); iwona.jarocka-karpowicz@umb.edu.pl (I.J.-K.)
* Correspondence: elzbieta.skrzydlewska@umb.edu.pl; Tel.: +48-857-485-882

Received: 19 November 2019; Accepted: 23 December 2019; Published: 25 December 2019

Abstract: Cannabidiol (CBD) is one of the main pharmacologically active phytocannabinoids of *Cannabis sativa* L. CBD is non-psychoactive but exerts a number of beneficial pharmacological effects, including anti-inflammatory and antioxidant properties. The chemistry and pharmacology of CBD, as well as various molecular targets, including cannabinoid receptors and other components of the endocannabinoid system with which it interacts, have been extensively studied. In addition, preclinical and clinical studies have contributed to our understanding of the therapeutic potential of CBD for many diseases, including diseases associated with oxidative stress. Here, we review the main biological effects of CBD, and its synthetic derivatives, focusing on the cellular, antioxidant, and anti-inflammatory properties of CBD.

Keywords: cannabidiol; cannabidiol synthetic derivatives; endocannabinoids; oxidative stress; lipid peroxidation; inflammation; membrane receptors

1. Introduction

The endocannabinoid system is an important molecular system responsible for controlling homeostasis and is becoming an increasingly popular target of pharmacotherapy. Endocannabinoids are ester, ether, and amide derivatives of long chain polyunsaturated fatty acids (PUFAs), such as arachidonic acid, and they act mainly as cannabinoid receptor ligands [1]. Endocannabinoids belong to a large group of compounds with a similar structure and biological activity called cannabinoids. Cannabinoids are chemical derivatives of dibenzopyrene or monoterpenoid, and to date over four hundred have been identified. The most important of these are Δ^9-tetrahydrocannabinol (Δ^9-THC), Δ^8-tetrahydrocannabinol (Δ^8-THC), cannabinol (CBN), and cannabidiol (CBD), and they are members of a large group of biologically active compounds found in *Cannabis sativa* L. [2]. The medical use of cannabinoids, in particular phytocannabinoids, has been one of the most interesting approaches to pharmacotherapy in recent years.

CBD is one of the main pharmacologically active phytocannabinoids [3]. It is non-psychoactive, but has many beneficial pharmacological effects, including anti-inflammatory and antioxidant effects [4]. In addition, it belongs to a group of compounds with anxiolytic, antidepressant, antipsychotic, and anticonvulsant properties, among others [5]. The biological effects of cannabidiol, including the various molecular targets, such as cannabinoid receptors and other components of the endocannabinoid system, with which it interacts, have been extensively studied. The therapeutic potential of CBD has been evaluated in cardiovascular, neurodegenerative, cancer, and metabolic diseases, which are usually accompanied by oxidative stress and inflammation [6]. One of the best studied uses of CBD is for therapeutic effect in diabetes and its complications in animal and human studies [7]. CBD, by activating the cannabinoid receptor, CB2, has been shown to induce vasodilatation in type 2 diabetic rats [8,9], and by activating 5-HT$_{1A}$ receptors, CBD showed a therapeutic effect in diabetic neuropathy [10].

Moreover, this phytocannabinoid accelerated wound healing in a diabetic rat model by protecting the endothelial growth factor (VEGF) [11]. In addition, by preventing the formation of oxidative stress in the retina neurons of diabetic animals, CBD counteracted tyrosine nitration, which can lead to glutamate accumulation and neuronal cell death [12].

This review summarizes the chemical and biological effects of CBD and its natural and synthetic derivatives. Particular attention was paid to the antioxidant and anti-inflammatory effects of CBD and its derivatives, bearing in mind the possibilities of using this phytocannabinoid to protect against oxidative stress and the consequences associated with oxidative modifications of proteins and lipids. Although CBD demonstrates safety and a good side effect profile in many clinical trials [4], all of the therapeutic options for CBD discussed in this review are limited in a concentration-dependent manner.

2. Molecular Structure of CBD

CBD is a terpenophenol compound containing twenty-one carbon atoms, with the formula $C_{21}H_{30}O_2$ and a molecular weight of 314.464 g/mol (Figure 1). The chemical structure of cannabidiol, 2-[1R-3-methyl-6R-(1-methylethenyl)-2-cyclohexen-1-yl]-5-pentyl-1,3-benzenediol, was determined in 1963 [13]. The current IUPAC preferred terminology is 2-[(1R,6R)-3-methyl-6-prop-1-en-2-ylcyclohex-2-en-1-yl]-5-pentylbenzene-1,3-diol. Naturally occurring CBD has a (−)-CBD structure [14]. The CBD molecule contains a cyclohexene ring (A), a phenolic ring (B) and a pentyl side chain. In addition, the terpenic ring (A) and the aromatic ring (B) are located in planes that are almost perpendicular to each other [15]. There are four known CBD side chain homologs, which are methyl, n-propyl, n-butyl, and n-pentyl [16]. All known CBD forms (Table 1) have absolute *trans* configuration in positions 1R and 6R [16].

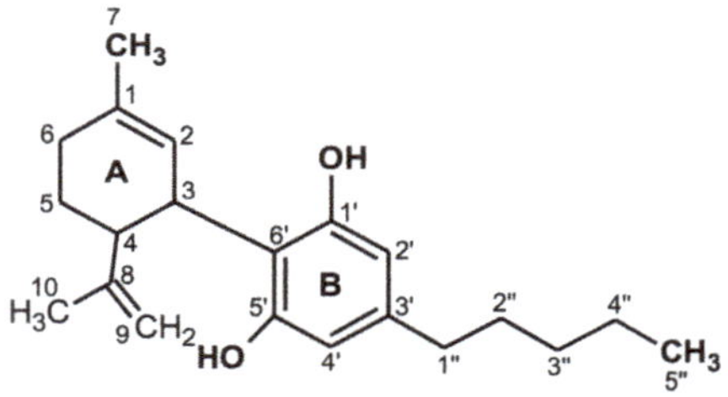

Figure 1. Chemical structure of cannabidiol (CBD) [16].

The CBD chemical activity is mainly due to the location and surroundings of the hydroxyl groups in the phenolic ring at the C-1′ and C-5′ positions (B), as well as the methyl group at the C-1 position of the cyclohexene ring (A) and the pentyl chain at the C-3′ of the phenolic ring (B). However, the open CBD ring in the C-4 position is inactive. Due to the hydroxyl groups (C-1′ and C-5′ in the B ring), CBD can also bind to amino acids such as threonine, tyrosine, glutamic acid, or glutamine by means of a hydrogen bond [17].

Table 1. Cannabidiol derivatives [16].

Compound	R^1	R^2	R^3
cannabidiolic acid (CBDA-C$_5$)	COOH	n-C$_5$H$_{11}$	H
(−)-cannabidiol (CBD-C$_5$)	H	n-C$_5$H$_{11}$	H
cannabidiol monomethyl ether (CBDM-C$_5$)	H	n-C$_5$H$_{11}$	Me
Cannabidiol-C$_4$ (CBD-C$_4$)	H	n-C$_4$H$_9$	H
cannabidivarinic acid (CBDVA-C$_3$)	COOH	n-C$_3$H$_7$	H
(−)-cannabidivarin (CBDV-C$_3$)	H	n-C$_3$H$_7$	H
cannabidiorcol (CBD-C$_1$)	H	CH$_3$	H

CBD has potential antioxidant properties because its free cationic radicals exhibit several resonance structures in which unpaired electrons are distributed mainly on ether and alkyl moieties, as well as on the benzene ring [18].

3. Biological Activity of CBD

CBD has a wide spectrum of biological activity, including antioxidant and anti-inflammatory activity, which is why its activity in the prevention and treatment of diseases whose development is associated with redox imbalance and inflammation has been tested [4,19,20]. Based on the current research results, the possibility of using CBD for the treatment of diabetes, diabetes-related cardiomyopathy, cardiovascular diseases (including stroke, arrhythmia, atherosclerosis, and hypertension), cancer, arthritis, anxiety, psychosis, epilepsy, neurodegenerative disease (i.e., Alzheimer's) and skin disease is being considered [20–22]. Analysis of CBD antioxidant activity showed that it can regulate the state of redox directly by affecting the components of the redox system and indirectly by interacting with other molecular targets associated with redox system components.

3.1. Direct Antioxidant Effects of CBD

CBD has been shown to affect redox balance by modifying the level and activity of both oxidants and antioxidants (Figures 2 and 3). CBD, like other antioxidants, interrupts free radical chain reactions, capturing free radicals or transforming them into less active forms. The free radicals produced in these reactions are characterized by many resonance structures in which unpaired electrons are mainly found on the phenolic structure, suggesting that the hydroxyl groups of the phenol ring are mainly responsible for CBD antioxidant activity [18].

CBD reduces oxidative conditions by preventing the formation of superoxide radicals, which are mainly generated by xanthine oxidase (XO) and NADPH oxidase (NOX1 and NOX4). This activity was shown in the renal nephropathy model using cisplatin-treated mice (C57BL/6J) [23] and in human coronary endothelial cells (HCAEC) [24]. In addition, CBD promoted a reduction in NO levels in the liver of doxorubicin-treated mice [25] and in the paw tissue of Wistar rats in a chronic inflammation model [26].

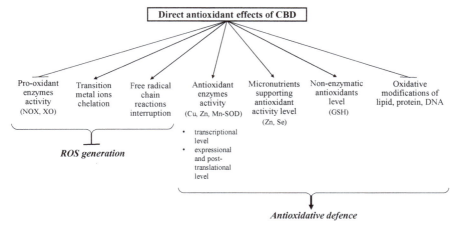

Figure 2. Direct antioxidant effects of CBD (closed arrows indicate reducing effects; opened arrows indicate inducing action).

CBD also reduces reactive oxygen species (ROS) production by chelating transition metal ions involved in the Fenton reaction to form extremely reactive hydroxyl radicals [27]. It was shown that CBD, acting similarly to the classic antioxidant butylated hydroxytoluene (BHT), prevents dihydrorodamine oxidation in the Fenton reaction [28]. In addition, CBD has been found to decrease β-amyloid formation in neurons by reducing the concentration of transition metal ions [29].

In addition to the direct reduction of oxidant levels, CBD also modifies the redox balance by changing the level and activity of antioxidants [19,26]. CBD antioxidant activity begins at the level of protein transcription by activating the redox-sensitive transcription factor referred to as the nuclear erythroid 2-related factor (Nrf2) [30], which is responsible for the transcription of cytoprotective genes, including antioxidant genes [31]. CBD was found to increase the mRNA level of superoxide dismutase (SOD) and the enzymatic activity of Cu, Zn- and Mn-SOD, which are responsible for the metabolism of superoxide radicals in the mouse model of diabetic cardiomyopathy type I and in human cardiomyocytes treated with 3-nitropropionic acid or streptozotocin [32]. Repeated doses of CBD in inflammatory conditions were found to increase the activity of glutathione peroxidase and reductase, resulting in a decrease in malonaldehyde (MDA) levels, which were six times higher in untreated controls [26]. Glutathione peroxidase activity (GSHPx) and glutathione level (GSH) were similarly changed after using CBD to treat UVB irradiated human keratinocytes. The high affinity of CBD for the cysteine and selenocysteine residues of these proteins is a possible explanation for this observation [33]. It is known that under oxidative conditions, alterations in enzymatic activity may be caused by oxidative modifications of proteins, mainly aromatic and sulfur amino acids [34]. It has also been suggested that the reactive CBD metabolite cannabidiol hydroxyquinone reacts covalently with cysteine, forming adducts with, for example, glutathione and cytochrome P450 3A11, and thereby inhibiting their biological activity [35]. In addition, CBD has been found to inhibit tryptophan degradation by reducing indoleamine-2,3-dioxygenase activity [36]. CBD also supports the action of antioxidant enzymes by preventing a reduction in the levels of microelements (e.g., Zn or Sn), which are usually lowered in pathological conditions. These elements are necessary for the biological activity of some proteins, especially enzymes such as superoxide dismutase or glutathione peroxidase [25].

By lowering ROS levels, CBD also protects non-enzymatic antioxidants, preventing their oxidation, as in the case of GSH in the myocardial tissue of C57BL/6J mice with diabetic cardiomyopathy [32] and doxorubicin-treated rats [25]. An increase in GSH levels after CBD treatment was also observed in mouse microglia cells [37] and in the liver of cadmium poisoned mice [25]. This is of great practical importance because GSH cooperates with other low molecular weight compounds in antioxidant action, mainly with vitamins such as A, E, and C [38]. CBD exhibits much more antioxidant activity (30–50%) than α-tocopherol or vitamin C [4].

3.2. The Consequences of Direct Antioxidant Action of CBD

The result of an imbalance between oxidants and antioxidants is oxidative stress, the consequences of which are oxidative modifications of lipids, nucleic acids, and proteins. This results in changes in the structure of the above molecules and, as a result, disrupts their molecular interactions and signal transduction pathways [39]. Oxidative modifications play an important role in the functioning of redox-sensitive transcription factors (including Nrf2 and the nuclear factor kappa B (NFκB). As a consequence, oxidative modifications play a role in the regulation of pathological conditions characterized by redox imbalances and inflammation, such as cancer, inflammatory diseases, and neurodegenerative diseases [40,41].

In this situation, one of the most important processes is lipid peroxidation, which results in the oxidation of polyunsaturated fatty acids (PUFA), such as arachidonic, linoleic, linolenic, eicosapentaenoic, and docosahexaenoic acids [42]. As a result of the ROS reaction with PUFAs, lipid hydroperoxides are formed, and as a result of oxidative fragmentation, unsaturated aldehydes are generated, including 4-hydroxynenenal (4-HNE), malonodialdehyde (MDA) or acrolein [43]. In addition, the propagation of oxidation chain reactions, especially with regard to docosahexaenoic acid, can lead to oxidative cyclization, resulting in production of isoprostanes or neuroprostanes [44]. The formation of lipid peroxidation products directly affects the physical properties and functioning of the cell membranes in which they are formed [42]. Due to their structure (the presence of a carbonyl groups and carbon-carbon double bonds) and electrophilic character, generated unsaturated aldehydes are chemically reactive molecules that can easily form adducts with the majority of the cell's nucleophilic components, including DNA, lipids, proteins, and GSH [45]. For example, 4-hydroxynonenal (4-HNE) has been identified as a stimulator of the cytoprotective transcription factor Nrf2, an inhibitor of antioxidant enzymes (e.g., catalase and thioredoxin reductase) and a pro-inflammatory factor acting through the NFκB pathway [46]. These reactions reduce the level of reactive lipid peroxidation products, while increasing the formation of adducts with proteins that promote cell signaling disorders, thus stimulating metabolic modifications that can lead to cellular dysfunction and apoptosis [47,48].

In addition to lipid peroxidation, oxidative conditions also favor the oxidative modification of proteins by ROS. The aromatic and sulfhydryl amino acid residues are particularly susceptible to modifications, and can result in production of levodopa (L-DOPA) from tyrosine, ortho-tyrosine from phenylalanine, sulfoxides and disulfides from cysteine, and kynurenine from tryptophan, among others [49]. The resulting changes in the protein structures cause disruption of their biological properties and, as in the case of lipid modification, affect cell metabolism, including signal transduction [46,50].

One of the most noticeable CBD antioxidant effects is the reduction in lipid and protein modifications [25,51]. CBD supplementation has been found to reduce lipid peroxidation, as measured by MDA levels, in mouse hippocampal (HT22) neuronal cells depleted of oxygen and glucose under reperfusion conditions [51]. A reduction in lipid peroxidation following CBD supplementation has also been shown in C57BL/6J mouse liver homogenates, assessed by 4-HNE levels [52]. CBD also protected the brain against oxidative protein damage caused by D-amphetamine in a rat model of mania [53]. On the other hand, CBD induced ubiquitination of the amyloid precursor protein (APP), an indicator of cellular changes in the brain of people with Alzheimer's disease, when evaluated in human neuroblastoma cells (SHSY5Y^{APP+}) [54]. In addition, CBD treatment has recently been shown to exhibit an unusual protective effect by transporting proteins including multidrug-1 resistance protein and cytosol transferases, such as S-glutathione-M1 transferase, prior to modification by lipid peroxidation products. This prevents elevation of 4-HNE and MDA adduct levels in fibroblast cell culture [55]. It was also shown that this phytocannabinoid reduced the level of small molecular $\alpha\beta$-unsaturated aldehydes in the myocardial tissue of Sprague-Dawley rats and mice with diabetic cardiomyopathy, and in the liver of mice from the acute alcohol intoxication model [21,25,32]. Additionally, CBD caused a reduction in the level of PUFA cyclization products, such as isoprostanes, in the cortex of transgenic mice (APPswe/PS1ΔE9) with Alzheimer's disease [56]. Thus, CBD protects lipids and proteins against oxidative damage by modulating the level of oxidative stress, which participates in cell signaling pathways.

3.3. Indirect Antioxidant Effects of CBD

Various cell metabolic systems, including the endocannabinoid system, are involved in the regulation of redox balance. Thus, the action of CBD as a phytocannabinoid may support the biological activity of the endocannabinoid system. CBD has recently been shown to modulate the endocannabinoid system activity by increasing anandamide (AEA) levels [5], which can affect cannabinoids signaling, including their interaction on cannabinoid receptors [57]. However, it is known that the peroxisome proliferator-activated receptor alpha (PPAR-α), for example, activated by endocannabinoids, directly regulates the expression of antioxidant enzymes such as superoxide dismutase by interacting with their promoter regions [58]. Therefore, it is believed that the most important antioxidant activity of CBD, like endocannabinoids, is associated with its effect on receptors. CBD, depending on the concentration, can activate, antagonize or inhibit cannabinoid receptors (CB1 and CB2), as well as ionotropic (TRP) and nuclear (PPAR) receptors (Figure 4) [52,59,60].

3.4. Cannabinoid Receptors

CBD has been shown to be a weak agonist of the human, mouse, and rat CB1 receptor [61]. The activation of the CB1 receptor increases ROS production and a pro-inflammatory response, including the downstream synthesis of tumor necrosis factor α (TNF-α) [62]. In addition, it was shown that CBD is a negative allosteric modulator of the CB1 receptor [63]. Regardless of the effect on the CB1 receptor, CBD is a weak agonist of the CB2 receptor [64], but it has also been suggested that it may demonstrate inverse agonism of the CB2 receptor [65]. Importantly, CB2 activation leads to a decrease in ROS and TNF-α levels, which reduces oxidative stress and inflammation [62]. Therefore, it has been suggested that CBD may indirectly improve anti-inflammatory effects. Clinical studies have confirmed that CBD reduces the levels of pro-inflammatory cytokines, inhibits T cell proliferation, induces T cell apoptosis and reduces migration and adhesion of immune cells [66]. In addition, CBD anti-inflammatory activity has been shown to be antagonized by both a selective CB2 antagonist and AEA, an endogenous CB2 receptor agonist [67].

CB1 and CB2 are receptors with strong expression in the central nervous system and the immune system primarily, but also occur in other tissues. CBD, acting on the above receptors, inhibits the activity of adenylyl cyclase and voltage gated calcium channels, activates potassium channels and activates mitogen activated protein kinase (MAPK), 3-phosphoinositol kinase (PI3K)/AKT, and the mammalian target of rapamycin (mTOR) signaling pathways [68]. The PI3K/AKT/mTOR pathway is one of the basic pathways necessary for physiological protein synthesis and induction of other intracellular pathways, such as the MAPK pathway, which plays an important role in regulating cell survival, proliferation, and apoptosis [69]. CBD was found to induce apoptosis in leukemia cells by reducing p38-MAPK levels [70]. However, CBD was also shown to inhibit apoptosis in human breast cancer cell lines (T-47D and MDA-MB-231) by inhibiting expression of oncogenic and pro-survival cyclin D1 and mTOR, and by increasing PPARγ receptor expression [71].

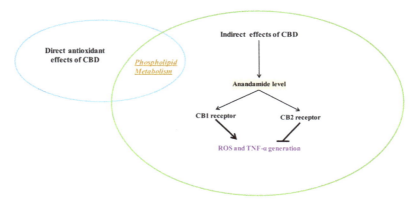

Figure 3. Indirect antioxidant and anti-inflammatory effects of CBD (closed arrows indicate inhibition; opened arrows indicate activation.

3.5. TRP Receptors

It has been shown that CBD can also affect redox balance and inflammation by modulating mammalian transient receptor potential (TRP) channels [72,73]. CBD activates vanilloid receptors (TRPV), directly or indirectly, by increasing the level of endogenous AEA, which is one of the endogenous TRPV1 agonists [64]. CBD, as a TRPV1 receptor agonist, binds to it and causes desensitization, leading to "paradoxical analgesic activity" similar to that of capsaicin [26]. It has been suggested that there is a relationship between molecular signaling of TRPV1 and oxidative stress [74] because ROS and lipid peroxidation products can regulate the physiological activity of TRPV1 by oxidizing its thiol groups [75]. Consequently, CBD not only activates TRP through a direct agonist-receptor interaction, but also by lowering the level of oxidative stress. In addition, CBD activates other vanilloid receptors such as TRPV2 and the potential ankarin protein 1 receptor subtype (TRPA1), while antagonizing the TRP-8 receptor (TRPM8) [72]. CBD has also been shown to stimulate calcium ions in transfected

HEK-293 cells via TRPV3 [76] and regulate calcium ion homeostasis in immune and inflammatory cells mainly via TRP channels, which is important for proliferation and pro-inflammatory related cytokine secretion [77]. In addition, Ca^{2+} ions control the activation of several transcription factors (e.g., NFAT) that regulate the expression of various cytokines, such as IL-2, IL-4 and IFNγ, which affect cellular inflammatory responses [78].

Regardless of the direct effect of CBD on TRP receptors, increasing the level of AEA, as a full TRPV1 agonist, also affects the activation of TRP receptors and negatively regulates the 2-arachidonoylglycerol (2-AG) metabolism [79]. It has been shown that both AEA and 2-AG can be synthesized in the plasma membrane. However, the degradation of phosphatidylinositol by phospholipase C results in the formation of a diacylglycerol precursor, whose hydrolysis (through diacylglycerol lipase activity, DAGL) allows the formation of 2-AG [80]. However, activation of DAGLα and DAGLβ requires GSH. Additionally, these enzymes are sensitive to Ca^{2+} ions [81]. TRPV1 agonists, such as capsaicin and AEA, have been shown to inhibit 2-AG synthesis in striatal neurons of C57BL/6 mice by glutathione-dependent pathways, since DAGL is stimulated by GSH [82]. In addition, the interaction between AEA and 2-AG has been shown to disappear after inactivation of TRPV1 channels. This suggests that the negative effect of AEA on 2-AG metabolism can be mimicked by stimulation of TRPV1 channels. Therefore, AEA and 2-AG interactions require redox balance, due to the participation of GSH in 2-AG synthesis. In summary, CBD modifies TRPV1 receptor activation through reducing oxidative stress as well as biosynthesis of 2-AG.

3.6. PPARγ Receptor

CBD is an agonist of the PPARγ receptor, which is a member of the nuclear receptor superfamily of ligand-inducible transcription factors [52]. PPARγ, an ubiquitin E3 ligase, has been shown to interact directly with NFκB. The interaction occurs between the ligand-binding domain of PPARγ and the Rel homology domain region of the p65 subunit of NFκB. Lys48-linked polyubiquitin of the ligand-binding domain of PPARγ is responsible for proteosomal degradation of p65 [83]. In this way, PPARγ participates in the modulation of inflammation by inducing ubiquitination proteosomal degradation of p65, which causes inhibition of pro-inflammatory gene expression, such as cyclooxygenase (COX2) and some pro-inflammatory mediators such as TNF-α, IL-1β, and IL-6, as well as inhibition of NFκB-mediated inflammatory signaling [84]. For this reason, PPARγ agonists can play an anti-inflammatory role by inhibiting the NFκB-mediated transcription of downstream genes [84]. This molecular mechanism is mediated by β-catenin and glycogen synthase kinase 3 beta (GSK-3β). β-catenin attenuates transcription of pro-inflammatory genes by inhibiting NFκB [85,86]. On the other hand, GSK-3β is decreased by PPARγ stimulation [87].

PPARγ cooperates also with another transcription factor, Nrf2, which controls the expression of genes encoding cytoprotective proteins, particularly antioxidant proteins [28,88]. PPARγ may bind to specific elements in the promoter region of genes it regulates, including Nrf2, catalase (CAT), glutathione S-transferase (GST), heme-oxygenase-1 (HO-1), and manganese-dependent superoxide dismutase (Mn-SOD). In contrast, Nrf2 can regulate PPARγ expression by binding to the PPARγ promoter in the sequence of antioxidant response elements (ARE) that are located in the -784/-764 and -916 regions of the PPARγ promoter [89,90]. The reduction in PPARγ expression in Nrf2 knockout mice provides confirmation of this regulation [91].

Acting through the PPARγ receptor, CBD demonstrates anti-inflammatory and antioxidant properties. In addition, direct CBD activity is enhanced by the action of AEA and 2-AG, which are also PPARγ agonists and whose levels are elevated by CBD [92]. It has been found that stimulation of PPARα and reduction of oxidative stress by CBD prevents amyloid β-induced neuronal death by increasing the levels of Wnt/β-catenin [84]. However, there are no data on the interaction between CBD and other PPAR subtypes (PPARα, β, δ). It is known that the endocannabinoids AEA (whose biosynthesis is stimulated by CBD) and 2-AG can activate PPARγ [92]. AEA activates PPARα, while the 2-AG derivative 15-hydroxyyeicosatetraenoic acid glyceryl ester increases the transcriptional activity of PPARα [92]. In summary, CBD demonstrates anti-inflammatory activity and antioxidant effects by activating PPARs, either directly or indirectly.

3.7. GPR Receptors

GPR55, which is strongly expressed in the nervous and immune systems as well as in other tissues, is a G-protein coupled receptor [93]. Activation of GPR55 increases the intracellular level of calcium ions [94]. CBD is a GPR55 antagonist and can modulate neuronal Ca^{2+} levels depending on the excitability of cells [95]. CBD antagonism is manifested as an anticonvulsant effect [96]. Because CBD increases endocannabinoid expression, it can also indirectly affect inflammation and redox balance via these molecules [58]. In addition, GPR55 knockout mice have been shown to have high levels of anti-inflammatory interleukins (IL-4, IL-10, and IFN-γ) [97], while high expression of GPR55 reduces ROS production [98]. Therefore, the organism's response to CBD depends on whether direct or indirect effects dominate.

CBD has also been shown to be an inverse agonist of other GPR receptors, including GPR3, GPR6 and GPR12. It reduces β-arestinin 2 levels and cAMP accumulation in amyloid plaque formation in the development of Alzheimer's disease, in a concentration-dependent manner [98]. In addition, one of the neuropharmacological effects of CBD is its reducing effect on hippocampal synaptosomes mediated by its interaction with GPR3 [99]. It has also been suggested that the effect of CBD on these orphan receptors represents a new therapeutic approach in diseases such as Alzheimer's disease, Parkinson's disease, cancer, and infertility [100].

3.8. 5-HT$_{1A}$ Receptor

CBD has direct affinity for the human 5-HT$_{1A}$ (serotonin) receptor [101]. In addition, CBD can induce the 5-HT$_{1A}$ receptor indirectly by increasing the level of AEA [102]. However, the activated 5-HT$_{1A}$ receptor can act as a membrane antioxidant by capturing ROS [103]. Therefore, through activation of 5-HT$_{1A}$, CBD can counteract peroxidation of phospholipids and thus participate in the protection of biomembranes against oxidative modifications. In addition, studies in Wistar rats have shown that CBD, by activating 5-HT$_{1A}$ receptors, can reduce physiological and behavioral responses to restrictive stress [104]. CBD has also been suggested as a therapeutic compound for the treatment of painful diabetic neuropathy due to its ability to activate 5-HT$_{1A}$ receptors [10].

3.9. Adenosine A$_{2A}$ Receptors

CBD is also an agonist of adenosine A$_{2A}$ receptors [61], which are G-protein coupled receptors. They are expressed in various cell types, participate in numerous physiological and pathological processes and also regulate inflammatory processes [105]. Adenosine and its agonists exhibit anti-inflammatory activity in vivo [106]. Therefore, adenosine release is one of the mechanisms of immunosuppression during inflammation [107], and adenosine receptor agonists reduce TNF-α levels [108,109]. It has been shown that CBD by activating A$_{2A}$ adenosine receptors can reduce the level of vascular cell adhesion molecule (VCAM-1) in endothelial cells in SJL/J mice, which may provide a new mechanism to control neuroinflammatory diseases such as multiple sclerosis (MS) [110].

In addition, it has been found that A$_{2A}$ activation can prevent reperfusion consequences and alleviate oxidative stress in mitochondria [111]. This suggests that CBD prevents oxidative stress by activating A$_{2A}$ receptors. It was also shown that A$_{2A}$ receptors can form heteromers with CB1 receptors in CA1 neurons and in the hippocampus of C57BL/6J mice [112]. Therefore, CBD can modify the functioning of the entire heteromer, and thus modulate the activation of two groups of receptors involved in the regulation of redox balance and inflammation.

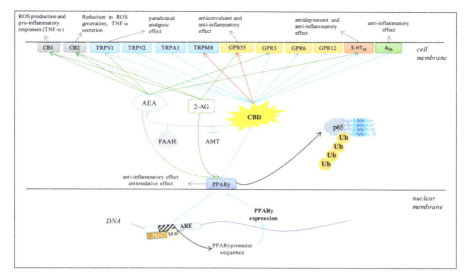

Figure 4. Major effects of CBD on several membrane receptors (AEA, anandamide; 2-AG, 2-arachidonoylglycerol; FAAH, fatty acid amide hydrolase; AMT, AEA membrane transporter; ROS, reactive oxygen species; Ub, ubiquitin; p65, transcription factor NF-κB; Nrf2, nuclear factor erythroid 2-related factor 2; ARE, antioxidant response elements. Blue arrows indicate agonist activity; red arrows indicate antagonist activity; dashed blue arrows indicate weakly agonistic activity; green arrows indicate endocannabinoid agonist activity; grey arrows indicate chemical and biological effects).

4. Effects of Natural Derivatives of CBD on Receptors

Due to the range of CBD metabolic effects known to date, interest in the possibility of using this phytocannabinoid is constantly growing. Considering the fact that modifications to the CBD structure may result in an improved therapeutic profile and biological activity, natural CBD derivatives are sought and their therapeutic utility is being evaluated. Therefore, known or potential effects of naturally occurring CBD derivatives are presented. Their activity through membrane receptors is emphasized, which are described in this review as those which under the influence of CBD show antioxidant and/or anti-inflammatory activities.

4.1. CB1/CB2 Receptors

Cannabidiolic acid (CBDA), being a C3′-carboxyl derivative of CBD (2,4-dihydroxy-3-[(1R, 6R)-3-methyl-6-prop-1-en-2-ylcyclohex-2-en-1-[alpha]-6-pentylbenzoic acid), acts as a selective COX2 and prostaglandin endoperoxide synthase inhibitor and exhibits anti-inflammatory properties in human breast cancer cells [113]. It has been suggested that its action may be due to a weak affinity for CB1 and CB2 receptors (Table 2) [114]. Similarly, other CBD derivatives, such as cannabidivarin (CBDV), which is a CBD analogue of C4′-propyl (2-[(1R, 6R)-3-methyl-6-prop-1-en-2)-ylcyclohex-2-en-1-yl]-5-propylbenzene-1,3-diol), 7-hydroxy-CBD (7-OH-CBD) and the hydroxylated CBD derivative of 7-carboxylic acid (7-COOH-CBD) have poor affinities for CB1 and CB2 (Table 2) [2]. There is no published data examining their impact on the redox balance.

4.2. GPR55 and TRPV1 Receptors

CBDV has been found to have antagonistic effects on GPR55 (Table 2), which probably leads to anticonvulsant effects [115]. Therefore, this compound is suggested for use when therapeutic antiepileptic activities are needed. On the other hand, CBDA has been suggested to be an effective

Antioxidants **2020**, *9*, 21

compound in analgesia and cancer through its agonistic action on TRPA1 and TRPV1 receptors (Table 2) and antagonistic action on TRPM8, similar to CBD [76,116].

Another natural phytocannabinoid is cannabimovone (1-[(1R, 2R, 3R, 4R)-3-(2,6-dihydroxy-4-pentylphenyl)-2-hydroxy-4-prop-1-en-2-ylcyclopentyl] ethanone), which has low affinity for the CB1 and CB2 receptors, but significant affinity for TRPV1 (Table 2) [117].

In contrast, cannabigivarin (a cannabigerol ropyl analogue) has been shown to stimulate and desensitize human TRPV1 (Table 2) [72]. It is also known that TRPV1 receptor activity is deeply involved in oxidative stress and inflammation [114]. Based on the understanding of the relationship between TRPV1 and oxidative stress described in Section 3.5, all of these derivatives may provide different therapeutic approaches in the case of inflammation and oxidative stress.

4.3. 5-HT$_{1A}$ and PPARγ Receptors

It has been shown that another CBD derivative, cannabigerol (CBG; (2-[(2E)-3,7-dimethylocta-2, 6-dienyl]-5-pentylbenzene-1,3-diol]), a naturally open analogue of cyclohexenyl CBD, activates TRPV1 as well as 5-HT$_{1A}$ (Table 2) and has antidepressant and anti-inflammatory effects in intestinal diseases [2,72,118]. CBG may also bind to PPARγ (Table 2) and increase its transcriptional activity [92]. Studies on the HEK293 cell line have shown that CBG, by activating PPARγ, significantly reduces the secretion of inflammatory mediators such as IL-6 and TNF-α [119].

5. Effects of Synthetic Derivatives of CBD on Receptors

Given the limitations in the biological activity of CBD itself and its natural derivatives and the fact that the biological properties of CBD derivatives depend on their structure, synthetic derivatives are produced that have been designed so that their structure allows direct interaction with components of the redox system or indirectly with molecular targets interacting with these components, including the cannabinoid receptors (Table 2). The derivatives with potential antioxidant and anti-inflammatory effects include, but are not limited to, (+)-CBD derivatives, dihydrocannabidiol and tetrahydrocannabidiol derivatives, and (+)-dihydro-7-hydroxy-CBD [2]. Promising synthetic derivatives that can modulate redox balance and/or inflammation are presented below.

5.1. CB1/CB2 Receptors

It has been shown that both the naturally occurring (−)-CBD enantiomer and its synthetic derivatives [(−)-7-hydroxy-5′-dimethylheptyl-CBD, and (−)-1-COOH-5′-dimethylheptyl-CBD] have weak affinity for the CB1 and CB2 cannabinoid receptors (Table 2). However, (+)-CBD and its derivatives [(+)-5′-dimethylheptyl-CBD and (+)-7-hydroxy-5′-dimethylheptyl-CBD] have high CB1 receptor affinity, slightly lower affinity for the CB2 receptor and inhibit AEA cellular uptake [120]. Similarly, (−)-7-hydroxy-dimethylheptyl-CBD can inhibit both AEA uptake and degradation through fatty acid amide hydrolase (FAAH) activity [121]. Recently, (−)-dimethylheptyl-CBD has been shown to be a CB1 receptor agonist (Table 2) in HEK-293A cells [122]. In addition, by reducing the expression of pro-inflammatory genes (IL-1b, IL-6, and TNF-α), it exhibits a dose-dependent anti-inflammatory effect on microglia BV-2 cells [30].

Furthermore, hydrogenated CBD derivatives such as (+)-dihydrocannabidiol and (+)-tetrahydrocannabidiol have CB1 receptor affinity (Table 2) and show anti-inflammatory effects on the peritoneal cells of C57BL/6 mice and a macrophage cell line. This behavior may suggest that the activation of pro-inflammatory mediators is not directly through the CB1 cannabinoid receptor [123]. Similarly, the (+)-8,9-dihydro-7-hydroxy-CBD derivative (HU-465), which has anti-inflammatory activity, especially at higher concentrations, binds to both CB1 and CB2 receptors, while its (−) enantiomer, (−)-8,9-dihydro-7-hydroxy-CBD (HU-446) has negligible affinity for both CB1 and CB2 receptors (Table 2). However, both HU-465 and HU-446 have been found to exhibit anti-inflammatory activity by inhibiting the release of IL-17 in mouse encephalitogenic T cells (TMOG) [124].

In addition, the pinene dimethoxy-dimethylheptyl-CBD derivative HU-308 [(3R, 4S, 6S)-2-[2,6-dimethoxy-4-(2-methyloctan-2-yl)phenyl]-7,7-dimethyl-4-bicyclo[3.1.1]hept-3-enyl]methanol] and its enantiomer HU-433 [(3S, 4R, 6R)-2-[2,6-dimethoxy-4-(2-methyloctan-2-yl)phenyl]-7,7-dimethyl-4-bicyclo[3.1.1]hept-3-enyl]methanol] were shown to have specific agonistic activity for the CB2 receptor (Table 2), and consequently, anti-inflammatory activity in cultured calvarial osteoblasts from C57BL/6J mice [125]. However, it has been found that HU-433 exhibits greater anti-inflammatory activity with poorer CB2 receptor binding affinity (Table 2) [125]. In contrast, HU-308, a CB2 agonist, was found to decrease TNF-α-induced expression of ICAM-1 and VCAM-1 in sinusoidal endothelial cells of human liver tissue [24]. Another CB2 receptor agonist, HU-910 ((1S,4R)-2-[2,6-dimethoxy-4-(2-methyloctan-2-yl)phenyl]-7,7-dimethyl-1-bicyclo[2.2.1]hept-2enyl]methanol)), significantly inhibits the effects of LPS that lead to increased inflammation (assessed by increased TNF-α expression) and increased oxidative stress (assessed by increased levels of 4-HNE and protein carbonyl groups) in mouse Kupffer cells [126]. This suggests that these effects are associated with CB2 receptor activation (Table 2).

5.2. GPR Receptors

Abnormal CBD (4-[(1R,6R)-3-methyl-6-prop-1-en-2-ylcyclohex-2-en-1-yl]-5-pentylbenzene-1,3-diol), named O-1602, is a synthetic CBD regioisomer and a selective GPR55 receptor agonist (Table 2), but not a CB1/CB2 receptor agonist. It causes vasodilation independent of the receptors as well [2,127]. Therefore, it has been suggested that O-1602 can be used to regulate ROS/RNS levels and modify the effect of oxidative stress on cellular metabolism by modulating GPR55 receptor activation [58]. In addition, O-1602, as a GPR18 agonist, mediates the reduction of cyclic adenosine monophosphate (cAMP) and the activation of the PI3K/AKT and ERK1/2 pathways in vitro [128]. A Sprague-Dawley rat study showed that GPR18 receptor activation via O-1602 leads to reduction of ROS levels, however inhibition of GPR18 receptor activation increases oxidative stress by increasing ROS production [129].

5.3. PPAR γ Receptor

Recently it has also been found that the synthetic quinoline derivatives of CBG, VCE-003 [(2-[(2E)-3,7-dimethylocta-2,6-dienyl]-3-hydroxy-5-pentylcyclohexa-2,5-dien-1,4-dione)] and HU-331 [(3-hydroxy-2-[(1R,6R)-3-methyl-6-prop-1-en-2-ylcyclohex-2-en-1-yl]-5-pentylcyclohexa-2,5-dien-1, 4-dione], activate the PPARγ receptor as CBG does (Table 2) [130]. HU-311 was shown to interfere with mitochondrial transmembrane potential and induce ROS generation, as well as activate the Nrf2 pathway [130]. However, another CBD derivative, VCE-003, induces PPARγ-mediated antioxidant and anti-inflammatory activity that prevents neuronal damage caused by inflammation in the Parkinson's mouse model (intravascular LPS injection). The same effect was seen in the in vitro cellular model of neurological inflammation (BV2 cells exposed to LPS and M-213 cells treated with media prepared from BV2 cells exposed to LPS) [131].

5.4. TRPV1, 5-HT$_{1A}$ and Adenosine A$_{2A}$ Receptors

Despite extensive research into the biological effects of synthetic CBD derivatives, they have not been evaluated for their interaction with TRPV1, 5-HT$_{1A}$ and adenosine A$_{2A}$ receptors in the context of anti-inflammatory and antioxidant activity.

Table 2. Influence of natural and synthetic CBD derivatives on receptor activation (X: agonist activation or Y: antagonist activation by related CBD derivative; * weak affinity; #: full name is in chapter 4.1) [2,24,72,76,114–116,120,122–127,130].

CBD Derivatives	Membrane Receptors								
Natural	CB1	CB2	Gpr55	Gpr18	TRPV1	TRPA1	TRPM8	5-HT$_{1A}$	PPARγ
cannabigerol (CBG)					X			X	
cannabigivarin (CBGV)					X				
cannabidiolic acid (CBDA)	X *	X *			X	X	Y		
cannabidivarin (CBDV)	X *	X *	Y						
cannabimovone	X *	X *			X				
7-OH-CBD	X *	X *							
7-COOH-CBD	X *	X *							
Synthetic	CB1	CB2	Gpr55	Gpr18	TRPV1	TRPA1	TRPM8	5-HT$_{1A}$	PPARγ
(−)-dimethylheptyl-CBD (DMH-CBD)	X								
(−)-7-hydroxy-5′-dimethylheptyl-CBD	X *	X *							
(−)-1-COOH-5′-dimethylheptyl-CBD	X *	X *							
(+)-5′-dimethylheptyl-CBD	X	X *							
(+)-7-hydroxy-5′-dimethylheptyl-CBD	X	X *							
(+)-dihydrocannabidiol (H2-CBD)	X								
(+)-tetrahydrocannabidiol (H4-CBD)	X								
(-)-8,9-dihydro-7-hydroxy-CBD (HU-446)	X *	X *							
(+)-8,9-dihydro-7-hydroxy-CBD (HU-465)	X	X							
pinene dimethoxy-dimethylheptyl-CBD derivative (HU-433)		X *							
pinene dimethoxy-dimethylheptyl-CBD derivative (HU-308)		X							
HU-910 #		X							
4′-fluorocannabidiol (HUF-101/4′-F-CBD)	X	X							
quinol derivative VCE-003									X
quinol derivative HU-331									X
abnormal-CBD			X	X					

6. Conclusions

Oxidative stress resulting from overproduction of ROS is a key element of the immune system's response to combat pathogens and initiates tissue repair. However, metabolic modifications resulting from overproduction of ROS also have many negative aspects and lead to the development and/or exacerbation of many diseases. It is believed that the endocannabinoid system, which includes G-protein coupled receptors and their endogenous lipid ligands, may be responsible for the therapeutic modulation of oxidative stress in various diseases. In this context, the phytocannabinoid cannabidiol, which was identified several decades ago and may interact with the cannabinoid system, is a promising molecule for pharmacotherapy.

Relatively recently, multidirectional biological effects have been demonstrated in various preclinical models, including the antioxidant and anti-inflammatory effects of cannabidiol [14,73]. In the context of the above data, CBD seems to be more preferred than other compounds from the phytocannabinoid group. Regardless of the beneficial pharmacological effects of CBD itself, if this compound is present in the Δ^9-THC environment, the undesirable effects of 99-THC are reduced, which improves its safety profile [132].

Important in CBD therapeutic applications is the lack of psychotropic effects. Furthermore, this phytocannabinoid is not teratogenic or mutagenic [133]. Until recently, CBD was thought to have only low toxicity to humans and other species [134], but recent studies indicate an increase in ALT and AST levels after CBD treatment, which disqualifies it as the drug of choice [135,136]. In addition, it has been found that CBD may interfere with the hepatic metabolism of some drugs by inactivating cytochrome P450 3A and P450 2C [137]. Such interactions should be considered when co-administering CBD with other drugs metabolized by above enzymes.

In order to find compounds with a greater therapeutic profile and activity than CBD, without any adverse effects, the biological properties of both natural and synthetic CBD derivatives were checked, with the hope of finding the perfect derivative that provides a close to ideal therapeutic effect.

Author Contributions: Conceptualization, E.S.; writing—original draft preparation, S.A. and I.J.-K.; writing—review and editing, E.S.; visualization, S.A.; supervision, E.S. All authors have read and agreed to the published version of the manuscript.

Funding: S.A.: co-author of the work, was supported by the project which has received funding from the European Union's Horizon 2020 research and innovation programme under the Marie Skłodowska-Curie grant agreement No 754432 and the Polish Ministry of Science and Higher Education, from financial resources for science in 2018-2023 granted for the implementation of an international co-financed project.

Conflicts of Interest: The authors declare no conflict of interest.

References

1. Battista, N.; Di Tommaso, M.; Bari, M.; Maccarrone, M. The endocannabinoid system: An overview. *Front. Behav. Neurosci.* **2012**, *6*, 9. [CrossRef] [PubMed]
2. Morales, P.; Reggio, P.H.; Jagerovic, N. An Overview on Medicinal Chemistry of Synthetic and Natural Derivatives of Cannabidiol. *Front. Pharmacol.* **2017**, *8*, 422. [CrossRef] [PubMed]
3. Rong, C.; Lee, Y.; Carmona, N.E.; Cha, D.S.; Ragguett, R.M.; Rosenblat, J.D.; Mansur, R.B.; Ho, R.C.; McIntyre, R.S. Cannabidiol in medical marijuana: Research vistas and potential opportunities. *Pharmacol. Res.* **2017**, *121*, 213–218. [CrossRef] [PubMed]
4. Iffland, K.; Grotenhermen, F. An Update on Safety and Side Effects of Cannabidiol: A Review of Clinical Data and Relevant Animal Studies. *Cannabis Cannabinoid Res.* **2017**, *2*, 139–154. [CrossRef]
5. Lim, K.; See, Y.M.; Lee, J. A Systematic Review of the Effectiveness of Medical Cannabis for Psychiatric, Movement and Neurodegenerative Disorders. *Clin. Psychopharmacol. Neurosci.* **2017**, *15*, 301–312. [CrossRef]
6. Oguntibeju, O.O. Type 2 diabetes mellitus, oxidative stress and inflammation: Examining the links. *Int. J. Physiol. Pathophysiol. Pharmacol.* **2019**, *11*, 45–63.

7. Smeriglio, A.; Giofrè, S.; Galati, E.M.; Monforte, M.T.; Cicero, N.; D'Angelo, V.; Grassi, G.; Circosta, C. Inhibition of aldose reductase activity by Cannabis sativa chemotypes extracts with high content of cannabidiol or cannabigerol. *Fitoterapia* **2018**, *127*, 101–108. [CrossRef]
8. Wheal, A.J.; Cipriano, M.; Fowler, C.J.; Randall, M.D.; O'Sullivan, S.E. Cannabidiol improves vasorelaxation in Zucker diabetic fatty rats through cyclooxygenase activation. *J. Pharmacol. Exp. Ther.* **2014**, *351*, 457–466. [CrossRef]
9. Wheal, A.J.; Jadoon, K.; Randall, M.D.; O'Sullivan, S.E. In Vivo Cannabidiol Treatment Improves Endothelium-Dependent Vasorelaxation in Mesenteric Arteries of Zucker Diabetic Fatty Rats. *Front. Pharmacol.* **2017**, *8*, 248. [CrossRef]
10. Jesus, C.H.A.; Redivo, D.D.B.; Gasparin, A.T.; Sotomaior, B.B.; de Carvalho, M.C.; Genaro, K.; Zuardi, A.W.; Hallak, J.E.C.; Crippa, J.A.; Zanoveli, J.M.; et al. Cannabidiol attenuates mechanical allodynia in streptozotocin-induced diabetic rats via serotonergic system activation through 5-HT1A receptors. *Brain Res.* **2019**, *1715*, 156–164. [CrossRef]
11. Yan, X.; Chen, B.; Lin, Y.; Li, Y.; Xiao, Z.; Hou, X.; Tan, Q.; Dai, J. Acceleration of diabetic wound healing by collagen-binding vascular endothelial growth factor in diabetic rat model. *Diabetes Res. Clin. Pract.* **2010**, *90*, 66–72. [CrossRef] [PubMed]
12. El-Remessy, A.B.; Khalifa, Y.; Ola, S.; Ibrahim, A.S.; Liou, G.I. Cannabidiol protects retinal neurons by preserving glutamine synthetase activity in diabetes. *Mol. Vis.* **2010**, *16*, 1487–1495.
13. Mechoulam, R.; Hanus, L. Cannabidiol: An overview of some chemical and pharmacological aspects. Part I: Chemical aspects. *Chem. Phys. Lipids* **2002**, *121*, 35–43. [CrossRef]
14. Burstein, S. Cannabidiol (CBD) and its analogs: A review of their effects on inflammation. *Bioorg. Med. Chem.* **2015**, *23*, 1377–1385. [CrossRef] [PubMed]
15. Jones, P.G.; Falvello, L.; Kennard, O.; Sheldrick, G.M.; Mechoulam, R. Cannabidiol. *Acta Crystallogr. B* **1977**, *33*, 3211–3214. [CrossRef]
16. Elsohly, M.A.; Slade, D. Chemical constituents of marijuana: The complex mixture of natural cannabinoids. *Life Sci.* **2005**, *78*, 539–548. [CrossRef] [PubMed]
17. Elmes, M.W.; Kaczocha, M.; Berger, W.T.; Leung, K.; Ralph, B.P.; Wang, L.; Sweeney, J.M.; Miyauchi, J.T.; Tsirka, S.E.; Ojima, I.; et al. Fatty acid-binding proteins (FABPs) are intracellular carriers for Δ9-tetrahydrocannabinol (THC) and cannabidiol (CBD). *J. Biol. Chem.* **2015**, *290*, 8711–8721. [CrossRef]
18. Borges, R.S.; Batista, J., Jr.; Viana, R.B.; Baetas, A.C.; Orestes, E.; Andrade, M.A.; Honório, K.M.; Da Silva, A.B. Understanding the Molecular Aspects of Tetrahydrocannabinol and Cannabidiol as Antioxidants. *Molecules* **2013**, *18*, 12663–12674. [CrossRef]
19. Peres, F.F.; Lima, A.C.; Hallak, J.E.C.; Crippa, J.A.; Silva, R.H.; Abílio, V.C. Cannabidiol as a Promising Strategy to Treat and Prevent Movement Disorders? *Front. Pharmacol.* **2018**, *9*, 482. [CrossRef]
20. Da Silva, V.K.; De Freitas, B.S.; Garcia, R.C.L.; Monteiro, R.T.; Hallak, J.E.; Zuardi, A.W.; Crippa, J.A.S.; Schröder, N. Antiapoptotic effects of cannabidiol in an experimental model of cognitive decline induced by brain iron overload. *Transl. Psychiatry.* **2018**, *8*, 176. [CrossRef]
21. Yang, L.; Rozenfeld, R.; Wu, D.; Devi, L.A.; Zhang, Z.; Cederbaum, A. Cannabidiol protects liver from binge alcohol-induced steatosis by mechanisms including inhibition of oxidative stress and increase in autophagy. *Free Radic. Biol. Med.* **2014**, *68*, 260–267. [CrossRef] [PubMed]
22. Hammell, D.C.; Zhang, L.P.; Ma, F.; Abshire, S.M.; McIlwrath, S.L.; Stinchcomb, A.L.; Westlund, K.N. Transdermal cannabidiol reduces inflammation and pain-related behaviours in a rat model of arthritis. *Eur. J. Pain* **2016**, *20*, 936–948. [CrossRef] [PubMed]
23. Pan, H.; Mukhopadhyay, P.; Rajesh, M.; Patel, V.; Mukhopadhyay, B.; Gao, B.; Haskó, G.; Pacher, P. Cannabidiol attenuates cisplatin-induced nephrotoxicity by decreasing oxidative/nitrosative stress, inflammation, and cell death. *J. Pharmacol. Exp. Ther.* **2008**, *328*, 708–714. [CrossRef] [PubMed]
24. Rajesh, M.; Mukhopadhyay, P.; Bátkai, S.; Haskó, G.; Liaudet, L.; Drel, V.R.; Obrosova, I.G.; Pacher, P. Cannabidiol attenuates high glucose-induced endothelial cell inflammatory response and barrier disruption. *Am. J. Physiol. Heart Circ. Physiol.* **2007**, *293*, 610–619. [CrossRef] [PubMed]
25. Fouad, A.A.; Albuali, W.H.; Al-Mulhim, A.S.; Jresat, I. Cardioprotective effect of cannabidiol in rats exposed to doxorubicin toxicity. *Environ. Toxicol. Pharmacol.* **2013**, *36*, 347–357. [CrossRef]

26. Costa, B.; Trovato, A.E.; Comelli, F.; Giagnoni, G.; Colleoni, M. The non-psychoactive cannabis constituent cannabidiol is an orally effective therapeutic agent in rat chronic inflammatory and neuropathic pain. *Eur. J. Pharmacol.* **2007**, *556*, 75–83. [CrossRef]
27. Hamelink, C.; Hampson, A.; Wink, D.A.; Eiden, L.E.; Eskay, R.L. Comparison of cannabidiol, antioxidants, and diuretics in reversing binge ethanol-induced neurotoxicity. *J. Pharmacol. Exp. Ther.* **2005**, *314*, 780–788. [CrossRef]
28. Campos, A.C.; Fogaça, M.V.; Sonego, A.B.; Guimarães, F.S. Cannabidiol, neuroprotection and neuropsychiatric disorders. *Pharmacol. Res.* **2016**, *112*, 119–127. [CrossRef]
29. Iuvone, T.; Esposito, G.; Esposito, R.; Santamaria, R.; Di Rosa, M.; Izzo, A.A. Neuroprotective effect of cannabidiol, a non-psychoactive component from Cannabis sativa, on beta-amyloid-induced toxicity in PC12 cells. *J. Neurochem.* **2004**, *89*, 134–141. [CrossRef]
30. Juknat, A.; Pietr, M.; Kozela, E.; Rimmerman, N.; Levy, R.; Gao, F.; Coppola, G.; Geschwind, D.; Vogel, Z. Microarray and pathway analysis reveal distinct mechanisms underlying cannabinoid-mediated modulation of LPS-induced activation of BV-2 microglial cells. *PLoS ONE* **2013**, *8*, e61462. [CrossRef]
31. Vomund, S.; Schäfer, A.; Parnham, M.J.; Brüne, B.; von Knethen, A. Nrf2, the Master Regulator of Anti-Oxidative Responses. *Int. J. Mol. Sci.* **2017**, *18*, 12. [CrossRef] [PubMed]
32. Rajesh, M.; Mukhopadhyay, P.; Bátkai, S.; Patel, V.; Saito, K.; Matsumoto, S.; Kashiwaya, Y.; Horváth, B.; Mukhopadhyay, B.; Becker, L.; et al. Cannabidiol attenuates cardiac dysfunction, oxidative stress, fibrosis, and inflammatory and cell death signaling pathways in diabetic cardiomyopathy. *J. Am. Coll. Cardiol.* **2010**, *56*, 2115–2125. [CrossRef] [PubMed]
33. Jastrząb, A.; Gęgotek, A.; Skrzydlewska, E. Cannabidiol Regulates the Expression of Keratinocyte Proteins Involved in the Inflammation Process through Transcriptional Regulation. *Cells* **2019**, *8*, 827. [CrossRef] [PubMed]
34. Wall, S.B.; Oh, J.Y.; Diers, A.R.; Landar, A. Oxidative modification of proteins: An emerging mechanism of cell signaling. *Front. Physiol.* **2012**, *3*, 369. [CrossRef] [PubMed]
35. Wu, H.Y.; Han, T.R. Cannabidiol hydroxyquinone-induced apoptosis of splenocytes is mediated predominantly by thiol depletion. *Toxicol. Lett.* **2010**, *195*, 68–74. [CrossRef]
36. Hornyák, L.; Dobos, N.; Koncz, G.; Karányi, Z.; Páll, D.; Szabó, Z.; Halmos, G.; Székvölgyi, L. The Role of Indoleamine-2,3-Dioxygenase in Cancer Development, Diagnostics, and Therapy. *Front. Immunol.* **2018**, *9*, 151. [CrossRef]
37. Wu, H.Y.; Goble, K.; Mecha, M.; Wang, C.C.; Huang, C.H.; Guaza, C.; Jan, T.R. Cannabidiol-induced apoptosis in murine microglial cells through lipid raft. *Glia* **2012**, *60*, 1182–1190. [CrossRef]
38. Gęgotek, A.; Ambrożewicz, E.; Jastrząb, A.; Jarocka-Karpowicz, I.; Skrzydlewska, E. Rutin and ascorbic acid cooperation in antioxidant and antiapoptotic effect on human skin keratinocytes and fibroblasts exposed to UVA and UVB radiation. *Arch. Dermatol. Res.* **2019**, *311*, 203–219. [CrossRef]
39. Kim, E.K.; Jang, M.; Song, M.J.; Kim, D.; Kim, Y.; Jang, H.H. Redox-Mediated Mechanism of Chemoresistance in Cancer Cells. *Antioxidants* **2019**, *8*, 471. [CrossRef]
40. Chio, I.I.C.; Tuveson, D.A. ROS in Cancer: The Burning Question. *Trends Mol. Med.* **2017**, *23*, 411–429. [CrossRef]
41. Pizzino, G.; Irrera, N.; Cucinotta, M.; Pallio, G.; Mannino, F.; Arcoraci, V.; Squadrito, F.; Altavilla, D.; Bitto, A. Oxidative Stress: Harms and Benefits for Human Health. *Oxidative Med. Cell. Longev.* **2017**, *2017*. [CrossRef] [PubMed]
42. Gaschler, M.M.; Stockwell, B.R. Lipid peroxidation in cell death. *Biochem. Biophys. Res. Commun.* **2017**, *482*, 419–425. [CrossRef] [PubMed]
43. Ayala, A.; Muñoz, M.F.; Argüelles, S. Lipid peroxidation: Production, metabolism, and signaling mechanisms of malondialdehyde and 4-hydroxy-2-nonenal. *Oxidative Med. Cell. Longev.* **2014**, *2014*. [CrossRef] [PubMed]
44. Milne, G.L.; Yin, H.; Hardy, K.D.; Davies, S.S.; Roberts, L.J. Isoprostane generation and function. *Chem. Rev.* **2011**, *111*, 5973–5996. [CrossRef] [PubMed]
45. Nam, T.G. Lipid Peroxidation and Its Toxicological Implications. *Toxicol. Res.* **2011**, *27*, 1–6. [CrossRef]
46. Łuczaj, W.; Gęgotek, A.; Skrzydlewska, E. Antioxidants and HNE in redox homeostasis. *Free Radic. Biol. Med.* **2017**, *111*, 87–101. [CrossRef]
47. Dalleau, S.; Baradat, M.; Guéraud, F.; Huc, L. Cell death and diseases related to oxidative stress: 4-hydroxynonenal (HNE) in the balance. *Cell Death Differ.* **2013**, *12*, 1615–1630. [CrossRef]

48. Środa-Pomianek, K.; Michalak, K.; Świątek, P.; Poła, A.; Palko-Łabuz, A.; Wesołowska, O. Increased lipid peroxidation, apoptosis and selective cytotoxicity in colon cancer cell line LoVo and its doxorubicin-resistant subline LoVo/Dx in the presence of newly synthesized phenothiazine derivatives. *Biomed. Pharmacother.* **2018**, *106*, 624–636. [CrossRef]

49. Gianazza, E.; Crawford, J.; Miller, I. Detecting oxidative post-translational modifications in proteins. *Amino Acids* **2007**, *33*, 51–56. [CrossRef]

50. Sottero, B.; Leonarduzzi, G.; Testa, G.; Gargiulo, S.; Poli, G.; Biasi, F. Lipid Oxidation Derived Aldehydes and Oxysterols Between Health and Disease. *Eur. J. Lipid Sci. Technol.* **2018**, *121*. [CrossRef]

51. Sun, S.; Hu, F.; Wu, J.; Zhanga, S. Cannabidiol attenuates OGD/R-induced damage by enhancing mitochondrial bioenergetics and modulating glucose metabolism via pentose-phosphate pathway in hippocampal neurons. *Redox Biol.* **2017**, *11*, 577–585. [CrossRef]

52. Wang, Y.; Mukhopadhyay, P.; Cao, Z.; Wang, H.; Feng, D.; Haskó, G.; Mechoulam, R.; Gao, B.; Pacher, P. Cannabidiol attenuates alcohol-induced liver steatosis, metabolic dysregulation, inflammation and neutrophil-mediated injury. *Sci. Rep.* **2017**, *7*, 12064. [CrossRef] [PubMed]

53. Valvassori, S.S.; Elias, G.; De Souza, B.; Petronilho, F.; Dal-Pizzol, F.; Kapczinski, F.; Trzesniak, C.; Tumas, V.; Dursun, S.; Chagas, M.H.; et al. Effects of cannabidiol on amphetamine-induced oxidative stress generation in an animal model of mania. *J. Psychopharmacol.* **2011**, *25*, 274–280. [CrossRef] [PubMed]

54. Scuderi, C.; Steardo, L.; Esposito, G. Cannabidiol promotes amyloid precursor protein ubiquitination and reduction of beta amyloid expression in SHSY5YAPP+ cells through PPARγ involvement. *Phytother. Res.* **2014**, *28*, 1007–1013. [CrossRef]

55. Gęgotek, A.; Atalay, S.; Domingues, P.; Skrzydlewska, E. The Differences in the Proteome Profile of Cannabidiol-Treated Skin Fibroblasts following UVA or UVB Irradiation in 2D and 3D Cell Cultures. *Cells* **2019**, *8*, 995. [CrossRef]

56. Cheng, D.; Low, J.K.; Logge, W.; Garner, B.; Karl, T. Chronic cannabidiol treatment improves social and object recognition in double transgenic APPswe/PS1ΔE9 mice. *Psychopharmacology* **2014**, *231*, 3009–3017. [CrossRef]

57. Bih, C.I.; Chen, T.; Nunn, A.V.W.; Bazelot, M.; Dallas, M.; Whalley, B.J. Molecular Targets of Cannabidiol in Neurological Disorders. *Neurotherapeutics* **2015**, *12*, 699–730. [CrossRef]

58. Gallelli, C.A.; Calcagnini, S.; Romano, A.; Koczwara, J.B.; De Ceglia, M.; Dante, D.; Villani, R.; Giudetti, A.M.; Cassano, T.; Gaetani, S. Modulation of the Oxidative Stress and Lipid Peroxidation by Endocannabinoids and Their Lipid Analogues. *Antioxidants* **2018**, *7*, 93. [CrossRef]

59. De Petrocellis, L.; Nabissi, M.; Santoni, G.; Ligresti, A. Actions and Regulation of Ionotropic Cannabinoid Receptors. *Adv. Pharmacol.* **2017**, *80*, 249–289. [CrossRef]

60. Ghovanloo, M.R.; Shuart, N.G.; Mezeyova, J.; Dean, R.A.; Ruben, P.C.; Goodchild, S.J. Inhibitory effects of cannabidiol on voltage-dependent sodium currents. *J. Biol. Chem.* **2018**, *293*, 16546–16558. [CrossRef] [PubMed]

61. McPartland, J.M.; Duncan, M.; Di Marzo, V.; Pertwee, R.G. Are cannabidiol and tetrahydrocannabivarin negative modulators of the endocannabinoid system? A systematic review. *Br. J. Pharmacol.* **2015**, *172*, 737–753. [CrossRef] [PubMed]

62. Han, K.H.; Lim, S.; Ryu, J.; Lee, C.W.; Kim, Y.; Kang, J.H.; Kang, S.S.; Ahn, Y.K.; Park, C.S.; Kim, J.J. CB1 and CB2 cannabinoid receptors differentially regulate the production of reactive oxygen species by macrophages. *Cardiovasc. Res.* **2009**, *84*, 378–386. [CrossRef] [PubMed]

63. Laprairie, R.B.; Bagher, A.M.; Kelly, M.E.; Denovan-Wright, E.M. Cannabidiol is a negative allosteric modulator of the cannabinoid CB1 receptor. *Br. J. Pharmacol.* **2015**, *172*, 4790–4805. [CrossRef] [PubMed]

64. Muller, C.; Morales, P.; Reggio, P.H. Cannabinoid Ligands Targeting TRP Channels. *Front. Mol. Neurosci.* **2018**, *11*, 487. [CrossRef]

65. Pertwee, R.G. The diverse CB1 and CB2 receptor pharmacology of three plant cannabinoids: Δ9-tetrahydrocannabinol, cannabidiol and Δ9-tetrahydrocannabivarin. *Br. J. Pharmacol.* **2008**, *153*, 199–215. [CrossRef]

66. Jean-Gilles, L.; Braitch, M.; Latif, M.L.; Aram, J.; Fahey, A.J.; Edwards, L.J.; Robins, R.A.; Tanasescu, R.; Tighe, P.J.; Gran, B.; et al. Effects of pro-inflammatory cytokines on cannabinoid CB1 and CB2 receptors in immune cells. *Acta Physiol.* **2015**, *214*, 63–74. [CrossRef]

67. Petrosino, S.; Verde, R.; Vaia, M.; Allaraà, M.; Iuvone, T.; Marzo, V.D. Anti-inflammatory properties of cannabidiol, a non-psychotropic cannabinoid, in experimental allergic contact dermatitis. *J. Pharmacol. Exp. Ther.* **2018**, *365*, 652–663. [CrossRef]

68. Callén, L.; Moreno, E.; Barroso-Chinea, P.; Moreno-Delgado, D.; Cortés, A.; Mallol, J.; Casadó, V.; Lanciego, J.S.; Franco, R.; Lluis, C.; et al. Cannabinoid Receptors CB1 and CB2 Form Functional Heteromers in Brain. *J. Biol. Chem.* **2012**, *287*, 20851–20865. [CrossRef]

69. Giacoppo, S.; Galuppo, M.; Pollastro, F.; Grassi, G.; Bramanti, P.; Mazzon, E. A new formulation of cannabidiol in cream shows therapeutic effects in a mouse model of experimental autoimmune encephalomyelitis. *Daru* **2015**, *23*, 48. [CrossRef]

70. McKallip, R.J.; Jia, W.; Schlomer, J.; Warren, J.W.; Nagarkatti, P.S.; Nagarkatti, M. Cannabidiol-induced apoptosis in human leukemia cells: A novel role of cannabidiol in the regulation of p22phox and Nox4 expression. *Mol. Pharmacol.* **2006**, *70*, 897–908. [CrossRef]

71. Sultan, A.S.; Marie, M.A.; Sheweita, S.A. Novel mechanism of cannabidiol-induced apoptosis in breast cancer cell lines. *Breast* **2018**, *41*, 34–41. [CrossRef] [PubMed]

72. De Petrocellis, L.; Ligresti, A.; Moriello, A.S.; Allarà, M.; Bisogno, T.; Petrosino, S.; Stott, C.G.; Di Marzo, V. Effects of cannabinoids and cannabinoid-enriched Cannabis extracts on TRP channels and endocannabinoid metabolic enzymes. *Br. J. Pharmacol.* **2011**, *163*, 1479–1494. [CrossRef] [PubMed]

73. Pellati, F.; Borgonetti, V.; Brighenti, V.; Biagi, M.; Benvenuti, S.; Corsi, L. *Cannabis sativa* L. and Nonpsychoactive Cannabinoids: Their Chemistry and Role against Oxidative Stress, Inflammation, and Cancer. *Biomed. Res. Int.* **2018**, *2018*. [CrossRef] [PubMed]

74. Miller, B.A.; Zhang, W. TRP channels as mediators of oxidative stress. *Adv. Exp. Med. Biol.* **2011**, *704*, 531–544. [CrossRef] [PubMed]

75. Ogawa, N.; Kurokawa, T.; Fujiwara, K.; Polat, O.K.; Badr, H.; Takahashi, N.; Mori, Y. Functional and Structural Divergence in Human TRPV1 Channel Subunits by Oxidative Cysteine Modification. *J. Biol. Chem.* **2016**, *291*, 4197–4210. [CrossRef] [PubMed]

76. De Petrocellis, L.; Orlando, P.; Moriello, A.S.; Aviello, G.; Stott, C.; Izzo, A.A.; Di Marzo, V. Cannabinoid actions at TRPV channels: Effects on TRPV3 and TRPV4 and their potential relevance to gastrointestinal inflammation. *Acta Physiol.* **2012**, *204*, 255–266. [CrossRef] [PubMed]

77. Bujak, J.K.; Kosmala, D.; Szopa, I.M.; Majchrzak, K.; Bednarczyk, P. Inflammation, Cancer and Immunity—Implication of TRPV1 Channel. *Front. Oncol.* **2019**, *9*, 1087. [CrossRef]

78. Minke, B. TRP channels and Ca^{2+} signalling. *Cell Calcium* **2006**, *40*, 261–275. [CrossRef]

79. Zou, S.; Kumar, U. Cannabinoid Receptors and the Endocannabinoid System: Signaling and Function in the Central Nervous System. *Int. J. Mol. Sci.* **2018**, *19*, 833. [CrossRef]

80. Lipina, C.; Hundal, H.S. Modulation of cellular redox homeostasis by the endocannabinoid system. *Open Biol.* **2016**, *6*. [CrossRef]

81. Di Marzo, V. Endocannabinoid signaling in the brain: Biosynthetic mechanisms in the limelight. *Nat. Neurosci.* **2011**, *14*, 9–15. [CrossRef] [PubMed]

82. Maccarrone, M.; Rossi, S.; Bari, M.; De Chiara, V.; Fezza, F.; Musella, A.; Gasperi, V.; Prosperetti, C.; Bernardi, G.; Finazzi-Agrò, A.; et al. Anandamide inhibits metabolism and physiological actions of 2-arachidonoylglycerol in the striatum. *Nat. Neurosci.* **2008**, *11*, 152–159. [CrossRef] [PubMed]

83. Hou, Y.; Moreau, F.; Chadee, K. PPARγ is an E3 ligase that induces the degradation of NFκB/p65. *Nat. Commun.* **2012**, *3*, 1300. [CrossRef] [PubMed]

84. Vallée, A.; Lecarpentier, Y.; Guillevin, R.; Vallée, J.N. Effects of cannabidiol interactions with Wnt/β-catenin pathway and PPARγ on oxidative stress and neuroinflammation in Alzheimer's disease. *Acta Biochim. Biophys. Sin.* **2017**, *49*, 853–866. [CrossRef] [PubMed]

85. Hoeflich, K.P.; Luo, J.; Rubie, E.A.; Tsao, M.S.; Jin, O.; Woodgett, J.R. Requirement for glycogen synthase kinase-3beta in cell survival and NFkappaB activation. *Nature* **2000**, *406*, 86–90. [CrossRef] [PubMed]

86. Beurel, E.; Michalek, S.M.; Jope, R.S. Innate and adaptive immune responses regulated by glycogen synthase kinase-3 (GSK3). *Trends Immunol.* **2010**, *31*, 24–31. [CrossRef] [PubMed]

87. Inestrosa, N.C.; Godoy, J.A.; Quintanilla, R.A.; Koenig, C.S.; Bronfman, M. Peroxisome proliferator-activated receptor gamma is expressed in hippocampal neurons and its activation prevents beta-amyloid neurodegeneration: Role of Wnt signaling. *Exp. Cell Res.* **2005**, *304*, 91–104. [CrossRef]

88. Paunkov, A.; Chartoumpekis, D.V.; Ziros, P.G.; Sykiotis, G.P. A Bibliometric Review of the Keap1/Nrf2 Pathway and its Related Antioxidant Compounds. *Antioxidants* **2019**, *8*, 353. [CrossRef]

89. Cho, H.Y.; Gladwell, W.; Wang, X.; Chorley, B.; Bell, D.; Reddy, S.P.; Kleeberger, S.R. Nrf2-regulated PPAR {gamma} expression is critical to protection against acute lung injury in mice. *Am. J. Respir. Crit. Care Med.* **2010**, *182*, 170–182. [CrossRef]

90. Lee, C. Collaborative Power of Nrf2 and PPARγ Activators against Metabolic and Drug-Induced Oxidative Injury. *Oxidative Med. Cell. Longev.* **2017**, *2017*. [CrossRef]

91. Haung, J.; Tabbi-Anneni, I.; Gunda, V.; Wang, L. Transcription factor Nrf2 regulates SHP and lipogenic gene expression in hepatic lipid metabolism. *Am. J. Physiol. Gastrointest. Liver Physiol.* **2010**, *299*, 1211–1221. [CrossRef] [PubMed]

92. O'Sullivan, S.E. An update on PPAR activation by cannabinoids. *Br. J. Pharmacol.* **2016**, *173*, 1899–1910. [CrossRef] [PubMed]

93. Zhou, J.; Burkovskiy, I.; Yang, H.; Sardinha, J.; Lehmann, C. CB2 and GPR55 Receptors as Therapeutic Targets for Systemic Immune Dysregulation. *Front. Pharmacol.* **2016**, *2016*, 264. [CrossRef] [PubMed]

94. Lauckner, J.E.; Jensen, J.B.; Chen, H.Y.; Lu, H.C.; Hille, B.; Mackie, K. GPR55 is a cannabinoid receptor that increases intracellular calcium and inhibits M current. *Proc. Natl. Acad. Sci. USA* **2008**, *105*, 2699–2704. [CrossRef]

95. Sylantyev, S.; Jensen, T.P.; Ross, R.A.; Rusakov, D.A. Cannabinoid- and lysophosphatidylinositol-sensitive receptor GPR55 boosts neurotransmitter release at central synapses. *Proc. Natl. Acad. Sci. USA* **2013**, *110*, 5193–5198. [CrossRef]

96. Marichal-Cancino, B.A.; Fajardo-Valdez, A.; Ruiz-Contreras, A.E.; Méndez-Díaz, M.; Prospéro-García, O. Advances in the Physiology of GPR55 in the Central Nervous System. *Curr. Neuropharmacol.* **2017**, *15*, 771–778. [CrossRef]

97. Staton, P.C.; Hatcher, J.P.; Walker, D.J.; Morrison, A.D.; Shapland, E.M.; Hughes, J.P.; Chong, E.; Mander, P.K.; Green, P.J.; Billinton, A.; et al. The putative cannabinoid receptor GPR55 plays a role in mechanical hyperalgesia associated with inflammatory and neuropathic pain. *Pain* **2008**, *139*, 225–236. [CrossRef]

98. Balenga, N.A.; Aflaki, E.; Kargl, J.; Platzer, W.; Schröder, R.; Blättermann, S.; Kostenis, E.; Brown, A.J.; Heinemann, A.; Waldhoer, M. GPR55 regulates cannabinoid 2 receptor-mediated responses in human neutrophils. *Cell Res.* **2011**, *21*, 1452–1469. [CrossRef]

99. Calpe-López, C.; García-Pardo, M.P.; Aguilar, M.A. Cannabidiol Treatment Might Promote Resilience to Cocaine and Methamphetamine Use Disorders: A Review of Possible Mechanisms. *Molecules* **2019**, *24*, 2583. [CrossRef]

100. Laun, A.S.; Shrader, S.H.; Brown, K.J.; Song, Z.H. GPR3, GPR6, and GPR12 as novel molecular targets: Their biological functions and interaction with cannabidiol. *Acta Pharmacol. Sin.* **2019**, *40*, 300–308. [CrossRef]

101. Russo, E.B.; Burnett, A.; Hall, B.; Parker, K.K. Agonistic Properties of Cannabidiol at 5-HT1A Receptors. *Neurochem. Res.* **2005**, *30*, 1037. [CrossRef] [PubMed]

102. Haj-Dahmane, S.; Roh-Yu Shen, R. Modulation of the Serotonin System by Endocannabinoid Signaling. *Neuropharmacology* **2011**, *61*, 414–420. [CrossRef] [PubMed]

103. Azouzi, S.; Santuz, H.; Morandat, S.; Pereira, C.; Côté, F.; Hermine, O.; El Kirat, K.; Colin, Y.; Le Van Kim, C.; Etchebest, C.; et al. Antioxidant and Membrane Binding Properties of Serotonin Protect Lipids from Oxidation. *Biophys. J.* **2017**, *112*, 1863–1873. [CrossRef] [PubMed]

104. Resstel, L.B.; Tavares, R.F.; Lisboa, S.F.; Joca, S.R.; Corrêa, F.M.; Guimarães, F.S. 5-HT$_{1A}$ receptors are involved in the cannabidiol-induced attenuation of behavioural and cardiovascular responses to acute restraint stress in rats. *Br. J. Pharmacol.* **2009**, *156*, 181–188. [CrossRef] [PubMed]

105. Guerrero, A. A2A Adenosine Receptor Agonists and their Potential Therapeutic Applications. An Update. *Curr. Med. Chem.* **2018**, *25*, 3597–3612. [CrossRef] [PubMed]

106. Noji, T.; Takayama, M.; Mizutani, M.; Okamura, Y.; Takai, H.; Karasawa, A.; Kusaka, H. KF24345, an adenosine uptake inhibitor, suppresses lipopolysaccharide-induced tumor necrosis factor-alpha production and leukopenia via endogenous adenosine in mice. *J. Pharmacol. Exp. Ther.* **2001**, *300*, 200–205. [CrossRef]

107. Haskó, G.; Cronstein, B.N. Adenosine: An endogenous regulator of innate immunity. *Trends Immunol.* **2004**, *25*, 33–39. [CrossRef]

108. Carrier, E.J.; Auchampach, J.A.; Hillard, C.J. Inhibition of an equilibrative nucleoside transporter by cannabidiol: A mechanism of cannabinoid immunosuppression. *Proc. Natl. Acad. Sci. USA* **2006**, *103*, 7895–7900. [CrossRef]

109. Ribeiro, A.; Ferraz-de-Paula, V.; Pinheiro, M.L.; Vitoretti, L.B.; Mariano-Souza, D.P.; Quinteiro-Filho, W.M.; Akamine, A.T.; Almeida, V.I.; Quevedo, J.; Dal-Pizzol, F.; et al. Cannabidiol, a non-psychotropic plant-derived cannabinoid, decreases inflammation in a murine model of acute lung injury: Role for the adenosine A(2A) receptor. *Eur. J. Pharmacol.* **2012**, *678*, 78–85. [CrossRef]

110. Mecha, M.; Feliú, A.; Iñigo, P.M.; Mestre, L.; Carrillo-Salinas, F.J.; Guaza, C. Cannabidiol provides long-lasting protection against the deleterious effects of inflammation in a viral model of multiple sclerosis: A role for A2A receptors. *Neurobiol. Dis.* **2013**, *59*, 141–150. [CrossRef]

111. Xu, J.; Bian, X.; Liu, Y.; Hong, L.; Teng, T.; Sun, Y.; Xu, Z. Adenosine A2 receptor activation ameliorates mitochondrial oxidative stress upon reperfusion through the posttranslational modification of NDUFV2 subunit of complex I in the heart. *Free Radic. Biol. Med.* **2017**, *106*, 208–218. [CrossRef] [PubMed]

112. Aso, E.; Fernández-Dueñas, V.; López-Cano, M.; Taura, J.; Watanabe, M.; Ferrer, I.; Luján, R.; Ciruela, F. Adenosine A2A-Cannabinoid CB1 Receptor Heteromers in the Hippocampus: Cannabidiol Blunts Δ9-Tetrahydrocannabinol-Induced Cognitive Impairment. *Mol. Neurobiol.* **2019**, *56*, 5382–5391. [CrossRef] [PubMed]

113. Takeda, S.; Okazaki, H.; Ikeda, E.; Abe, S.; Yoshioka, Y.; Watanabe, K.; Aramaki, H. Down-regulation of cyclooxygenase-2 (COX-2) by cannabidiolic acid in human breast cancer cells. *J. Toxicol. Sci.* **2014**, *39*, 711–716. [CrossRef] [PubMed]

114. Nahler, G.; Jones, T.M.; Russo, E.B. Cannabidiol and Contributions of Major Hemp Phytocompounds to the "Entourage Effect"; Possible Mechanisms. *J. Altern. Complementary Integr. Med.* **2019**, *5*. [CrossRef]

115. Vigli, D.; Cosentino, L.; Raggi, C.; Laviola, G.; Woolley-Roberts, M.; De Filippis, B. Chronic treatment with the phytocannabinoid Cannabidivarin (CBDV) rescues behavioural alterations and brain atrophy in a mouse model of Rett syndrome. *Neuropharmacology* **2018**, *140*, 121–129. [CrossRef] [PubMed]

116. Anavi-Goffer, S.; Baillie, G.; Irving, A.; Gertsch, J.; Greig, I.R.; Pertwee, R.G.; Ross, R.A. Modulation of L-α-lysophosphatidylinositol/GPR55 mitogen-activated protein kinase (MAPK) signaling by cannabinoids. *J. Biol. Chem.* **2012**, *287*, 91–104. [CrossRef]

117. Taglialatela-Scafati, O.; Pagani, A.; Scala, F.; De Petrocellis, L.; Di Marzo, V.; Grassi, G.; Appendino, G. Cannabimovone, a cannabinoid with a rearranged terpenoid skeleton from hemp. *Eur. J. Org. Chem.* **2010**, *2010*, 2067–2072. [CrossRef]

118. Borrelli, F.; Fasolino, I.; Romano, B.; Capasso, R.; Maiello, F.; Coppola, D.; Orlando, P.; Battista, G.; Pagano, E.; Di Marzo, V.; et al. Beneficial effect of the non-psychotropic plant cannabinoid cannabigerol on experimental inflammatory bowel disease. *Biochem. Pharmacol.* **2013**, *85*, 1306–1316. [CrossRef]

119. Granja, A.G.; Carrillo-Salinas, F.; Pagani, A.; Gómez-Cañas, M.; Negri, R.; Navarrete, C.; Mecha, M.; Mestre, L.; Fiebich, B.L.; Cantarero, I.; et al. A cannabigerol quinone alleviates neuroinflammation in a chronic model of multiple sclerosis. *J. Neuroimmune Pharmacol.* **2012**, *7*, 1002–1016. [CrossRef]

120. Bisogno, T.; Hanus, L.; De Petrocellis, L.; Tchilibon, S.; Ponde, D.E.; Brandi, I.; Moriello, A.S.; Davis, J.B.; Mechoulam, R.; Di Marzo, V. Molecular targets for cannabidiol and its synthetic analogues: Effect on vanilloid VR1 receptors and on the cellular uptake and enzymatic hydrolysis of anandamide. *Br. J. Pharmacol.* **2001**, *134*, 845–852. [CrossRef]

121. Fride, E.; Ponde, D.; Breuer, A.; Hanus, L. Peripheral, but not central effects of cannabidiol derivatives: Mediation by CB (1) and unidentified receptors. *Neuropharmacology* **2005**, *48*, 1117–1129. [CrossRef] [PubMed]

122. Tham, M.; Yilmaz, O.; Alaverdashvili, M.; Kelly, M.E.M.; Denovan-Wright, E.M.; Laprairie, R.B. Allosteric and orthosteric pharmacology of cannabidiol and cannabidiol-dimethylheptyl at the type 1 and type 2 cannabinoid receptors. *Br. J. Pharmacol.* **2019**, *176*, 1455–1469. [CrossRef] [PubMed]

123. Ben-Shabat, S.; Hanus, L.O.; Katzavian, G.; Gallily, R. New cannabidiol derivatives: Synthesis, binding to cannabinoid receptor, and evaluation of their antiinflammatory activity. *J. Med. Chem.* **2006**, *49*, 1113–1117. [CrossRef] [PubMed]

124. Kozela, E.; Haj, C.; Hanuš, L.; Chourasia, M.; Shurki, A.; Juknat, A.; Kaushansky, N.; Mechoulam, R.; Vogel, Z. HU-446 and HU-465, Derivatives of the Non-psychoactive Cannabinoid Cannabidiol, Decrease the Activation of Encephalitogenic T Cells. *Chem. Biol. Drug Des.* **2016**, *87*, 143–153. [CrossRef] [PubMed]

125. Smoum, R.; Baraghithy, S.; Chourasia, M.; Breuer, A.; Mussai, N.; Attar-Namdar, M.; Kogan, N.M.; Raphael, B.; Bolognini, D.; Cascio, M.G.; et al. CB2 cannabinoid receptor agonist enantiomers HU-433 and HU-308: An inverse relationship between binding affinity and biological potency. *Proc. Natl. Acad. Sci. USA* **2015**, *112*, 8774–8779. [CrossRef]
126. Horváth, B.; Magid, L.; Mukhopadhyay, P.; Bátkai, S.; Rajesh, M.; Park, O.; Tanchian, G.; Gao, R.Y.; Goodfellow, C.E.; Glass, M.; et al. A new cannabinoid CB2 receptor agonist HU-910 attenuates oxidative stress, inflammation and cell death associated with hepatic ischaemia/reperfusion injury. *Br. J. Pharmacol.* **2012**, *165*, 2462–2478. [CrossRef]
127. McKillop, A.M.; Moran, B.M.; Abdel-Wahab, Y.H.A.; Flatt, P.R. Evaluation of the insulin releasing and antihyperglycaemic activities of GPR55 lipid agonists using clonal beta-cells, isolated pancreatic islets and mice. *Br. J. Pharmacol.* **2013**, *170*, 978–990. [CrossRef]
128. Matouk, A.I.; Taye, A.; El-Moselhy, M.A.; Heeba, G.H.; Abdel-Rahman, A.A. The Effect of Chronic Activation of the Novel Endocannabinoid Receptor GPR18 on Myocardial Function and Blood Pressure in Conscious Rats. *J. Cardiovasc. Pharmacol.* **2017**, *69*, 23–33. [CrossRef]
129. Penumarti, A.; Abdel-Rahman, A.A. The novel endocannabinoid receptor GPR18 is expressed in the rostral ventrolateral medulla and exerts tonic restraining influence on blood pressure. *J. Pharmacol. Exp. Ther.* **2014**, *349*, 29–38. [CrossRef]
130. Del Río, C.; Navarrete, C.; Collado, J.A.; Bellido, M.L.; Gómez-Cañas, M.; Pazos, M.R.; Fernández-Ruiz, J.; Pollastro, F.; Appendino, G.; Calzado, M.A.; et al. The cannabinoid quinol VCE-004.8 alleviates bleomycin-induced scleroderma and exerts potent antifibrotic effects through peroxisome proliferator-activated receptor-γ and CB2 pathways. *Sci. Rep.* **2016**, *6*. [CrossRef]
131. García, C.; Gómez-Cañas, M.; Burgaz, S.; Palomares, B.; Gómez-Gálvez, Y.; Palomo-Garo, C.; Campo, S.; Ferrer-Hernández, J.; Pavicic, C.; Navarrete, C.; et al. Benefits of VCE-003.2, a cannabigerol quinone derivative, against inflammation-driven neuronal deterioration in experimental Parkinson's disease: Possible involvement of different binding sites at the PPARγ receptor. *J. Neuroinflamm.* **2018**, *15*, 19. [CrossRef] [PubMed]
132. Zhornitsky, S.; Potvin, S. Cannabidiol in Humans-The Quest for Therapeutic Targets. *Pharmaceuticals* **2012**, *5*, 529–552. [CrossRef] [PubMed]
133. Rosenkrantz, H.; Hayden, D.W. Acute and subacute inhalation toxicity of Turkish marihuana, cannabichromene, and cannabidiol in rats. *Toxicol. Appl. Pharmacol.* **1979**, *48*, 375–386. [CrossRef]
134. Rosenkrantz, H.; Fleischman, R.W.; Grant, R.J. Toxicity of short-term administration of cannabinoids to rhesus monkeys. *Toxicol. Appl. Pharmacol.* **1981**, *58*, 118–131. [CrossRef]
135. Thiele, E.; Marsh, E.; Mazurkiewicz-Beldzinska, M.; Halford, J.J.; Gunning, B.; Devinsky, O.; Checketts, D.; Roberts, C. Cannabidiol in patients with Lennox-Gastaut syndrome: Interim analysis of an open-label extension study. *Epilepsia* **2019**, *60*, 419–428. [CrossRef]
136. Devinsky, O.; Patel, A.D.; Cross, J.H.; Villanueva, V.; Wirrell, E.C.; Privitera, M.; Greenwood, S.M.; Roberts, C.; Checketts, D.; VanLandingham, K.E.; et al. Effect of cannabidiol on drop seizures in the Lennox-Gastaut syndrome. *N. Engl. J. Med.* **2018**, *378*, 1888–1897. [CrossRef]
137. Samara, E.; Brown, N.K.; Harvey, D.J. Microsomal metabolism of the 1″,1″-dimethylheptyl analogue of cannabidiol: Relative percentage of monohydroxy metabolites in four species. *Drug Metab. Dispos.* **1990**, *18*, 548–549.

© 2019 by the authors. Licensee MDPI, Basel, Switzerland. This article is an open access article distributed under the terms and conditions of the Creative Commons Attribution (CC BY) license (http://creativecommons.org/licenses/by/4.0/).

MDPI
St. Alban-Anlage 66
4052 Basel
Switzerland
Tel. +41 61 683 77 34
Fax +41 61 302 89 18
www.mdpi.com

Antioxidants Editorial Office
E-mail: antioxidants@mdpi.com
www.mdpi.com/journal/antioxidants

Lightning Source UK Ltd.
Milton Keynes UK
UKHW050725190920
370074UK00007B/185